Star-hopping is a technique that uses the brighter stars as markers on a celestial path that leads to fainter stars, star clusters, and the distant galaxies. All new stargazers need to learn these skills, but few books give real guidance on how to relate the baffling view at the telescope to a map in a star atlas. With star-hopping you first locate an easy-to-find bright star. Then you move the telescope in a series of overlapping steps until you reach the target object. You can start with star-hopping any time, because the heart of this book is a series of twelve monthly star-hops. Two or more tours are given for each month of the year, but many more can be tried at any time of course. One chapter is devoted to the popular Messier Marathon, which involves trying to see in a single night more than 100 bright galaxies, star clusters, and gas clouds.

Star-Hopping

Your Visa to Viewing the Universe

Star-Hopping

Your Visa to Viewing the Universe

ROBERT A. GARFINKLE

Foreword by Richard Berry

CAMBRIDGE UNIVERSITY PRESS

Published by the Press Syndicate of the University of Cambridge
The Pitt Building, Trumpington Street, Cambridge CB2 1RP
40 West 20th Street New York, NY 10011-4211, USA
10 Stamford Road, Oakleigh, Melbourne 3166, Australia

First published 1994
Reprinted 1995 , 1997
First paperback edition 1997

Printed in the United States of America

A catalogue record for this book is available from the British Library

Library of Congress cataloguing in publication data

Garfinkle, Robert A.
Star-hopping: your visa to viewing the universe / Robert A. Garfinkle; foreword by Richard Berry.
   p. cm.
Includes bibliographical references and index.
ISBN 0 521 41590 X
1. Stars – Charts, diagrams, etc.   2. Stars – Observers' manuals.
3. Stars – Amateurs' manuals.   4. Astronomy – Amateurs' manuals.
I. Title.
QB65.G37   1994
523.8'02'2 – dc20   93-8204 CIP

ISBN 0 521 41590 X hardback
ISBN 0 521 59889 3 paperback

To my parents, Wil and Wilma Garfinkle,
who taught me to look up and appreciate the night sky.

Contents

Contents

Figures

Tables

Photocredits

Eyepiece impressions by Rob Toebe

Foreword

Astronomy is full of star-hoppers. I am a star-hopper, Bob Garfinkle is a star-hopper, and soon you will be a star-hopper, too. This book tells you about star-hopping. Its purpose is to teach you how to star-hop, and it teaches by inviting you to take 27 tours of the night sky, one or more tour each month of the year. You can start at any time, and by the end of the year you will be a well-seasoned star-hop traveler, your celestial visa bearing the stamps of the constellations you have explored.

No skill is more basic to observing with a telescope than star-hopping. Once you have mastered star-hopping, you will have the skills to use any telescope, any time, any place. Armed with nothing more than a star chart, a red-filtered flashlight, and your sense of the sky, you will be able to guide your friends to celestial sights that you have enjoyed.

Without doubt you will eventually meet astronomers with telescopes that have digital setting circles and tiny buzzing electric motors that flit from star to star. There is nothing wrong with that – in fact, you may some day want to join the high-tech crowd. In the meantime, your star-hopping skills will stand you in good stead as you hop your way to the celestial sights you want to see.

The essence of star-hopping is to develop a feel for how to interpret the myriad dots on a star chart as stars in the sky. As you follow Bob's 27 sky tours, bit-by-bit you will discover how the view through an eyepiece relates to the dots on the star chart. You will also develop a sense of how to move the telescope, and how those movements relate to a star chart.

When I was a teenager just getting started in astronomy, I found it hard to remember which way was which. Once I realized that all I needed was to move my telescope a short distance toward Polaris to identify north, or push it slightly toward the eastern horizon to figure out which way was east in the eyepiece, my troubles were over. On every star chart in this book, Bob has thoughtfully provided an arrow pointing toward Polaris to help you keep your celestial directions clear.

In developing this book, Bob has worked out 27 tours of the night sky. These tours are the heart of the book. In each tour, he has woven the wealth of information that you would naturally expect from a good tour guide. Before you go star-hopping though, it helps to read through each tour to absorb some of the rich background that Bob has assembled. Read through quickly; ignore the details. Just get a sense for what you are going to see.

Under the stars, take Bob's tours. At first you may want to follow him every step of the way, but later you may wish to improvise your own tours with the aid of a star atlas. Be daring! Hop off and get yourself lost in the trackless depths of space. Nothing can happen to you – at a moment's notice, you can return safely to Earth.

After your star-hop, reread the tour and once again listen to your guide. Bob evokes mythic traditions that are thousands of years old, and recounts hundreds of years of modern astronomy in a few short pages. It is a lot to absorb, but rest assured that understanding will come.

Amateur astronomy is for enjoyment, and star-hopping is one of the skills that makes astronomy more enjoyable. Let the memories and pleasures of the perfect nights carry you through those occasional nights that will, in retrospect, seem funny. After all, what do spilled cans of soda, forgotten star atlases, dead batteries, and lost eyepieces really matter when stacked against the glory of the firmament? Draw three deep breaths. Relax. Look around the sky and you will see your old friends, the stars, still shining overhead. You had better get ready for me, stars, because I am hopping!

Richard Berry
Former Editor-in-Chief of *Astronomy* magazine
Cedar Grove, WI

Preface

This book will teach you the techniques of star-hopping. The old proverb says that every journey starts with the first step; your journey up the mountain of stargazing experience must also start with a first step – learning how to "star-hop."

As an observer of the night sky, you will be joining the millions of people throughout history who have enjoyed the twinkling points of light overhead. We were meant to be stargazers. During the creation of the world, according to the ancient Greeks, the Titan Prometheus took some mud from the Earth, mixed it with drops of Oceanus's water, and kneaded the mixture into man. To separate man from all of the other animals, Prometheus created us upright so that, whereas all of the other animals look downward toward the ground, we gaze upward toward the stars.

Whether you observe the wonders of the night with only your eyes, with binoculars, or with telescopes, you can use the skills covered in this book. I have employed star-hopping to get around the sky when using only my eyes, and when using various amateur telescopes. Even when looking through the 30-inch telescope at Fremont Peak in California, I have to rely on star-hopping to locate an object, because the telescope is not equipped with celestial coordinates setting circles.

Sometimes the most basic techniques of any hobby or scientific field can get lost in the shuffle by people once they progress beyond the beginner level. Without thinking about it, the more advanced observer moves his scope from object to object just as naturally as walking. As a newcomer to stargazing, no one led me through the starry maze. After many years of observing and studying the night sky, I began to think back to how I had learned to find my way to the "faint fuzzies." My decision to write this book so as to help you get started was based on ideas similar to those sentiments expressed by the Spanish-Arab astronomer Nūr al-Dīn, Abu Isḥāk, al-Biṭrūdjī (Latinized as Alpetragius) (11??–1217?) in the introduction to his treatise *On the Principles of Astronomy*:

> ... my intention is to inform you of a matter which has preoccupied my mind and to tell you the plain meaning of some of the noble secrets. This has come to my mind after much thought and diligence for most of my life.

I used to be the type of stargazer who swept the sky with binoculars and stopped at anything that looked interesting. Lacking any real guidance, I had no idea what I was looking at nor how much I was missing. At the time, I did not even own a sky chart. I bought a small telescope and was ready to conquer the universe. The instructions in my first guide book were, to me, utterly useless. I soon returned to my "peek-a-boo" method of locating the objects I had read about.

One night at a star party, I had the chance to look at the Ring Nebula (M57) in Lyra through a telescope for the first time. As I looked through the eyepiece and then at the point in the sky where the telescope seemed to be pointing, I asked the owner of the telescope where exactly in the sky the nebula was. I wanted to be able to find it myself. My friend pointed to Vega. He showed me how to find Beta and Lambda Lyrae, and then to center the telescope's field of view halfway between them. Bingo! In one simple lesson I had been introduced to naked-eye and telescopic star-hopping. After that, I went in search of objects I had so many times before passed over. My friend had set me on a new course of discovery among the stars.

The night sky consists of billions of objects that can be observed with the right equipment. Unfortunately, the vast majority of these would require that you have access to professional optical or radio telescopes. I am going to present a sampling of what the sky contains by taking you to many of the beautiful and key celestial objects; those studied by astronomers to further our understanding of the universe and our place in it. With each successful completion of the star-hops in this book, you will steadily progress up the mountain of experience.

My goal is for you to be able to find and view faint stellar objects by using easy to locate stars as your guideposts and to do the planning to get you from the guideposts to the faint fuzzies. When you finish this book, you will feel very comfortable with the night sky, and should easily be able to find each of the beautiful objects that I have taken you to.

I will assume you have looked skyward at least a few times before you put this book in your hands. I am also going to assume you may know absolutely nothing about stargazing, or maybe you own binoculars or a telescope, but have never used them to view the heavens, except possibly to observe the Moon and the planets.

Astronomy is the one science that links the present with all who have come before us. I have included at the commencement of each chapter words of wisdom or poetry from different times and lands to remind us of our forebears and the lessons about the sky they have passed on to us.

Like my friend who stamped my visa and opened my vista to the universe, I hope that with this book I can stamp your visa to a lifetime of exciting universal exploration. As al-Biṭrūdjī wrote: "Now I ask you, and those who study what I have written, to think well of me...."

Acknowledgements

This book could not have been completed without the encouragement and assistance of my family, many friends, and experts in various aspects of astronomy. I am indebted first to the staff members of *Astronomy* magazine, in particular Richard Berry and David Eicher. Richard took the chance on this then new writer by publishing my first book reviews and encouraged me to write my first star-hopping article. Those phone conversations were the starter yeast from which this book has risen. After he left *Astronomy* Richard graciously accepted my offer to write the foreword. Thank you Richard for this and your valuable tips. David became my book review editor in 1990, and put me in touch with Dr. Simon Mitton at Cambridge University Press. Thanks go to Dr. Mitton and the staff at CUP for all of their help in the production of this book.

My friend and renowned amateur astronomer, Ben Mayer also encouraged me to write this book and implanted many ideas and thoughts on how to make it easier for you to learn from it.

University of Texas astronomy professor Gérard de Vaucouleurs reviewed and commented on the section on galaxy classification. He developed the system, so I went to the source to be sure I had the correct information. His colleague at UOT, Professor Robert Robbins, assisted with material about the Mimbres Indian pottery that depicts the supernova explosion of 1054. The Anthropology Department of the University of Minnesota supplied the photograph of the particular piece of pottery. Dr. William Morgan, at the Yerkes Observatory, reviewed the section on his MKK luminosity classification system.

I am indebted to the following friends who reviewed various portions of the manuscript and gave of their time for this project – Chabot College Astronomy instructor Scott Hildreth; fellow technical writer Jerry Taylor; and expert amateur astronomer Jack Zieders. Comet-hunter and another expert amateur astronomer, Don Machholtz, has led the San Jose Astronomical Association on many Messier Marathons so I asked him to review that chapter. Dr. Jack Marling is the developer of some of the special filters covered in this work and he reviewed and commented on that portion of the book.

I am also very grateful to Rob Toebe for the splendid eyepiece impressions. In the "its a small world" category, after Rob began to help me on this project, we discovered we were delivered by the same doctor, though 18 years apart.

The astrophotographs were supplied by Lee C. Coombs, Martin C. Germano, Tony Hallas and his wife Daphne Mount, and Dr. Jack Marling. Alan Hale, President of Celestron Corporation, supplied the photographs of the telescopes in chapter 2. Dr. Martha Hazen, Curator of Astronomical Photographs at the Harvard College Observatory, located the historic spectrogram of Vega. Thanks to all of you for your time, talents, and assistance.

Dr. Don Markos, Professor of English at California State University, Hayward, assisted with suggestions for some of the poets whose work I have incorporated in the book.

I also want to acknowledge the assistance of those poets, writers, scientists, and astronomers who came before me and helped by providing the reference source material referred to throughout this book.

The artwork was prepared on my friend Elin Thomas's Macintosh. She gave freely of her computer and I want to thank her for allowing me to spend many days at her machine.

There is no way I can fully express my deep gratitude to my wife and two daughters for putting up with all the hassles and headaches that must be overcome along with the undone chores as I sequestered myself away in my office over the course of about a year. Thank you Kathy, Kim, and my budding astronomer Annmarie.

Robert A. Garfinkle

Great indeed are the things which in this brief
treatise I propose for observation and
consideration by all students of nature. I say great,
because of the excellence of the subject itself, the
entirely unexpected and novel character of these
things, and finally because of the instrument by
means of which they have been revealed to our
senses.

Galileo Galilei (1546–1642),
Sidereus Nuncius
("The Starry Messenger"),
1610

1 How to use this book and what you are going to see

For what could be more beautiful than the heavens which contain all beautiful things? Their names make this clear: *Caelum* (heavens) by naming that which is beautifully carved.... So if the worth of the arts were measured by the matter with which they deal, this art — which some call astronomy,... would be by far the most outstanding.

Nicolas Copernicus (1473–1543),
De revolutionibus orbium coelestium
("On the Revolutions of the Celestial Spheres,"
Book One),
1543

How to use this book

When planning a vacation you usually get out a map of the place you want to visit and decide how best to get there. You will also read guidebooks and pamphlets so you can determine what sights to see and where they are. Would you hop on a train or airplane and set out for a distant land without knowing in which direction it lies or the best route to get you there? Would you arrive there with no idea of what to see or the significance of the places? Would you set out on your trip without luggage? The same holds true for the night sky; you need to be able to find your way among the stars, understand what you are looking at, and have the equipment and guidebooks necessary for a successful and enjoyable journey.

Astronomy is laden with baffling jargon and technical information. Under the heading of "What you are going to see", this chapter has brief definition paragraphs for your background information on astronomical terms and the types of objects we are going to hop to. A classification system for each type of object exists and these are given in tables in Appendix A. Refer to the glossary for additional definitions on the topics and terms covered in this section and in the balance of the book.

In chapter 2, we will learn how the sky moves, how to read a star chart, how to determine the size of your field of view, how to observe various kinds of objects, and how to star-hop.

Like preparing for any trip, you need to do some planning, have an itinerary, and pack the necessary items. Planning a star-hop begins with having a basic understanding of the night sky. Though not entirely necessary, you should consider acquiring a good star atlas, circle templates or wire rings, a variety of guidebooks or catalogues, a

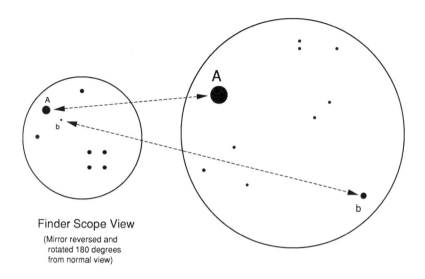

Finder Scope View

(Mirror reversed and
rotated 180 degrees
from normal view)

Figure 1.1
Comparing finder
scope and eyepiece
fields of view.

Telescopic View of the Same Field

dim flashlight with a red lens, and a logbook. Refer to the bibliography for astronomical books and supplies you can obtain to supplement this book.

There are two ways to star-hop: naked-eye and telescopic (telescopic also means the use of binoculars). You will employ both methods as you view the heavens and become more experienced.

To naked-eye star-hop, you only have to look up and be able to recognize the star groupings called constellations and to know who is neighbor to whom. Several prominent constellations (Orion, Leo, Pegasus, Scorpius, Cassiopeia, Ursa Major, and Cygnus) serve as convenient major guideposts. As evening twilight changes to night, your eyes will naturally search out these constellations, since they contain stars that are usually the first to appear. We use their neighborly relationships to hop from the brighter constellations to fainter ones nearby. Naked-eye hopping gets you to the general area of the sky where your target object is. You then switch to telescopic star-hopping.

To telescopic star-hop, you use the diameter of the circle of the sky (field of view) visible in your optical instrument as a distance gauge. Most of the time, you use the field of view in your finder scope rather than in your eyepiece, because you can see a larger area of the sky with the finder, as shown in Figure 1.1.

No matter how big or powerful your telescope is, the amount of sky you can see through it at any one time is very small. Knowing how many field of view diameters it takes to get from one object to the next, you can then move your scope the correct number of hops to reach the vicinity of the object. Once you are near the object, you may have to tweak the scope's aim slightly to center the object in your eyepiece.

Following chapter 2, we will take interesting star-hop adventures each month. Besides the actual act of going outside and stargazing, it can be very pleasurable to read about the objects you are going to observe. In the 12 monthly star-hop chapters, I will present more celestial details and weave the objects we view into their astronomical, mythological, and historical places. Along the way we will learn what the objects are and how they are classified. I am also going to discuss some of the trailblazing astronomers and what their discoveries mean to us.

The last chapter will assist you in taking one of the most popular and challenging star-hops: a sundown-to-sunup Messier Marathon. In this one night of viewing, you will search for and attempt to observe the 110 objects French astronomer Charles Messier (1730–1817) and his colleague Pierre François André Méchain (1744–1804) took 24 years to locate and catalogue in the late 1700's. You will have no problem venturing off to find more faint fuzzies on your own after you finish this book.

What you are going to see

Astronomy, like all sciences, relies on facts and precise terminology. Some of the numerical data given in this book are absolute and finite whereas others are astronomers' best guesses or are as accurate as modern technology can detect. In researching the literature for material for this book, I found wide ranges in the values for some of the data for some of the objects.

Most of the terminology used in astronomy is interlocked like the pieces of a celestial puzzle. In the following section, we will begin to put the edges of the puzzle together. The rest of the pieces will fall into place as we proceed through the book.

What are stars and what do we know about them?

In the simplest sense, a star is a ball of hydrogen atoms, so compressed by its own gravitational force that its internal temperature has risen to the point where nuclear fusion takes place. Fusion is the act of combining atomic particles from two or more lighter atoms to create a new heavier one, releasing energy in the process. In stars, the first level of fusion is the creation of helium from hydrogen atoms; heavier elements are then formed from the helium.

With stars being so hot and so far away that we cannot get close to them, how do we know what they are? Astronomers use a vast array of instruments combined with the knowledge of several sciences to study the composition of the universe. Four main stellar characteristics used to classify stars are: *spectral type, surface temperature, magnitude,* and *luminosity*. The age and mass of a star are also considered.

Spectral type

All energy can be measured either by its wavelength or frequency on the electromagnetic spectrum. A single component of the electromag-

Table 1.1. *The electromagnetic spectrum.*

Type of radiation	Wavelength range (cm)	Frequency range (Hz)
Gamma rays	$< 10^{-8}$	$> 3 \times 10^{19}$
X-rays	$1{-}200 \times 10^{-8}$	$1.5 \times 10^{16}{-}3 \times 10^{19}$
Extreme ultraviolet	$200{-}900 \times 10^{-8}$	$3.3 \times 10^{15}{-}1.5 \times 10^{16}$
Ultraviolet	$900{-}4{,}000 \times 10^{-8}$	$7.5 \times 10^{14}{-}3.3 \times 10^{15}$
Visible	$4{,}000{-}7{,}000 \times 10^{-8}$	$4.3 \times 10^{14}{-}7.5 \times 10^{14}$
Near infrared	$0.7{-}20 \times 10^{-4}$	$1.5 \times 10^{13}{-}4.3 \times 10^{14}$
Far infrared	$20{-}100 \times 10^{-4}$	$3.0 \times 10^{12}{-}1.5 \times 10^{13}$
Radio	> 0.01	$< 3.0 \times 10^{12}$
(Radar)	$(2{-}20)$	$(1.5{-}15 \times 10^{9})$
(FM radio)	$(250{-}350)$	$(85{-}110 \times 10^{6})$
(AM radio)	$(18{,}000{-}38{,}000)$	$(800{-}1{,}600 \times 10^{3})$

netic spectrum is called a spectra. When energy is dispersed, such as by a prism, its component parts can be seen as they arrange themselves in the order of their wavelengths. Spectra is actually the plural form of the Latin word spectrum. You will see the singular and plural forms of this word commonly incorrectly used to refer to one and sometimes more than one component of the electromagnetic spectrum.

Classifying a star by its spectral type relies on the fact stars produce a tremendous amount of energy over all wavelengths and frequencies of the electromagnetic spectrum. As visual observers of the sky, we are mainly interested in the spectrum's narrow visible portion, which consists of the colors red, orange, yellow, green, blue, indigo, and violet. To remember the colors in order, use the mnemonic: Roy G. Biv.

The ranges for electromagnetic wavelengths of gamma rays, X-rays, ultraviolet light, visible light, infrared light, and radio, are listed in Table 1.1 and shown in Figure 1.2. As energy moves, it travels in alternating electric and magnetic waves. The distance between two successive crests of the waves is called the wavelength. The frequency at which the wave crests occur is measured in hertz (Hz), where 1 cycle per second = 1 Hz.

Sir Isaac Newton (1642–1727) experimented with sunlight passing through a prism. He discovered that as the light beam passes through the glass the speed of each color is reduced causing the beam to bend (refract) into the component parts we see as a rainbow. By studying the refracted light, scientists are able to determine the chemical composition and temperature of its source.

In 1802, the British chemist and physicist, William Hyde Wollaston (1766–1828) noticed dark lines in the spectrum of the Sun. He thought these were simply the boundaries between the colors. The German optician and physicist Joseph von Fraunhofer (1787–1826) found 576 of these dark lines and established a map of them, in 1818. He labeled the prominent lines A through G. He also found that their relative placement on the solar spectrum is constant.

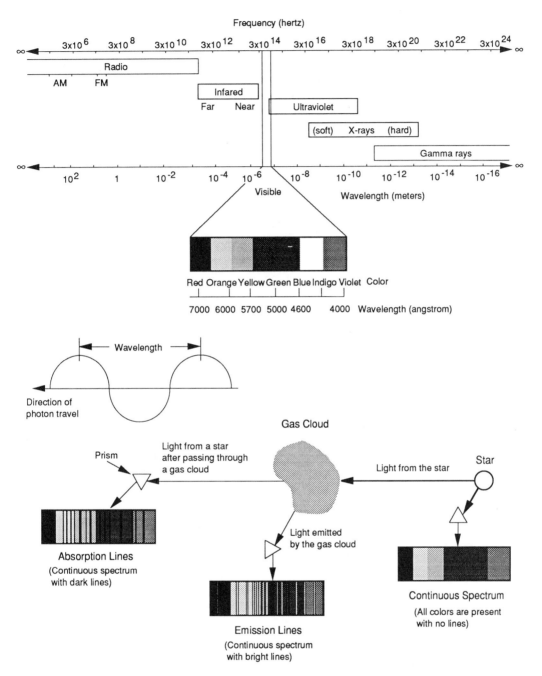

Figure 1.2 The electromagnetic spectrum.

The German chemist Robert Wilhelm Bunsen (1811–99) and physicist Gustav Robert Kirchhoff (1824–87) discovered that light passing through colored gases or the light emitted by a flame left different dark or bright lines on a spectrum.

The distances between individual spectra are measured in increments called angstroms (Å). This unit of length is equal to one ten-billionth of a meter (10^{-10} m) (100,000,000 Å = 1 cm). The angstrom was named for Swedish physicist and founder of spec-

Photo 1.1   An historical spectrogram of Vega (α Lyrae). This spectrogram was taken on August 31, 1896, using the 13-inch Boyden Refractor, located at the Boyden Station of the Harvard College Observatory in Arequipa, Peru. This is one of four spectra used by Anne J. Cannon to classify the spectrum of this star, under the Henry Draper Memorial. The Balmer lines of hydrogen are labeled, as is the K line of calcium.

troscopy, Anders Jonas Ångstrom (1814–74). He developed his system of identifying the signature of the elements in a beam of light and published his findings beginning in 1853. In 1863, he published his widely used map of the normal solar spectrum.

A *spectroscope* is a device attached to a telescope, or a specially designed telescope, used to spread out a beam of light into its spectrum so it can be recorded on a spectrogram (a photograph of the spectra) like those shown in Photo 1.1. The first spectrogram of a star (Vega, 3 Alpha Lyrae) was made, in 1872, by American amateur astronomer Henry Draper (1837–82). Two basic types of spectroscopes can be used: the transmission (prism) or reflecting (grating) spectroscope. The transmission spectroscope uses a glass prism to refract and separate the starlight into its spectrum. A reflecting spectroscope uses a piece of very flat glass with thousands of grooves per inch to reflect and break up the light beam into its spectrum.

Surface temperature

Three types of visual spectrum are used by astronomers to determine the makeup of a star – continuous, absorption, and emission. Each of these spectrum types reveals something different about the makeup and temperature of the object producing the radiation energy. A hot dense gas, like a star, producing radiation covering all of the visual spectrum would produce a continuous spectrum. On a spectrogram of a continuous spectrum, all of the colors are present with no gaps.

If a particular gas obscures a portion of the light coming from a more distant object, black absorption bands appear on a spectrogram. These dark bands or lines are caused by the absorption of radiation in electrons in the intervening gas. This happens only when the energy of the incoming radiation precisely equals the difference between the electron's ground (undisturbed) state and an excited state in the gas. The absorption spectra reveal the presence of the atom that absorbed the energy.

High temperatures can cause the electrons in atoms of a gas to jump from one excited state to another and then to give off radiation as they return to a lower state. This radiation appears on a spectrogram as colored emission lines. Each line is from a distinct transition of an electron between energy levels in a particular atom. Variations in the temperature of a gas cause it to emit different emission spectrum lines. A lower case "e" is added to a star's spectral classification to indicate the presence of emission lines in the spectra.

A system of classifying stars based on their spectra was developed at the Harvard College Observatory under the direction of Edward Charles Pickering (1846–1919), and published in the *Henry Draper Catalogue* so named in memory of his friend. The catalogue was published in nine volumes of the *Annals of the Harvard College Observatory* from 1890 to 1924. Most of the painstaking classification work on the stars was accomplished by a group of women "computers" nicknamed "Pickering's Harem." These women looked at thousands of individual spectrograms and classified each star on the plates. Scottish-born Williamina Paton Fleming (1857–1911) was the first of Pickering's female assistants and led the team of women who devised the original system that classified stars based upon their chemical composition and placed them in spectral classes labeled A through N.

Annie Jump Cannon (1863–1941) joined the Harvard Observatory staff in 1896. She spent about four years (1911–14) classifying 250,000 stars by studying their spectra. She discovered the smooth transition of strong absorption lines from one type of star to the next and rearranged the original alphabetical ordering into the O, B, A, F, G, K, M arrangement, as listed in Table A.1 in Appendix A. Cannon also established the ten subclasses for each type. An easy way to remember the order of classes is the famous mnemonic device attributed to American astronomer Henry Norris Russell (1877–1957):

> *Oh Be A Fine Girl* (Guy) *Kiss Me.*

The Harvard system has been modified with the addition of three minor classes N, R, and S and a subclass called Wolf–Rayet (W) stars. The N, R, and S classes are rare and are noted by their deep red tint. A few years ago, the N and R classes were replaced by a single new class. This new C class of stars shows intense bands of carbon molecules and the stars are cool at 2,500 kelvin (K). The W subclass consists of extremely hot white stars with expanding gas shells. They fit within the O class.

Though at the time the *Henry Draper Catalogue* was published its compilers did not know their system was actually based upon the differences in surface temperatures of the stars, this factor is now considered a part of the system. The presence or absence of key spectral lines determines what spectral type the star is.

The early sky watchers assumed everything in the sky, except for the wandering stars (planets) and the Sun and Moon, was fixed in place. Austrian physicist Christian Doppler (1803–53) noticed that as any object moves, the placement of its emission or absorption lines on a spectrogram shifts in relationship to the movement. In 1842, Doppler published his findings in his book *Concerning the Colored Light of Double Stars*. As the relative motion and distance between an object and observer increases, the spectral lines shift toward the red end of the spectrum, and as the distance between them decreases, they shift toward the blue end. Stellar motion was also confirmed by comparing the different placement of some stars on photographic plates of the same area of the sky taken many years apart.

Stellar magnitude

The earliest system of classifying stars uses the differences in their apparent brightness. The system was developed around 130 BC by Turkish astronomer Hipparchus of Rhodes (flourished between 146 and 126 B.C.). He separated the stars into six groups with the brightest being first magnitude and the faintest as sixth magnitude. A person with normal sight can generally see, on a night of good seeing at a dark site, stars down to about sixth magnitude.

The farther away you are from a glowing object, the dimmer it appears. A standard of absolute magnitude (also absolute brightness) was established and refers to how bright an object would be if viewed from the distance of 10 parsec (32.58 light-years).

The logarithmic magnitude scale was devised, in 1856, by British astronomer Norman Robert Pogson (1829–91). His scale has each magnitude being 2.512 times brighter or dimmer than the magnitude next to it. This logarithmic scale has negative numbers representing the brightest objects (-26.7 for the Sun) and increasing positive numbers to indicate the relative brightness of dimmer objects. The faintest objects have the highest positive numbers. The standard reference for the scale is the 0.03-magnitude star Vega. The apparent brightness of Vega is used to calibrate the magnitude scale on photoelectric equipment.

Using the new techniques of electronic imagery, scientists are able to study extremely faint objects. As I am writing this book, the faintest celestial objects detected are two galaxies of magnitude +29. This magnitude has been compared to being how bright a lit cigar on the Moon would appear as seen from Earth.

Photographic classification

The majority of stars are too faint to be studied visually. With the advent of astrophotography, the system of classifying stars based upon how bright they appear on a photographic plate was devised. A drawback to photographic magnitudes is that most photographic emulsions are more sensitive to the blue – violet portion of the visual spectrum, whereas the human eye is more sensitive to the yellow – green area. Photographic magnitudes, therefore, are somewhat misleading as to the actual magnitude of a star.

To overcome the problems associated with photographic magnitudes, the technique of *photovisual* magnitude is used. By using emulsions on plates or film that are sensitive to green and yellow along with a yellow filter on the camera, the photovisual magnitude of a stellar object can be determined. This system gives a magnitude rating that is close to how bright an object appears to your eyes.

In 1953, American astronomers William Wilson Morgan (1906–94) and Harold Lester Johnson (1921–) established the UBV system of stellar magnitude using photoelectric photometry in three broad wavelength bands: (U) ultraviolet, (B) blue, and (V) visual spectral regions. By comparing the starlight in these three bands and making corrections for interstellar dust absorption (which reddens a star's B-V magnitude and is called color excess), temperature (which affects a star's B and V magnitudes to a different degree than color excess), and spectral type, the UBV system gives a more accurate magnitude for a star than any other technique. It can also be used on stars which are too faint for spectroscopic study. The V magnitude scale is indexed to the visual magnitudes established, in 1859–62, with the publication of the *Bonner Durchmusterung* ("Bonn Survey") by German astronomer Friedrich Wilhelm August Argelander (1799–1875). This catalogue and atlas contained all 324,198 of the northern sky stars, down to ninth magnitude. The U-B and B-V indices have their zero-magnitude point set as that for spectral-type A, subclass 0, stars of normal luminosity.

Bolometric magnitude is used to determine the brightness of a star based on the measurement of its electromagnetic radiation at all wavelengths. This magnitude is measured with a *spectrum bolometer* invented, in 1878, by American astronomer and physicist Samuel Pierpoint Langley (1834–1906).

Luminosity

The last major key to understanding the composition of stars is knowing how much energy a star is emitting. The measure of this energy output is the star's absolute luminosity. The more mass a star contains, the hotter its surface temperature and the brighter it shines. The largest normal stars are the hottest and brightest: the blue supergiants. The MK (also known as MKK) luminosity classification system was introduced at Yerkes Observatory, in 1943, by astronomers W. W. Morgan and Philip Childs Keenan (1908–), and their darkroom assistant Edith Kellman (1911–), in *An Atlas of Stellar Spectra With an Outline of Spectral Classification*. The MK system, as listed in Table A.2 in Appendix A, uses Roman numerals that are appended to the Henry Draper spectral classification designation. For each luminosity class and subclass, a standard star for that class has been selected. All stars in the same class are compared to the standard star.

For stars that show strong emission lines for some elements, the symbol for that element is appended to the star's spectrum classification. For example, the spectrum for the normal giant star 77 Theta[1] Tauri is K0 III: Fe-1, which shows this is a spectral class K, subclass 0, normal giant star that is rich in ionized iron (Fe-1).

Stellar evolution

As astronomers developed the techniques and equipment that allows for the classification of celestial objects it was possible to begin to understand celestial evolution. Working independently, then combining forces, Danish astronomer Ejnar Hertzsprung (1873–1967) and Henry Russell developed their absolute luminosity/absolute magnitude/spectral type/surface temperature diagram, in 1914. Their Hertzsprung–Russell (H–R) diagram, as shown in Figure 1.3, graphically plots stars according to these four components. Luminosity, mass, temperature, visual color, and age are interdependent and affect each other. As stars age, these stellar components gradually change, thereby causing a shift in a star's H–R diagram position. The diagram therefore shows the evolution of a star from its birth as a protostar to its lifetime as a normal star fusing hydrogen to helium, to its final sputters of life as an old-age red dwarf or perhaps a more cataclysmal end as an exploding star.

About 90 percent of the stars near us have luminosity and spectral classes that place them on the downward sloping line called the *main-sequence* on the H–R diagram. The line shows a sequence of stars arranged in the order of decreasing mass, from the most massive stars in the upper left down to the least massive stars at the lower right end of the line. All of these stars are in the normal process of converting their hydrogen fuel core to helium. Above the line are the giants and below are the dwarfs. The giants and dwarfs have abnormal relationships between their brightness and temperature.

Stars, like our Sun, gradually move upward and to the right on the H–R diagram. Depending on its mass and the rate its nuclear fuel is consumed, a star can move away from the main-sequence becoming a bright giant or a faint dwarf.

Stars can have either a short life or live for hundreds of billions (10^9) of Earth years. The longevity of a star and its path on the H–R diagram is initially based on the amount of material in the star as its nuclear furnace begins to create helium. The Sun is near the middle of the main-sequence. The faintest known red stars have masses about 1/25 that of the Sun. At the opposite end of the mass scale, the bright blue stars can be as massive as 100 times that of the Sun.

Whether a star becomes a giant or a dwarf depends on its ability to maintain an equilibrium between the forces of gravity and pressure. If a star's internal radiation pressure is greater than its gravity, the star expands into a giant. Too much gravity and the star shrinks into a dwarf. A rapid imbalance in the gravity/pressure relationship causes the star to pulsate as it tries to maintain stability. Failure to maintain stability can lead to the explosive destruction of the star.

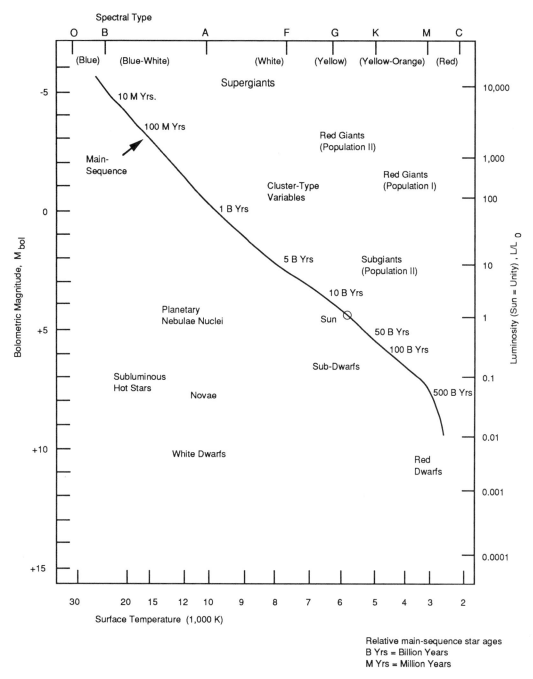

Relative main-sequence star ages
B Yrs = Billion Years
M Yrs = Million Years

Figure 1.3
Hertzsprung–Russell
diagram.

British astronomer Sir Arthur Stanley Eddington (1882–1944) developed the equation of stellar equilibrium in 1914. He also predicted the source of a star's energy is the fusion of atoms. Eddington discovered that hydrogen is the predominant material in the universe. Dispersed throughout the universe are great clouds of hydrogen atoms and dust particles. The gravitational forces within these clouds cause atoms to slowly clump together to create a protostar.

Stellar death

A star can also end its life in the explosive flash of a supernova. The remnant of a supernova can become one of several types of stars, or if the explosion was complete the original star can cease to exist. The material thrown into space by the explosion can gradually combine with other interstellar material to form a new star, rich in heavy elements (a Population II star). The shell of material thrown into space by a nova can glow from its own radiation along with that of the remaining central star and become a glowing planetary nebula.

The last stages a star can pass through depends on what happened earlier in its life. If the core of a star remains after a supernova explosion, it can form a superdense neutron star where a cubic inch of material weighs several tons. Initially spinning very rapidly, neutron stars can emit flashes of high-energy radio wave radiation and are known as *pulsars*. Current theory holds that pulsars are surrounded by a massive magnetic field having a greater quantity of energy escaping along the axis of the field. As this axis sweeps past our line of sight to the star, we perceive a flash or pulse of energy.

Though known only in theory, a supernova remnant can become a black hole. This type of object is so dense and has such an extremely high gravitational force it prevents all material near it, including light, from escaping its pull.

Types of objects: from gas clouds to superclusters of galaxies

Nebulae

Space is not an empty vacuum as many people believe. Uncountable hydrogen atoms and dust particles are scattered throughout the universe. The gravitational pull from each object slowly pulls these atoms into great clouds of gas and dust. These clouds are the birthplaces for stars. As clumps of dust and gas grow into spheres, the gravitational pressure increases until these dark bodies begin to radiate at very low temperatures. These unseen objects are known theoretically as *brown dwarfs*. Brown dwarfs have accumulated too much mass and are too dense to be a planet, but are still too small to ignite into stars. These objects are believed to be the link between a gas cloud and its protostars.

As a protostar begins to shine, its energy stimulates and causes the cloud (nebula) it formed from to emit its own diffuse light. We can observe many of these glowing clouds, which are called *emission nebulae*. If stars are near a cloud, but not a part of it, the star's light can be reflected off the cloud. This type of nebula is a *reflection nebula*. Several examples are known where the nebula is both emitting light from internal stars and reflecting light from neighboring stars. Usually, an emission nebula contains hot blue stars, indicating its youthful age. They also can contain cool red stars, but these low luminosity stars are difficult to see within the cloud.

Photo 1.2
Emission/reflection
nebulae NGC
1973/75/77 in Orion.

Related to these bright nebulae are the dark nebulae. Dark nebulae are clouds of hydrogen gas and dust which are neither emitting nor reflecting light. These dark clouds lack internal stars that stimulate the hydrogen atoms of the cloud to emit their own radiation in the visible portion of the electromagnetic spectrum, like an emission nebula. Though they appear to be surrounded by millions of stars, these dark patches are actually quite distant from any stars. This means they cannot reflect light from nearby stars, as in a reflection nebula. Thin spots in the clouds allow some light to pass through them, but other places in the clouds are so thick with interstellar matter they block all of the background light and look like ink blots on the Milky Way. We see the dark nebulae simply because they obscure the light coming from objects farther from us.

Variable stars

We know that many stars do not remain constant in brightness. We can detect these changes and have classified these variable stars in six main classes of cataclysmic, eclipsing binary, eruptive, pulsating, rotating, and X-ray, as listed in Table A.3 in Appendix A.

Novae and dwarf novae make up the cataclysmic class as these stars may be losing matter to a nearby larger star. They may also be undergoing violent outbursts of energy.

The apparent magnitude of an eclipsing binary varies when the orbit of one of the stars takes it in front of or behind the other star.

The combined brightness of the pair increases when we see both stars and dims when one of the stars is hidden.

Eruptive variables are caused by the instability in a star's outer atmosphere. This instability causes the brightness of the stars to change, generally in short periods.

A star is able to pulsate as a result of instability in its pressure/gravity relationship. The heat and pressure of radiation causes a star to expand while gravity works to contract a star's size. When these forces are in balance, a star is stable as to its size and luminosity. When the forces are out of balance, the star undergoes luminosity changes depending on which force is dominant: it brightens as it expands and dims during contraction. In a pulsating variable, these two forces exchange their dominant positions back and forth on a regular and steady basis.

Not all stars are round nor are they uniform in brightness over their entire surface. Like the Sun, many stars are known to have dark spots. On some stars, these spots are massive enough to dim the star's overall brightness, so as the star rotates its brightness varies.

X-ray variables occur in a close binary system that consists of a hot compact star (like a neutron star, a white dwarf, or a black hole) and a second star which is cooler than its companion. Matter can be pulled from the cooler star, and, as it flows into the hotter star, high-energy X-rays are produced by this exchange of matter. These X-rays are reflected off the cooler star as high-temperature optical radiation, which causes a detectable variation in the brightness of the entire system.

Double stars

About half of the stars we can observe are actually two or more stars so close together they appear to be just one. Double stars can be actual companions rotating around a common mass point forming an *actual* binary or multiple system. Visual doubles appear close, but in reality they are very far apart and totally unrelated (an *apparent* binary system).

Many companion stars are much fainter than their associated brighter primary. These unseen stars are revealed by their spectra or by the wobble their gravitational pull induces in the proper motion of the primary star. The orbital movement of the secondary (also known as a *comes*) can be detected by movement of its lines on the spectrogram of the primary star. As the star moves, its lines merge and separate at regular intervals. This type of binary system is known as a spectroscopic binary system.

When we look at double stars we are concerned with how far apart the primary star and its secondary star(s) appear to be. This angular distance between the stars is measured in arcseconds between the centers of their Airy disks. The Airy disk is the smallest image that a telescope can make of the point of the light from a star. Doubles are classified in terms of their separation as listed in Table A.4 in Appendix A. To know where to look for the comes in most cases requires knowledge of its *position angle*. The position angle is the

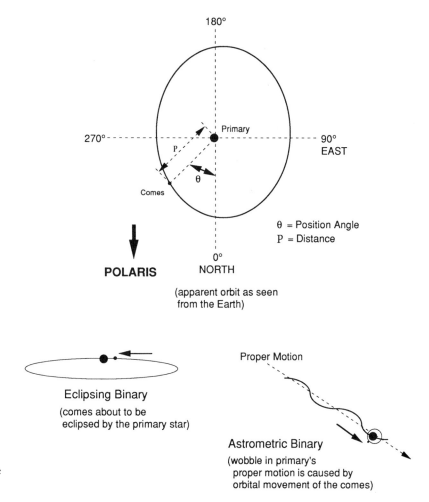

Figure 1.4 Double stars.

angular distance of the star along its orbit, with 0° being at the north pole of the primary as we see it. The distance is measured counterclockwise from the north pole, as shown in Figure 1.4.

Planetary nebulae

A planetary nebula is not a planet, but the remains of a star which has either expelled large amounts of material or has exploded in a violent outburst such as a supernova. The star is enveloped in a shroud of expanding glowing gas which is speeding away from the star. This envelope gives the star either a halo appearance or leaves an irregular patch of glowing gas where the star was. Planetary nebulae are classified by a system introduced in 1934 by Russian astronomer Boris Aleksandrovich Vorontsov-Vel'iaminov (1904–) and is listed in Table A.5 in Appendix A.

Asterisms

Scattered throughout the billions of stars in the sky, we can see pat-

15

Photo 1.3 Planetary nebula IC 4406 (PK 319+15.1) in Lupus.

terns and geometrical shapes created by groups of stars called asterisms. These asterisms can be groupings like the Big Dipper, the bright stars that make up Orion's belt and sword, or the Summer Triangle. In many cases, the stars making up the shape are far from each other in space. Their appearance as a group is the result of their accidental line of sight from Earth. We will also use asterisms consisting of a few stars in the shapes of triangles, rectangles, trapezoids, arcs, and other geometric shapes to guide us to our target objects.

Clusters

When several stars are bound together by their mutual gravity and exhibit as a group a similar proper motion they are referred to as a cluster. A large grouping of these stars is also known as a moving group. There are two main types of clusters: open and globular. Any cluster located in our Galaxy is also known as a galactic or open cluster.

The stars in open clusters are somewhat scattered, yet all members of the cluster have approximately the same actual motion. The simple system for classifying open clusters was devised, in 1930, by Swiss-American astronomer Robert Julius Trumpler (1886–1956). The system bases open clusters according to their size, magnitude, and richness of stars. The system, as shown in Table A.6 in Appendix A, consists of three components which are combined to classify the cluster.

Globular clusters look like dense balls of stars with a few stragglers nearby. Globular clusters are known to exist in other galaxies and are more numerous than open clusters in these other galaxies. Globular clusters are classified according to how concentrated the stars are

Photo 1.4 Open cluster NGC 1960 (M36) in Auriga.

toward the center of the cluster. The system was devised in 1927 by Harvard Observatory astronomers Harlow Shapley (1885–1972) and Helen Battles Sawyer Hogg (1905–93). Their system values range from 1 to 12 with 1 representing the most compact and 12 being the least compact clusters.

Galaxies, clusters of galaxies, and superclusters

The largest structures in the universe are the galaxies, clusters of galaxies, and superclusters. A galaxy is a massive assemblage of stars, nebulae, clusters, and interstellar matter. Galaxies are classified according to their appearance as we view them. The classification system used today was first devised in 1925 by American astronomer Edwin Powell Hubble (1889–1953). His system has been expanded by other astronomers, in particular French-American astronomer Gérard Henri de Vaucouleurs (1918–95). Hubble's system is shown in Figure 1.5, and his system, as expanded by De Vaucouleurs, is listed in Table A.7 in Appendix A. Whereas Hubble kept his system simple, De Vaucouleurs tried to classify each of the almost infinite varieties and mixtures of galaxies. He also expanded the system to make room for varieties recognized after Hubble's death. Galaxies can also be members of a group or cluster consisting of a few galaxies or several thousands of gravitationally bound galaxies. Our Milky Way Galaxy and the Magellanic Clouds are members of the Local Group consisting of 20 galaxies that are within about 1 megaparsec (three million

17

Photo 1.5 Globular cluster NGC 7089 (M2) in Aquarius.

light-years) from our Galaxy. We also know of several superclusters of galaxies. As a member of the Local Group, we are also members of the Local Supercluster of Galaxies, which includes many galaxies in addition to the local one, clouds of galaxies, and the Virgo Cluster.

About two percent of galaxies fall into a class of active galaxies known as Seyfert galaxies. These galaxies have very bright nuclei caused by light emissions induced by charged particles being accelerated in strong magnetic fields within the galaxy's core. This synchrotron radiation is polarized and requires very large total energies from the center of the galaxy to operate. This energy drives hot gases within the nuclei at speeds of several thousand kilometers per second (km/s). These galaxies were first studied and classified, beginning in 1943, by American astronomer Carl Keenan Seyfert (1911–60).

Photo 1.6 Spiral galaxy NGC 4258 (M106) in Canes Venatici.

Star coordinate systems

When ancient man looked up at the night sky, he imagined the stars were all at the same distance above the Earth and were attached to a rotating crystal celestial sphere. With the advent of map making, the stars as well as terrestrial features were plotted on stone walls, globes,

19

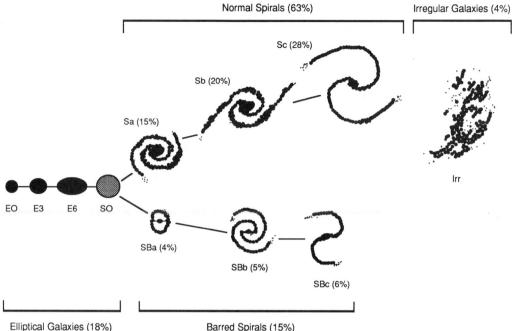

Normal Spirals (63%)

Irregular Galaxies (4%)

Sc (28%)

Sb (20%)

Sa (15%)

EO   E3   E6   SO

Irr

SBa (4%)

SBb (5%)

SBc (6%)

Elliptical Galaxies (18%)

Barred Spirals (15%)

Figure 1.5 Hubble's "Tuning Fork" classification of galaxies.

parchment, and paper. Over the aeons, many systems of plotting these stars were devised then discarded as improved systems were created.

Today, several coordinate systems abound in astronomy, such as the *galactic coordinate system* (based upon the axial poles and equator of our Galaxy), the *horizon system of coordinates* (based on your location on Earth and your horizon), and the *equatorial coordinate system*. We will be using the equatorial coordinate system, as shown in Figure 1.6, so being able to locate an object on a star chart and then being able to find it in the sky requires a fundamental understanding of this system.

The equatorial coordinate system uses the plane of the Earth's equator and axial poles, as if projected out into space, as its main reference points. On the sky, the Earth's equator becomes the *celestial equator*, the north pole is the *north celestial pole*, and the south pole is the *south celestial pole*. The positions along the celestial equator are marked in 24 segments known as *right ascension* (R.A.) These 24 segments are divided into hours, minutes, and seconds of arc. Positions above (north) and below (south) of the celestial equator are measured in degrees, minutes, and seconds of arc, known as *declination* (dec).

Meridian right ascension lines run across the sky from the celestial north pole to the celestial south pole. These lines are similar to the lines of longitude on a terrestrial map.

The starting point for the hours of right ascension is known as the *prime meridian of the sky*. This celestial meridian runs from the north celestial pole, bisects the ecliptic and celestial equator at 0 hours of right ascension and 0 degrees of declination, then ends at the south celestial pole.

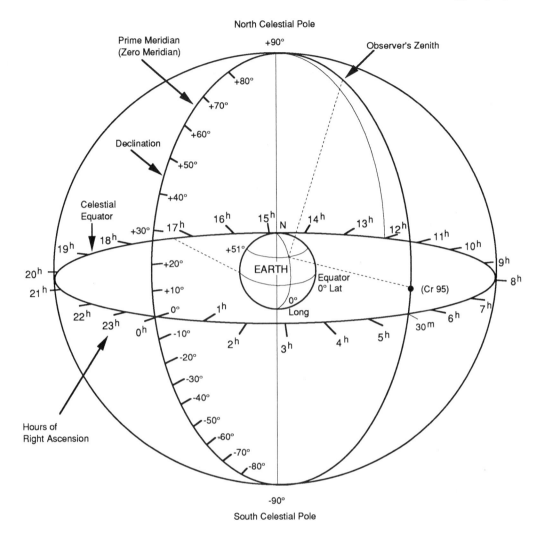

North Celestial Pole

Observer at Greenwich, UK (51° latitude, 0° longitude)
looking at open cluster Cr 95 at 6 hours 30 minutes
right ascension, +10° declination

Figure 1.6
Equatorial
coordinate system.

Star positions are referenced by the star's distance above or below this celestial equator (declination), its position from the prime meridian (right ascension), and the *epoch* for the catalogue or map. Once you know a star's coordinates, the star can be located either on a map or in the night sky. In chapter 2, we will learn how to use the equatorial coordinate system to locate our target objects.

Two other key factors in knowing how to locate an object on your star charts involves dealing with the problems of *precession* and the object's own motion in space. Precession, which is the product of the slow wobble of the Earth on its axis, as shown in Figure 1.7, causes the coordinates of the stars to slowly shift in relation to your charts. The Earth takes about 25,725 years to complete one revolution. Like everything else in the universe, stars are speeding in all directions.

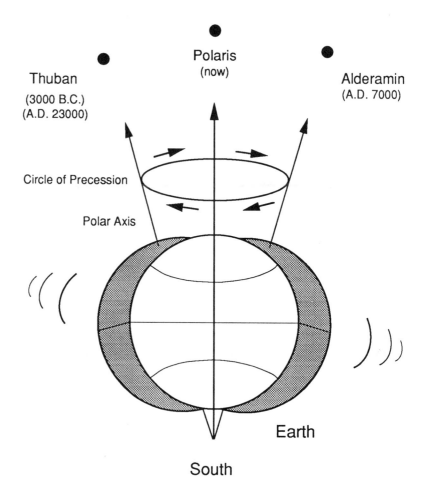

North Polar Stars

Figure 1.7
Precession.

Observations conducted over many years (centuries for some bright stars) show what direction a star is moving in. Proper motion is really just the direction in which the star *appears* to be moving, as shown in Figure 1.8. True space motion is the direction the star is *actually* moving in through three-dimensional space. The radial velocity of an object tells us how fast an object is approaching or receding from us. By studying the Doppler shift in the object's spectra, we can determine this speed and direction.

To keep things straight, sky maps are keyed to the date on which the coordinates are correct as plotted. The date becomes the chart's epoch or equinox. Generally, maps are plotted as of January 1 in 50-year intervals (1900, 1950, 2000). Most of the charts and catalogues you will come across are for either standard epoch 1950.0 or 2000.0. Personal computer software is available that will allow you to create your own sky charts and catalogues for any epoch; past, present, or future.

As an example of precession, along with proper motion, Arcturus

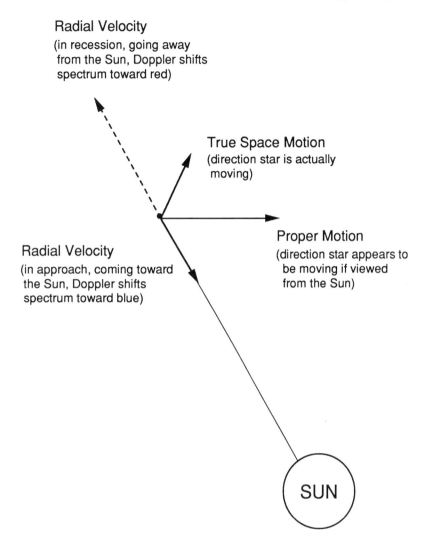

Radial Velocity
(in recession, going away
from the Sun, Doppler shifts
spectrum toward red)

True Space Motion
(direction star is actually
moving)

Radial Velocity
(in approach, coming toward
the Sun, Doppler shifts
spectrum toward blue)

Proper Motion
(direction star appears to
be moving if viewed
from the Sun)

SUN

Figure 1.8 Proper motion and radial velocity.

(16 Alpha Boötis) has shifted from its epoch 1950.0 right ascension position of 14 hours, 13 minutes, 40 seconds, to be at 14 hours, 15 minutes, and 39 seconds for epoch 2000.0. Its declination has shifted from +19 degrees, 27 minutes for epoch 1950.0 to +19 degrees, 10 minutes for epoch 2000.0. The change in minutes in both coordinates means that Arcturus has moved celestially southeast during this 50-year period. You only have to worry about precession and proper motion when you are looking for an object listed by coordinates of one epoch and you are using a star chart or catalogue for another epoch. If you are using epochs that are close together, there is little problem, but if the charts or catalogues are 100 years or more apart in age, the stars will have made significant shifts from one epoch to the other. The epoch 2000.0 equatorial coordinates will be given after a star's identification, such as [14:15.39;+19d10'] for Arcturus.

One of the most often asked questions is "How far away is an object?" Celestial distances are exceedingly vast and hard to comprehend, since most of us think of distances as measured in the astro-

23

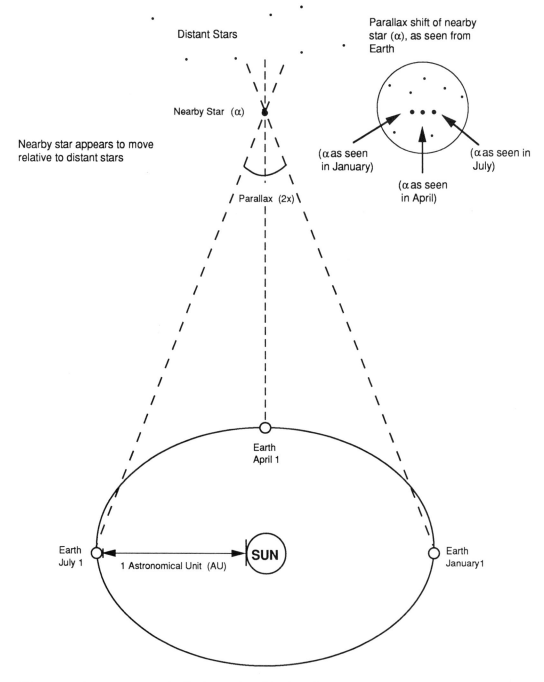

Distant Stars

Parallax shift of nearby
star (α), as seen from
Earth

Nearby Star (α)

Nearby star appears to move
relative to distant stars

(α as seen
in January)

(α as seen in
July)

(α as seen
in April)

Parallax (2x)

Earth
April 1

Earth
July 1

1 Astronomical Unit (AU)

SUN

Earth
January 1

Figure 1.9 Stellar parallax.

nomically tiny units of inches, feet, miles, or centimeters, meters, and kilometers. The way most amateurs think of space distances is in units based on how far light will travel in one Earth year (at the speed of light of 186,000 miles per second in a vacuum). One light-year equals about 5,878,797,613,000 miles (9,460,536,000,000 km).

The method preferred by professionals, which is gaining acceptance with advanced amateurs, is the *parsec* (pc). A parsec is the dis-

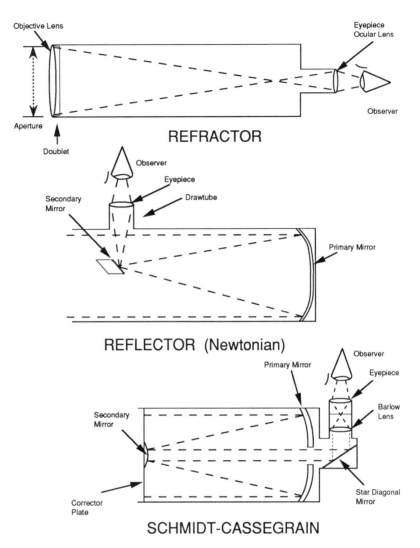

Figure 1.10
Telescope designs.

tance at which a star would have an annual apparent shift, or *parallax*, of one second of arc and equals 3.261631 light-years or about 19,177,611,000,000 miles (30,856,777,000,000 km).

As shown in Figure 1.9, stellar parallax can be used to determine the distance to a celestial body. The two points used are opposite positions along the Earth's orbit. Knowing the distance between these points (2 astronomical units) the angle of the shift (parallax) can be determined, then the distance to the star can be calculated.

Instruments

Telescopes

Until the Italian scientist Galileo Galilei aimed his crude "spyglass" skyward in 1609, all stargazing had been done naked-eye. Over the years, many telescope designs have been created, but they all are only

Photo 1.7 A refractor on an altazimuth mount on a tripod.

modifications of the two basic types: refractor and reflector. The word "telescope" was coined in 1648 from the Greek word *teleskopos* meaning "farseeing."

A refractor, as shown in Figure 1.10 and Photo 1.7, passes light through a glass objective lens or combination of lenses at the front opening (aperture) of the telescope tube. The light then passes through the focusing lens called an eyepiece (also called an ocular lens). Refractors excel at lunar, solar, and planetary views. Small refractors (under 4-inch aperture) are easy to transport and will give you decent views of nebulae. The larger the aperture, the longer the refractor has to be to work properly. Refractors with apertures larger than 100mm (four inches) require very long tubes which in turn makes them very expensive and bulky. The bending of the light through the objective also causes the various light wavelengths to come into focus at different distances from the objective. This can be overcome to a certain extent by the use of different configurations of lenses, but not entirely. Except for the most expensive models, small refractors generally suffer from some amount of *achromatism*, which results in colored haloes and cross-like patterns around bright stars. For deep-sky work, you would need a large high-quality refractor, but their size and bulk make them difficult to transport to a dark site. Focusing is accomplished by moving the eyepiece closer to or farther away from the objective lens. Refractors require very little mainte-nance as the lenses are the only optical components exposed to dust and moisture. Good refractors with an aperture over four inches are

Photo 1.8 A reflector on a german equatorial mount on a tripod.

very expensive. You can get more telescope power for your money by purchasing a telescope with mirrors.

A reflector, as shown in Figure 1.10 and Photo 1.8, bounces and focuses the light off a curved mirror and passes the light through an eyepiece. Reflectors are best at deep-sky work, are generally easy to transport, and are the cheapest per inch of aperture to manufacture. Reflectors are better for deep-sky observing than refractors, because they give sharper star images and are more compact than a similar power refractor. The size of a reflector is determined by the diameter of its primary mirror. The most popular reflector style is based on the design created by Scottish mathematician James Gregory (1638–75) in 1663, then improved five years later by Sir Isaac Newton. The Newtonian design employs a secondary mirror, angled at 45 degrees to the primary, mounted in the light path to direct the image to the eyepiece. Focusing is accomplished by moving the eyepiece drawtube in and out of the telescope until the sharpest image is obtained. To align (*collimate*) the optics, the primary mirror is adjusted to be in alignment with the secondary mirror. One problem with reflectors is that their mirrors are exposed to dust and moisture and are hard to keep clean. Their reflectiveness slowly degrades as the mirrors tarnish.

A popular telescope design is the Schmidt–Cassegrain, as shown in Figure 1.10 and Photo 1.9. In this *catadioptric* type of instrument, the light passes through a glass corrector plate, bounces off a primary and

Photo 1.9 A
Schmidt–Cassegrain
on an equatorial fork
mount on a Tripod.

secondary convex mirror mounted on the corrector plate, passes
through a hole in the primary, then through the eyepiece. This design
allows for a long focal length telescope in a compact package that
is easy to transport and set up. To collimate the optics in this type
of telescope, the position of the secondary is adjusted by moving
adjustment screws on the secondary mirror mounting cell. These
telescopes are good overall astronomical units. They give excellent
views of solar system and deep-sky objects. The closed design also
protects the mirrors from dust and tarnishing moisture. The design
for this type of reflector was conceived, in 1672, by the French doctor
Jacques Cassegrain (also known as Guillaume Cassegrain and N.
Cassegrain) (1652–1712). Little is known of him and it is believed
that he never constructed a telescope of this design. James Short
(1710–68) was the first person known to build a Cassegrain-type
telescope. Estonian astronomer Bernhard Voldemar Schmidt
(1879–1935) developed the concept of passing the light beam
through a curved corrector plate as it entered a reflecting camera.

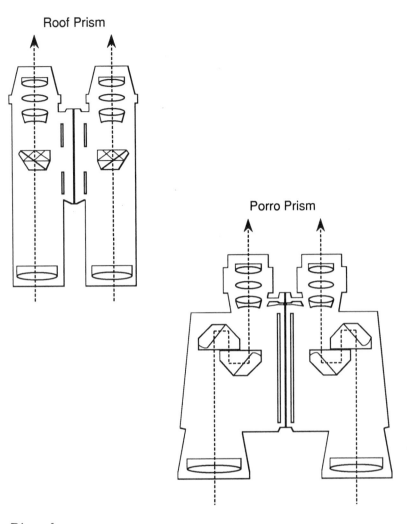

Figure 1.11
Binocular designs.

Binoculars

Binoculars are essentially a matched pair of low-power refractors which are mounted together. They are used mainly for terrestrial work, but give great views of star fields, the Moon, and bright celestial objects such as clusters. I find it very refreshing to just sit back with binoculars and browse through star fields, looking for nothing in particular. Binoculars come in two basic configurations: porro and pent roof, as shown in Figure 1.11. Porro prisms are the most common type and use two prisms mounted back to back to reflect the light four times as it passes from the objective to the eyepiece. Pent roof binoculars are usually more compact and cheaper than porro types and use two angled prisms mounted back to back to cause the light to be reflected off at least one aluminized surface. Pent roof binoculars are not as good for astronomical work as porro types.

To keep this book from becoming cumbersome or confusing, when I use the term telescope, I am also referring to the use of binoculars. Depending on the diameter of their aperture and power of magnification, binoculars work like twin finder scopes. Binoculars and finder scopes with the same aperture size and magnification power will give you the same field of view. Therefore directions given for using a finder scope are generally applicable to the use of binoculars. Finder scope, finder, and guide scope are interchangeable labels for the same piece of optical equipment.

Magnification and power

Two questions almost always asked at a star party are "How bright an object can you see?" and "How powerful is your scope?" All telescopes have a limit to how dim an object can be and still be seen. A telescope's limiting magnitude is the faintest level of magnitude that can theoretically be seen in it. Local seeing and transparency can limit the visible magnitude to objects brighter than the theoretical limiting magnitude for your scope. As shown in Figure 1.12, the theoretical limiting magnitude of a scope is a function of its aperture. Use this figure to determine if on a good "seeing" night an object will be detectable in your scope.

Magnification and power refer to a telescope's ability to increase in size the image of an object. The magnifying lens allows you to focus your eye closer to an object than you normally can, which in turn makes the object appear larger to you. The focal length of an eyepiece is given in millimeters. To determine the power of a particular eyepiece/telescope combination, divide the focal length of the telescope by the eyepiece focal length:

$$\text{magnification} = \frac{\text{telescope focal length (in millimeters)}}{\text{eyepiece focal length (in millimeters)}}$$

The resulting answer is given as the magnification power, such as 100x. As you increase magnification, you lose brightness and sharp focusing becomes more critical and harder to obtain.

All stellar images suffer some blurring due to atmospheric distortion, so you will not see pinpoint sharp images. A "good" telescope will keep the blur image to a minimum (about 1 arcsec) and give you crisp images on nights of excellent seeing. Another measure of a telescope's resolution capabilities is called the *Dawes limit*. Famous double star observer, British astronomer William Rutter Dawes (1799–1868) calculated the ability of a telescope to resolve close double stars depended on a factor of 115 multiplied by the aperture, in millimeters, of the scope. The theoretical Dawes limits for telescopes is shown in Figure 1.12.

The useful magnification of an optical instrument is limited by its ability to increase the size of an image while keeping that image sharp and bright. At too high a magnification, the resolving power of an eyepiece reaches a point where the image is too dim to see and almost impossible to focus. The minimum useful power is 3.5 times the aperture in millimeters. The maximum useful power is 60 times the aperture in millimeters. When you try to use too high a power eyepiece for your scope, you will discover it is very hard to focus and the

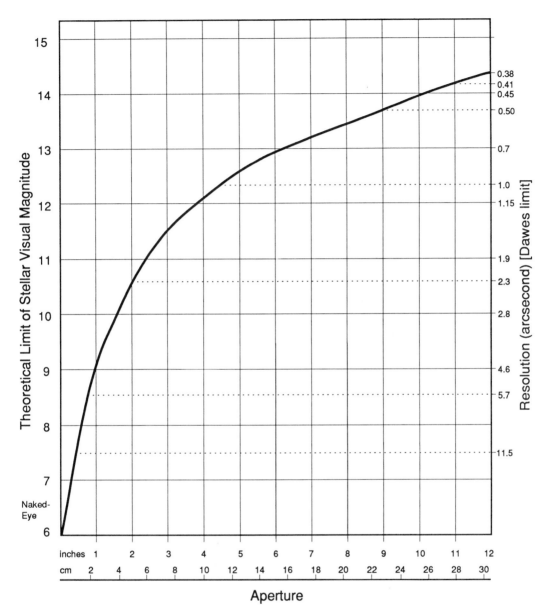

Figure 1.12
Limiting magnitude
and Dawes limit for
various scope sizes.

object will become very dim. Use the right power that gives you the best view.

Telescope mounts

When you go looking for a telescope, you will find two basic mountings available to choose from: altazimuth and equatorial. An altazimuth mount does not track the sky as the stars move. It is set parallel to the ground so you have to constantly adjust both your telescope's horizontal and vertical positions as the sky moves.

An equatorial-type mount is set so it rotates in parallel with the Earth's polar axis. The mount's horizontal axis is set perpendicular to

31

the polar axis and is also set equal to your latitude. This allows the telescope to follow the stars during the night. On a properly polar-aligned unit with a motor drive, only slight adjustments in right ascension and declination are required for the telescope to follow a single star during the night. The motor drive will rotate the telescope at the sidereal rate so it moves at the same speed as the stars across the sky. Some motor drives are adjustable so you can also track the Sun or Moon as these objects have different rates of celestial motion than the sidereal rate of the stars.

Refer to the bibliography for additional books you may want to consult for more information. The monthly astronomical publications carry telescope buyers' guides and astronomical club information.

2 How the sky works, determining your field of view, observing tips, and how to navigate in the night sky

We have next to speak of the stars, as they are called, of their composition, shape, and movements.

Aristotle (384–322 B.C.),
De caelo
("On the Heavens," Book II, Chapter 7)

How the sky works

In the first chapter, I covered the technical aspects of the components of the universe. In this chapter we will get our equipment together and begin to observe these objects.

One of the trickiest concepts for the beginning star-hopper is understanding celestial directions. Astronomers think of sky object locations in points of celestial north, south, east, and west: not up, down, right, or left. Once you get used to thinking in celestial directions, you will easily be able to find your way when going from your charts to the night sky.

On star charts, north is *always toward Polaris*, the North Star. Star-hopping is done without the aid of setting circles (if your telescope is so equipped with this direction finding accessory we will not be using them). The first time I mention an object, I will refer to its epoch 2000.0 equatorial coordinate system position of right ascension hours and declination degrees, but only to assist you in finding it on a star chart. This information will appear in brackets after the object's name or catalogue designations, such as Polaris [2:31;+89d15']. The 2:31 is Polaris's right ascension position of 2 hours and 31 minutes. The + indicates the star is north of the celestial equator at 89 degrees and 15 minutes of declination. A minus (–) sign is used for objects south of the celestial equator.

You can easily determine if an object is celestially east or west of another by comparing their right ascension positions on a star chart or catalogue listing. The hours of right ascension encircle the Earth at the celestial equator and increase from 0 up to 23 hours, 59 minutes going east, as shown in Figure 1.6. The celestial equator represents the Earth's equator as if it were projected up into the sky. Due to the Earth's daily rotation and orbital motion around the Sun, a star on

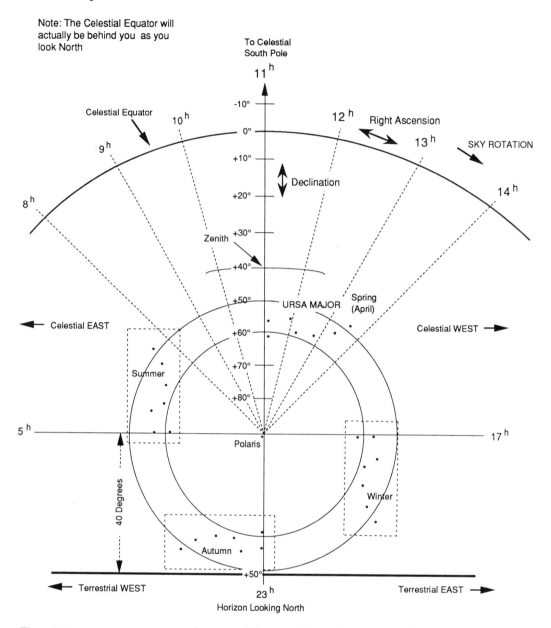

Note: The Celestial Equator will actually be behind you as you look North

To Celestial South Pole

11^h

-10°

Celestial Equator

10^h

0°

12^h Right Ascension

9^h

13^h

+10°

SKY ROTATION

8^h

+20°

Declination

14^h

+30°

Zenith

+40°

+50°   URSA MAJOR   Spring (April)

Celestial EAST

+60°

Celestial WEST

Summer

+70°

+80°

5^h

Polaris

17^h

40 Degrees

Winter

+50°

Terrestrial WEST   23^h Terrestrial EAST

Horizon Looking North

Figure 2.1 Equatorial coordinate system as viewed looking north. (From +40° latitude; in April; 9:00 p.m).

your southern meridian tonight at 8:00 p.m. will reach the meridian tomorrow night at 7:56, and four minutes earlier the next evening. One year later, the same star will return to the meridian at precisely the same time (if you disregard the fact a year is not exactly 365 days long).

To determine if an object is north or south of another, compare their declination positions. Positive degrees of declination increase going north from 0° at the celestial equator up to +90° at the north celestial pole. Polaris (pronounced poe-LAH-ris) is approximately one degree south of the north celestial pole, as shown in Figure 2.1. Negative degrees represent positions south of the celestial equator and go from 0° at the equator to -90° at the south celestial pole.

Figure 2.2 shows how the equatorial coordinate system works when

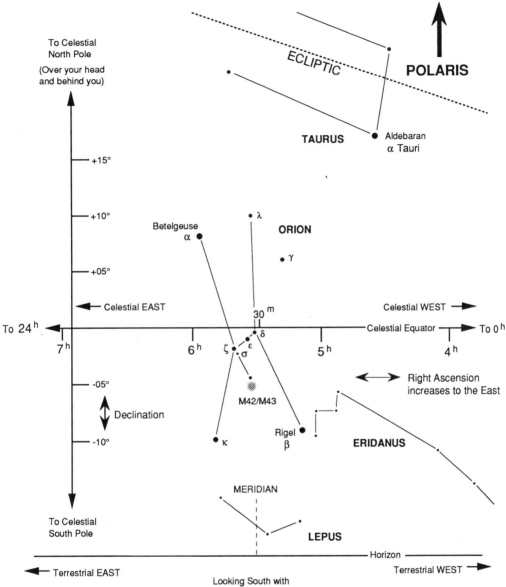

Figure 2.2
Equatorial
coordinate system as
viewed looking
south.

you are facing south and Orion is on your southern meridian. When you compare Figures 2.1 and 2.2, notice that terrestrial east and west have changed places whereas celestial east and west are constant.

Figure 2.1 shows the night sky for a person located at about 40° north latitude looking toward Polaris, and shows the position of the Big Dipper during evening hours in April. The celestial equator will be behind you. If you are located at a higher Earth latitude (farther north) Polaris will be higher above your horizon. If you are located at a latitude less than 40°, then Polaris will be closer to your horizon. An observer at the equator would see the celestial equator directly overhead and Polaris will be very close to your horizon.

35

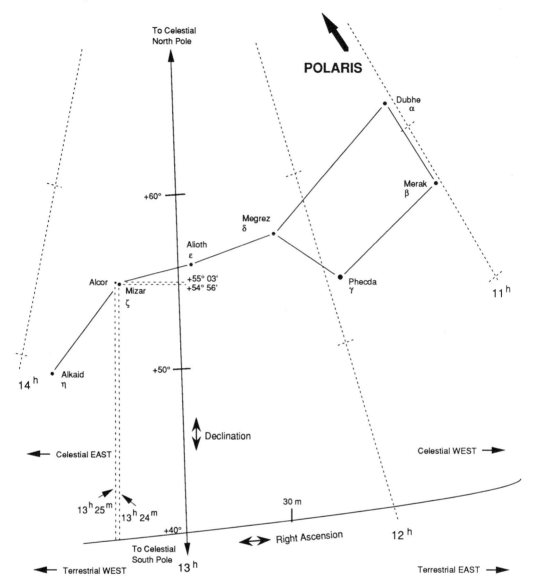

Figure 2.3 Mizar and Alcor and their equatorial coordinates.

Go outside tonight and stand with south behind you. Find the Big Dipper then Polaris. Stick your right hand out away from your side. Now sweep your outstretched hand to be in front of you. Raise it up to Polaris. As you moved your hand, you went higher in both right ascension and declination. Right hand ascending means higher right ascension and declination.

At sunset during January, the opening of the bowl of the Big Dipper in Ursa Major (UMa) appears to be facing up (in relation to the ground), as shown in Figure 2.3. Where the handle is bent is a pair of stars: Alcor and Mizar. Alcor (pronounced AL-kor) sits above (north) and to the left (west) of blue-white Mizar (79 Zeta Ursae Majoris) [13:24;+54d56'], and as shown in position 1 of Figure 2.4. But Alcor (80 UMa) [13:25;+55d03'] is celestially northeast of Mizar (pronounced MY-zar). Alcor's epoch 2000.0 coordinates of right

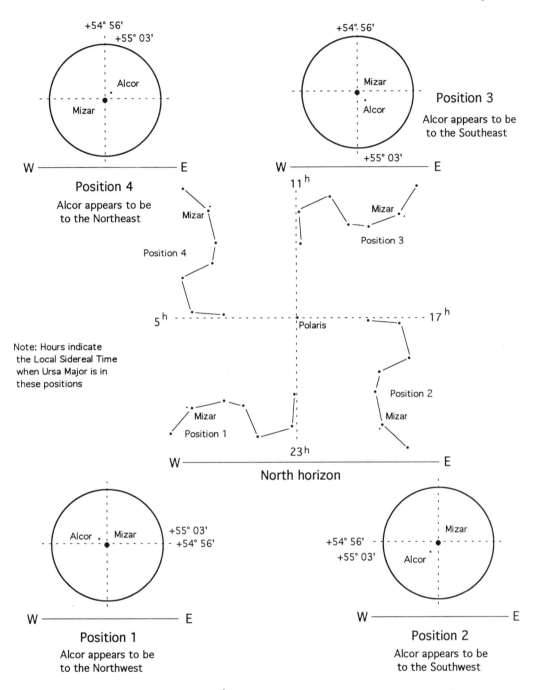

Note: Hours indicate
the Local Sidereal Time
when Ursa Major is in
these positions

North horizon

Position 1

Alcor appears to be
to the Northwest

Position 2

Alcor appears to be
to the Southwest

Figure 2.4 Positions
of Mizar and Alcor.

ascension at 13^h 25^m and declination at +55° 03' are slightly higher
numerically than Mizar's 13^h 24^m right ascension and +54° 56' declination positions.

The Big Dipper appears differently in relation to the horizon at different times of the night. As you watch through the night, the Dipper
will rotate in a counterclockwise motion around Polaris, as shown in
positions 2, 3, and 4 in Figure 2.4. The Dipper will be in position 1 at
about 8:00 p.m. on October 30. Around 2:00 a.m. it will appear near

37

position 2. Though the Sun will have risen by 8:00 a.m., the Dipper will be in position 3 then and in position 4 around 2:00 p.m.

During this rotation of the Big Dipper, the double stars Mizar and Alcor will change their relative positions to the horizon, but their celestial coordinates remain constant. The celestial coordinate system appears to move with the sky, but in reality we on Earth rotate beneath the celestial gridwork.

Do not be dismayed if you are unable to see Alcor with the naked eye. Lots of people are unable to do so. Many American Indian tribes used the ability to see these stars as separate points of light as a test of their warriors' eyesight.

Star names and catalogues

Stars and deep-sky objects are labeled on star charts in what may seem as a very confusing and apparently haphazard manner. Over the centuries many astronomers created their own systems for naming these objects. Some systems were accepted whereas many others were left to wither in obscurity.

As civilizations progressed, time keeping became an important aspect of daily life. Farmers needed to know when to plant and harvest their crops. The rhythm of the skies established the periods and seasons for the first time-keeping systems. Because of the importance of the night sky, ancient peoples gave the brighter stars names and created tales to explain the hows and whys of the heavens. Many of these popular names are still in use and their lore retold. In his 1725 book, *Principi Di Scienza Nuova Di Giambattista Vico D'Intorno Alla Comune Natura Delle Nazioni...* ("Principles of New Science of Giambattista Vico Concerning the Common Nature of the Nations..."), Italian philosopher Giambattista Vico (1668–1744) wrote:

> Thus, beginning with vulgar astronomy, the first peoples wrote in the skies the history of their gods and their heroes. Thence there remained this eternal property, that the memories of men full of divinity or of heroism are matter worthy of history.... And poetic history gave learned astronomers occasion for depicting the heroes and the heroic hieroglyphs in the sky with one group of stars rather than another....

Though several ancient cultures gave the stars names, few of these names have survived to still be in use. The earliest lists of the stellar names we still use were compiled by the Turkish astronomer Hipparchus in the second century BC. He built upon the foundations laid by earlier observers, especially that of Timocharis of Alexandria (*c.* 300 B.C.).

Unfortunately, much of Hipparchus's work has been lost, but Alexandrian astronomer and mathematician Ptolemy (also known as Claudius Ptolemaeus) (*c.* A.D. 83–161) (flourished between A.D. 127 and 151) used Hipparchus's catalogue and the works of others, then added his own observations to their efforts. Ptolemy's monumental work was originally entitled *Hē Mathēmatikē Syntaxis* ("The

Mathematical Collection"). Many years after his death, the title was changed to *Ho Megas Astronomos* ("The Great Astronomer") and transliterated into the Greek "Megistē" from which the Arabs in the ninth century translated it into "Al megisti," from which *Almagest* the name we know it by today is derived. This 13-volume book contained astronomical formulae as well as a listing of 1,022 stars and 48 constellations. About a dozen of these stars Ptolemy listed by their then common Arabic name. The rest of the names he gave refer to the stars' positions within the constellations.

In the 13th century, Spanish King Alfonso X (The Wise) (1221–84) had many Arabic texts, in particular scientific works, translated into Latin. The *Almagest* was among those translated. In the translation process from the Arabic alphabet to the Roman, the spelling of many of the names were changed to be phonetic representations of the original words. Due mainly to our European bias, we use the Latin rather than the Arabic spelling, but the names retain their original Arabic meaning.

The stars were known by their cumbersome Arabic names, until 1540, when Italian astronomer Alessandro Piccolomini (1508–78) published his inaccurate stellar atlas *De le Stelle Fisse*. He assigned Roman letters to the brighter stars. This system was not well received and was supplanted in 1603, when German lawyer and avid astronomer Johann Bayer (1572–1625) published his celestial atlas, entitled *Uranometria*. Bayer introduced the system of labeling the stars employing the lower-case Greek alphabet and Roman letters. Bayer generally, but not always, labeled the *lucida* (brightest) star in each of Ptolemy's 48 constellations with the Greek letter "Alpha" (α). The next brightest became known as "Beta" (β) and so on to the end of the alphabet with "Omega" (ω). In Bayer's system, the Greek letter is followed by the genitive (possessive) form of the constellation's Latin name.

A star numbering system attributed to British astronomer John Flamsteed (1646–1719) was introduced in 1712, in an unauthorized and error-filled version of his catalogue, entitled *Historia Coelestis Britannica*. This system consecutively numbers the stars in each constellation in the order of their right ascension and declination. This catalogue was edited by Edmond Halley (1656–1742) and contains the numbering system whereas Flamsteed's own version, published in 1725 and entitled *Stellarum Inerrantium Catalogus Britannicus*, does not contain the numbers.

Thus, a star like Mizar can be known by its Arabic name, Bayer designation of Zeta (ζ), and Flamsteed number 79. Mizar's official designation is 79 Zeta Ursae Majoris. Being a double star, Mizar is also known as Σ 1744 (or STF 1744) in the *Dorpat Catalogue* of double stars, published in 1837, by German astronomer Friedrich Georg Wilhelm von Struve (1793–1864). In 1932, Lick Observatory director, Robert Grant Aitken (1864–1951) published his *New General Catalogue of Double Stars Within 120 Degrees of the North Pole*; this is known as the ADS. About 150 lesser catalogues list double stars, with most of these stars also included in the ADS. Mizar is listed as number

Photo 2.1 NGC 6720 (M57); the Ring Nebula in Lyra.

8891 in the ADS. Catalogue designations used in this book are listed in Table A.8 in Appendix A.

We also have the popular designations for the 110 nebulae and star clusters in the catalogue compiled by Charles Messier. He published his list in stages between 1771 and 1781. These objects will be found on your star charts with a capital M and a number (like M57). The objects are numbered generally in the order Messier and his colleague Pierre Méchain discovered or rediscovered them, and therefore the numbers are scattered around the sky in what appears to be a random order. Messier discovered 38 of the objects, Méchain found 28, and the rest were first observed by others prior to Messier commencing his list. Messier viewed them in telescopes that ranged in power from 44x to 138x. Several of his objects have received a popular name, such as the Ring Nebula (M57) in Lyra. Messier actually compiled this list so he would avoid confusing these objects with real comets in his quest for fame and fortune as a comet hunter.

German-British astronomer Sir Frederick William Herschel (1738–1822), his sister Caroline Lucretia Herschel (1750–1848), and William's son Sir John Herschel (1792–1871) added to Messier's list of nebulae in a series of catalogues and papers. In 1864, John published the results of his family's extensive collaborative work in a catalogue containing 5,079 objects. In 1888, Danish-British astronomer John Ludvig Emil Dreyer (1852–1926) published his *New General*

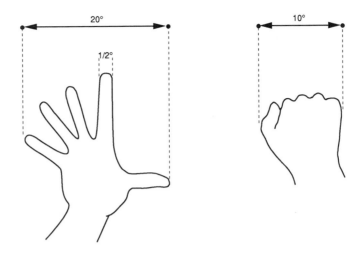

Figure 2.5
Naked-eye
star-hopping by
hand.

Your Hand Will Equal These Approximate Angular
Distances With Your Arm Fully Extended

Catalogue of Nebulae and Clusters of Stars which included 7,840
objects. These objects may be depicted on your charts as NGC numbers or just the number alone. The NGC catalogue was revised and
expanded in 1895 and 1908 with the publication of Dreyer's two
Index Catalogues. Objects first listed in these supplemental catalogues
are labeled as IC numbers. You will usually find these objects charted
with only an "I" preceding the object's catalogue number.

To add to the profusion of labels, many other astronomers and
observatories have published catalogues of their new discoveries.
Today's star charts are littered with abbreviations and numbers desig-
nating these additions to the collective knowledge of the heavens.

Determining your field of view

Star-hopping is like following the links of a chain to get from one end
of the chain to the other. To follow the trails of stellar objects, your
field of view becomes the single chain link you move to get around the
night sky. Whether your star-hopping measuring tool is your hand or
an eyepiece, you first need to learn how much of the sky you can see
at any one time.

You can naked-eye star-hop by using your hand or fist as your distance gauge. As shown in Figure 2.5, the distance across your outstretched fist equals about 10° of the sky. The distance from the tip of
your thumb to the end of your little finger is about 20°, and your
index finger is about $\frac{1}{2}$° wide. This rough scale works when you hold
your hand at arm's length and sight past it. This method works for
people of all ages. Younger viewers may have shorter arms, but their
small hands stay in proportion.

As you become the "night sky expert" and want to show others
where in the sky things are, you can quickly teach these simple mea-

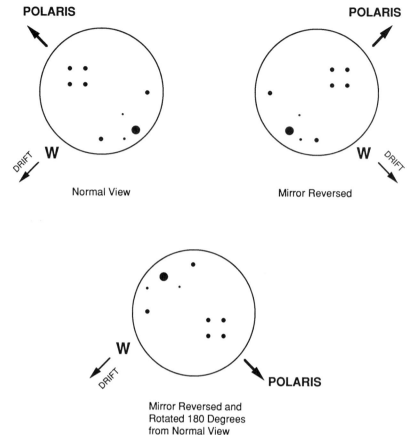

Figure 2.6 Eyepiece and finder scope views.
Note: The stars will drift to the west with a clock drive turned off.

surements to your friends. Using fists and fingers is a lot easier than trying to tell someone an object is so many inches from another object. You will find it is almost impossible to point at a dim object and have your friend locate it. The line of sight between your pointing finger and eye will not line up to the same place as your friend's eye and finger. Star-hopping with fists and fingers will easily get you and your friend to look at the same naked-eye object.

The sky is 90° from a flat horizon to a point directly over your head (the *zenith*). At the celestial equator, the sky moves from east to west at a rate of 15° ($1\frac{1}{2}$ fists) per hour. You can quickly determine how long it will take for an object to move to another position (say out from behind a tree) so you can see it better.

Using Polaris, you can easily determine your latitude by counting how many fists Polaris is above your horizon and multiplying the number of fists by ten. I live near latitude 37° north. This places Polaris about 37° above my horizon, or almost four fists high. If you use an equatorially mounted telescope, you will need to know your rough latitude in order to begin to set up your scope. To properly track a star, you will have to use additional steps for a more precise polar alignment of your instrument.

Before you could star-hop with the unaided eye, you had to learn how much of the sky your measuring tools cover; to star-hop with an

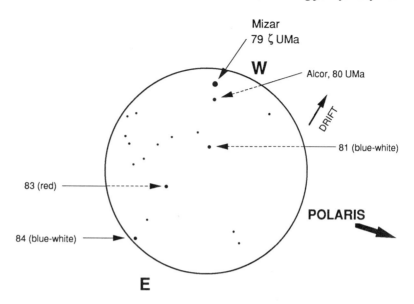

Figure 2.7 Alcor and Mizar, finder scope view, 9x60mm, with a star diagonal mirror.

optical aid you also have to find out know how much of the sky you can see in your finder and eyepieces. But, before you turn your finder scope toward the night sky, you need to know what view to expect in order to figure out the star patterns you will see in the finder and how to convert those patterns to the ones for the same stars on your charts. As shown in Figure 2.6, the image in your finder can either be inverted, mirror-image reversed, or right side up.

In daylight, aim your finder at a terrestrial object. Is the object upside down or right side up? Rotate the scope and see if the image rotates and in what direction. If your scope has a star diagonal, rotate it and see what happens to the image. Can you read letters and numbers or are they backwards? Make notes in your logbook.

If you become confused when looking at the night sky through a telescope, and you are not sure which way to move it, you can easily regain your bearings if you bump the telescope north toward Polaris. Stars will enter the view at the north side of the field. Move the scope celestially east. Stars will enter from the east. This works with all telescopes and mounts. Since the stars are moving toward the western horizon, they will leave on the western side of your eyepiece, if you turn off your clock drive (if you have one). Note the direction of the drift.

Go back to Alcor and Mizar as we will use them to determine the field of view in your finder or binoculars. At magnitude 2.3, Mizar is the brighter of the two stars. Fourth-magnitude Alcor is 11.8 arcminute northeast of Mizar. Some star charts will label these two stars by name and some charts give only their assigned numbers or the Greek letter ζ for Mizar.

Place Mizar at the 6 or 12 o'clock position in your finder with Alcor in line toward the center of the field, as shown in Figure 2.7. Using 6 or 12 o'clock will depend on whether your scope inverts the image or not. A line of three fifth-magnitude stars cut diagonally across the field. These stars are numbered 81, 83, and 84. Stars 81 and 84 appear bluish-white (81 is a spectrum A0 star and 84 is a B9) and 83 is a bright red spectrum M2 IIIa giant. South of 83 UMa you might also be able to see the cluster of sixth- and seventh-magnitude stars around 5.46-magnitude, spectrum A3V, 82 UMa. Keeping Mizar at the edge of your field, gently move your scope to place 82 UMa and (if your field is large enough) 84 UMa along another edge of the field. You now know how large your finder's field of view is.

Using the sky around Mizar to determine your field of view lets you see how large this field is as depicted on your charts. If you do not have a star chart yet, use the Big Dipper as shown in Figure 2.3. To help determine how many hops to make to get from one object to the next, use wire rings or a circle template with your star chart. Compare your circle to your star chart and find the circle which closely approximates your finder's field as well as all of your eyepieces and any combination of them and a magnifying Barlow lens. Unless otherwise indicated, the circles on the star-hop diagrams in this book represent the field of view on each individual chart, as seen in my 9x60mm finder. I suggest you acquire a translucent plastic kind of circle template which has a variety of circle diameters up to about three inches or make rings from a wire coathanger. You can buy a circle template at almost any office or artists' supply store. Since there is no universal scale size for star charts, a variety of circles will allow you to use the template with any of your guide materials.

After determining which circle roughly matches a field of view, mark the circle for future use. I marked my template with "SA2000fin" near the $1\frac{1}{4}$-inch circle that represents the field in my 9x60mm finder when compared to Wil Tirion's *Sky Atlas 2000.0*. "SA26" represents my 26mm eyepiece on the same atlas. The 3-inch circle closely delineates my finder field as compared to the Tirion, Barry Rappaport (1960–), and George Lovi *Uranometria 2000.0*. I marked "RMM26X2" for that eyepiece with a 2x Barlow used on a Rand McNally wall map of the Moon. I wrote this information on the inside cover of my logbook, next to a photocopy of the Greek alphabet.

You should also know how many minutes and seconds of arc you can see with your eyepieces. You will find the size of stellar objects, other than stars, given in arcseconds in catalogues and books. Place any star which is near the celestial equator, such as 34 Delta Orionis [5:32;–0d17'], as shown in Figure 2.2, on the eastern edge of your eyepiece. Time how long, in seconds, it takes for the star to cross through the center of the field and disappear at the western edge. Divide the time by four. Your answer is the field diameter of your eyepiece in arcminutes (arcmin). You can then multiply that number by 60 to get your field diameter in arcseconds (arcsec). Say a star takes three minutes (180 s) to cross the field. Dividing 180 by four yields a field diameter of 45 arcmin. Multiplied by 60, this means your eyepiece will also allow you to see a field 2,700 arcsec across.

Good, easy to locate, bright stars to use for determining your field diameter are Delta Orionis in the winter, 79 Zeta Virginis [13:34;–0d35'] in the spring, 65 Theta Aquilae [20:11;–0d49'] in the summer, and 34 Alpha Aquarii [22:05;–0d19'] in the autumn, but any other star within 10 degrees of the celestial equator will work.

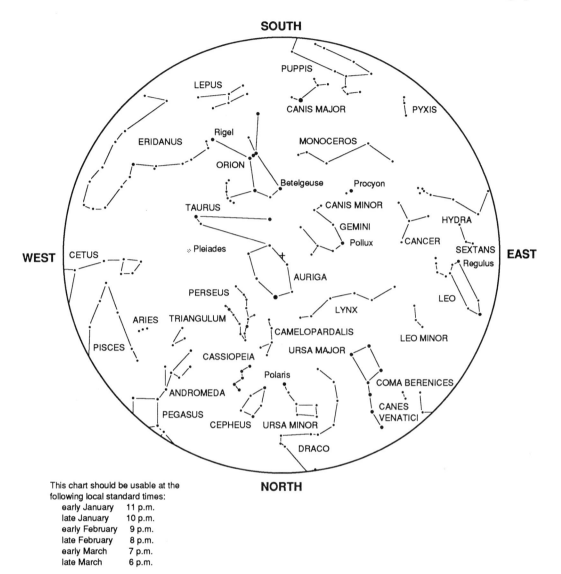

early January	11 p.m.
late January	10 p.m.
early February	9 p.m.
late February	8 p.m.
early March	7 p.m.
late March	6 p.m.

Figure 2.8
Midwinter evening
sky, mid-northern
latitudes. (January –
March)

Observing tips

The night sky is easy to get around in, if you know the constellations
and which ones are next to each other. As the Earth rotates on its axis
and orbits around the Sun, the sky appears to move from terrestrial
east to west, and objects visible in the evening winter sky set earlier
each evening into the western twilight and begin to reappear in the
east during the early pre-dawn hours in autumn. The procession of
the stars through each night and the seasons complicates the task of
learning how to navigate in the night sky.

About the only real tip on learning the constellations is to go out
and look at the sky with a sky chart, like those shown in Figures 2.8 to
2.11, or with a planisphere in your hands, and begin to memorize the
night skies. Every time I go out at night or even when I go out to pick

45

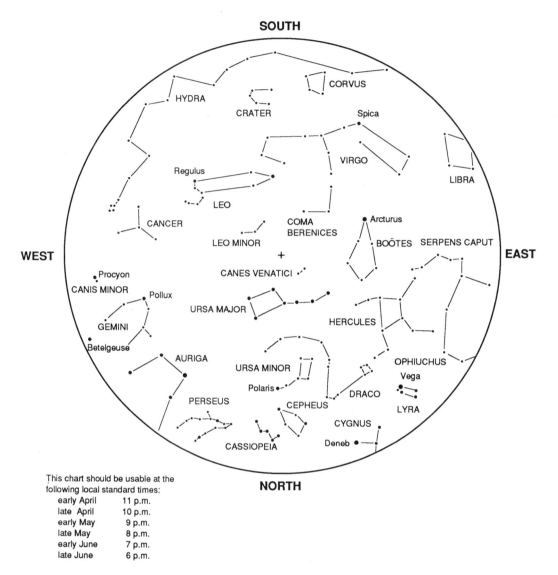

SOUTH

WEST   EAST

NORTH

This chart should be usable at the
following local standard times:

early April	11 p.m.
late April	10 p.m.
early May	9 p.m.
late May	8 p.m.
early June	7 p.m.
late June	6 p.m.

Figure 2.9
Mid-spring evening
sky, mid-northern
latitudes. (April –
June)

up the newspaper before dawn, I glance up and check out what stars
are visible. I pick out the bright constellations and try to see the
fainter ones. If you do not have your sky chart handy, try to remember the star patterns and check your charts later. See if you identified
any of them correctly.

As you star-hop, you will become more familiar with the constellations. With practice, you should be able to pick out even the obscure
and faint constellations such as Lacerta or Norma.

Figures 2.8 to 2.11 depict the sky, with the constellations marked,
as viewed in the early evening hours during the middle of each season.
Hold the appropriate sky figure vertically with its terrestrial compass
point toward the ground, for the direction you are looking in.

A handy tool you should consider buying is a planisphere. This
inexpensive device works in the same way as the sky figures, but
allows you to select the sky as seen on any given date and time. The

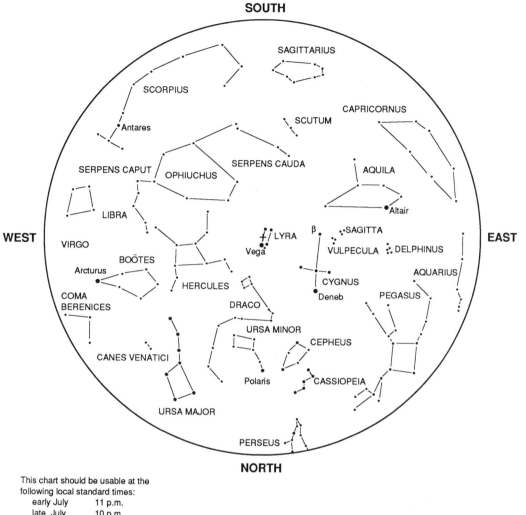

SOUTH

SAGITTARIUS

SCORPIUS

CAPRICORNUS

•Antares

SCUTUM

SERPENS CAPUT   OPHIUCHUS   SERPENS CAUDA

AQUILA

•Altair

LIBRA

WEST

VIRGO

β

SAGITTA

LYRA

VULPECULA

DELPHINUS

EAST

Vega

BOÖTES

Arcturus

HERCULES

CYGNUS

AQUARIUS

Deneb

PEGASUS

COMA
BERENICES

DRACO

CANES VENATICI

URSA MINOR

CEPHEUS

Polaris

CASSIOPEIA

URSA MAJOR

PERSEUS

NORTH

This chart should be usable at the
following local standard times:

early July	11 p.m.
late July	10 p.m.
early August	9 p.m.
late August	8 p.m.
early September	7 p.m.
late September	6 p.m.

Figure 2.10
Midsummer evening
sky, mid-northern
latitudes. (July –
September)

planisphere has the entire yearly night sky drawn on a disk. You
rotate the disk to match a right ascension hour with a date or civil
hour on the outer rim of the device. In the viewing window, you will
see the sky depicted for the time you have chosen. Planispheres are
designed to work at a specific latitude, plus and minus a few degrees.
They are usually labeled with the constellation names and many of
the brighter objects. Some of the constellation shapes are slightly distorted, because the spherical sky has been depicted on a plane, which
gives the device its name plani-sphere. Be sure to purchase one that
will work at your location. Monthly sky charts can be found in the
monthly publications listed in the bibliography.

Some other tools you will need are: a flashlight with a red lens; a
folding table to hold your charts and supplies; a sturdy case (a fishing
tackle box works well) to keep your accessories in; something to sit on

47

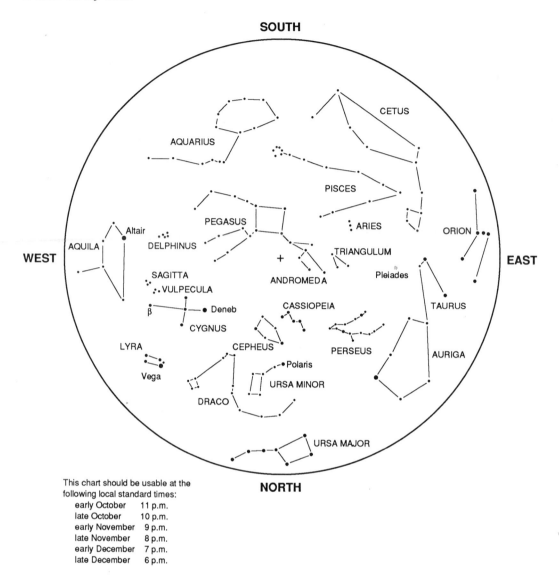

SOUTH

CETUS

AQUARIUS

PISCES

PEGASUS

Altair

AQUILA

DELPHINUS

ARIES

ORION

WEST

TRIANGULUM

EAST

SAGITTA

ANDROMEDA

Pleiades

VULPECULA

β ● Deneb

CASSIOPEIA

TAURUS

CYGNUS

LYRA

CEPHEUS

PERSEUS

AURIGA

Vega

Polaris

URSA MINOR

DRACO

URSA MAJOR

This chart should be usable at the following local standard times:

early October	11 p.m.
late October	10 p.m.
early November	9 p.m.
late November	8 p.m.
early December	7 p.m.
late December	6 p.m.

NORTH

Figure 2.11
Mid-autumn evening sky, mid-northern latitudes. (October – December)

at the telescope; and whatever it takes to make yourself comfortable (food and appropriate clothes) for long, sometimes cold, viewing sessions. If you become chilled, you will not want to stay out long and may miss the chance to see something.

I use several kinds of flashlights, each with red translucent plastic over the lens. Our eyes are not very sensitive to *dim* red light. This allows you to have some light to work with during an observing session. I use a regular flashlight, with a red lens, for setting up, and a red-lighted magnifying glass for reading the small print on star charts. Be sure to take extra batteries with you. In lieu of red plastic, you can use layers of brown paper bag placed under the regular lens. You want the light to be as dim as you can get it and still be able to see when holding the flashlight four inches above your chart. Anything brighter will affect your *dark adaptation*.

Refer to the bibliography for additional references and star charts to supplement this book. No one source has all the facts, so before you go out and buy the whole list, try to attend a local star party and see what materials other amateurs are using, and ask questions. Find out what products will suit your tastes and needs, then get them and use them. Check with the astronomy department at a local college or science center, or in the popular astronomy magazines for a listing of any astronomy clubs in your area. Attend their meetings and go to their star parties. Star parties are a great place to try out different equipment and see how the equipment really works. I have yet to meet an amateur who would not allow me to see through his or her telescope. Everyone shares the sky. Ask questions, be enthusiastic, courteous, and try not to monopolize another observer's evening.

Probably our worst nighttime enemy happens also to be one of the most interesting objects in the sky: the Moon. Being so bright from first quarter phase to full then to last quarter, our celestial companion makes it virtually impossible to see faint objects. Plan your observing sessions around the Moon's phases. The crescent Moon causes little problem as it sets early or rises just before dawn. Dark moonless nights at a dark site without light pollution will give you the best chance to see many of the objects covered in this book and the ones you will want to search for later.

How to navigate in the night sky

Star-hop Mizar to M51

Now that you have your equipment ready, it is time to venture off on our first trip. Our "get a feel for star-hopping" tour will be the telescopic star-hop shown in Figure 2.12. This star-hop will take us from Mizar to the double star Σ 1770, Alkaid (85 Eta UMa), and 24 Canum Venaticorum. We will end at the Whirlpool Galaxy, which is catalogued as NGC 5194 and M51. This beautiful galaxy is located in the constellation Canes Venatici (CVn); the Hunting Dogs.

Locate Mizar and place it in the crosshairs of your finder. Using a low-power eyepiece (about 100x), put Mizar (the brightest star in your field) near the southwestern edge of your eyepiece field, as shown in Figure 2.13. If this is the first time you have looked at Alcor and Mizar through a telescope, you might think Alcor is the closest star to Mizar, but actually Alcor will be near the center of the field, 708.6 arcsec away from Mizar. The blue-white, magnitude 3.95, spectrum A1m IV-V, star near Mizar A is Mizar B. Italian astronomer Giovanni Battista Riccioli (1598–1671) discovered Mizar B, in 1650. Though Mizar A and B appear to be only two stars, spectroscopic study shows them to actually be as many as six stars orbiting each other so closely that only their spectral signatures reveal their total number. In 1889, Edward Pickering noticed periodic shifts and doubling in Mizar A's spectral lines. He suspected the shifting and doubling was caused by the 20.5-day period of a star orbiting around Mizar A. Mizar was the first star to be known as a binary star system

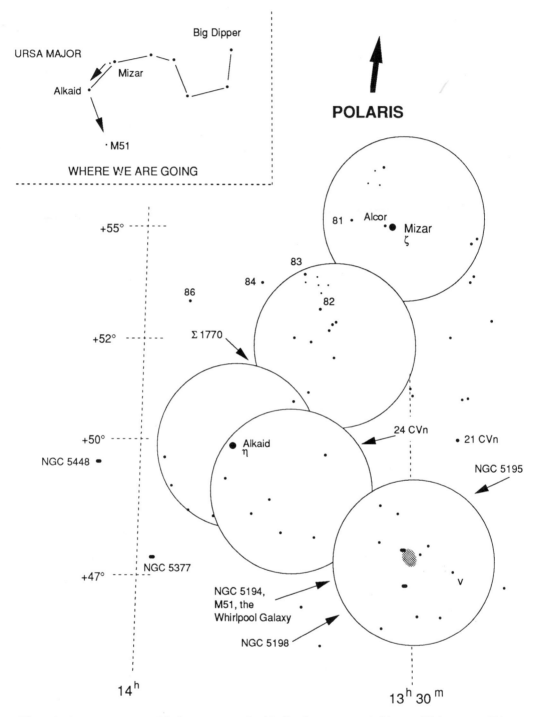

Figure 2.12
Star-hop, Mizar to
M51 (the Whirlpool
Galaxy).

and is known as a *double-lined spectroscopic binary*. This type of binary occurs when the two stars are very close and have approximately the same magnitude. A doubling of spectra is seen on a spectrogram, thereby revealing the presence of the secondary star. A *single-lined spectroscopic binary* occurs when one star is more than two magnitudes

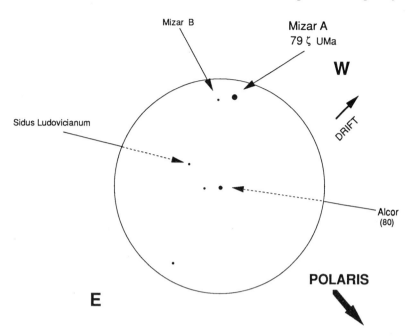

Figure 2.13 Alcor and Mizar, eyepiece view, 26mm (96x).

fainter than the other star, the fainter star's spectra will not be visible. The periodic shifting of the spectra reveals the secondary star. The faint star southeast of Alcor is Sidus Ludovicianum. This star forms a shallow triangle with Mizar and Alcor and was named, in 1723, for Ernst Ludwig (1667–1739), the Landgrave of Hasse-Darmstadt.

You can get a feel for how small an arcsecond is in your field of view when you realize Mizar B is only 14.4 arcsec from Mizar A and is at position angle 152°. In observations made 72 years apart, the comes, Mizar B, had moved one degree northwest in its orbit around its primary star, Mizar A.

Use your template to determine your field distance and how many hops you have to make to magnitude 1.86, spectrum B3 V, blue-white, main-sequence star Alkaid (85 Eta CMa) [13:47;+49d18']; the last star in the handle of the Big Dipper. Use the string of stars between Mizar and Alkaid (pronounced al-KADE) to guide you to Alkaid.

One of these stars is the very close double star system Σ1770 (ADS 8979) [13:37;+50d43']. The components of Σ1770 are a K5 spectral-type, 6.8-magnitude primary and its 8.3-magnitude comes located 1.8 arcsec away at position angle 121°. Theoretically, you should be able to separate these stars into individual points of light and see black space between them in a scope with at least a 2-inch aperture.

Place Alkaid on the northeastern edge of your field. Depending on the diameter of your field, you might have M51 [13:29;+47d12'] on the opposite side. If you do not see the very faint glow of M51, you can use the orangish-yellow, 6.8-magnitude star 24 CVn [13:34;+49d00'] to guide you to M51. You will find the faint hazy glow of M51 near the northern edge of a trapezoidal asterism of sixth- and seventh-magnitude stars. M51 is a face-on Sc-type galaxy with its

51

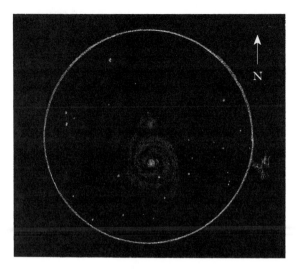

*Eyepiece impression
2.1   NGC 5194/95
(M51); the
Whirlpool Galaxy.*

irregular companion NGC 5195 [13:30;+47d16'] just to the north, as shown in Eyepiece Impression 2.1. These are two of the few galaxies you can easily view in binoculars, but you need to be at a dark site on a dark night to do it. You will only perceive one faint patch of light in normal binoculars.

Do not be surprised to see only the two glowing blobs of the nuclei of NGC 5194 and NGC 5195 surrounded by a peanut-shaped hazy gray glow. On a night of good seeing, and using averted vision in a telescope of at least 6-inch aperture, you might be able to see the spiral structure of M51, but on the best night, you probably will not be able to see the wispy connecting arm of stars between the smaller tenth-magnitude NGC 5195 and eighth-magnitude NGC 5194. Try to observe the arm anyway. In 8- to 10-inch scopes, the arms become more distinct in dark skies. Larger scopes reveal the spiral structure of the curving black dust lanes between the arms.

Older catalogues list M51 as being located about 11.5 Mpc (37 million light-years) from us. Recent studies utilizing spectrographic techniques have reduced this distance to 7.25 to 7.7 Mpc (23–25 million light-years). M51 is moving away from us at a combined radial velocity of 485 km/s (325 km/s + 160 km/s, which is our motion away from M51). The total mass of NGC 5194 has been calculated to be about 160 billion solar masses and its absolute magnitude is -20.75. Its diameter is about 100,000 light-years. This magnificent face-on galaxy was discovered by Messier, on October 13, 1773. Méchain discovered the smaller NGC 5195, in 1781.

In 1854, William Parsons, 3rd Earl of Rosse (1800–67) used his 72-inch reflector at Parsonstown, Southern Ireland, to discover the sweeping spiral structure of NGC 5194. Not until the 1920's was this nebula proven to be a system of stars beyond the limits of the Milky Way.

NGC 5195 is an irregular P-type galaxy that visually appears to be connected to the larger galaxy. Computer modeling studies reveal the two galaxies are probably gravitationally connected. The modeling of gravitational effects of two interacting galaxies maps closely to what we observe in photographs. They are not physically connected as the northern arm and dust lanes of M51 cross in front of NGC 5195. Though much smaller in apparent size, NGC 5195 contains almost twice the mass of NGC 5194. This dense galaxy glows at absolute magnitude -19.32 and shows a radial velocity in recession of 552 km/s.

The component galaxies of M51 are probably the finest examples of interacting galaxies which you can easily view. In 1959, Vorontsov-Vel'iaminov published a system of classifying interacting galaxies in his *Atlas of Interacting Galaxies*. In this photographic atlas, he used 396 specimens to establish a means of identifying over 60 types of interacting galaxies. These types range from almost normal galaxies showing little gravitational distortion effects to galaxies ripped apart and distorted by the gravitational pull of a nearby galaxy. The smaller NGC 5195 is rotating around NGC 5194 and is pulling the larger galaxy's northern arm with it. It will take NGC 5195 several hundred million years to make one orbit around NGC 5194.

One of the things about astronomy which still captures my imagination, is contemplating all that has happened on Earth in the 23 to 37 million years since the faint light seen tonight first left M51. These photons of light energy that strike the retina of your eye started on their journey long before man became a reality on Earth. After traveling all those years, the instant each photon hits your retina and stimulates a nerve ending, it ceases to exist, except in your memory. (Technically its light energy is converted to heat energy in your body and is radiated back into space as you stand out in the cold night air.)

Imagine you are a space traveler and you left the Earth about the time the dinosaurs died out (estimated to have been 65 million years ago). If you could travel at the speed of light, you would just be getting back home from your round trip to M51. Think about it and wonder at the magnificence of it all.

One other thing about the study of astronomy you will find utterly confusing is the inconsistency in facts about an object. Though astronomy is supposed to be an exact science, it cannot be until we actually venture out to distant objects, like M51, and physically measure the distance to them and give them close scrutiny or we vastly improve our measuring devices. The astronomical literature is laden with information on all of the objects covered in this book, and rarely do these sources give the exact same data on an object. Techniques change and new information overlays, modifies, and shatters older theories and accepted truths. Until the advent of astrophotography, in the late 1800's, we had no idea exactly what most of the faint fuzzies were. Not until the early years of the 20th century was the concept of island universes beyond the boundaries of the Milky Way proven. Much has happened in the field in the last few years, and it has been a struggle to locate the most current published data and verify them as best as I can. Therefore, I will sometimes give you multiple values for the same object when I know the facts are inconclusive or that new information has been produced that contradicts older

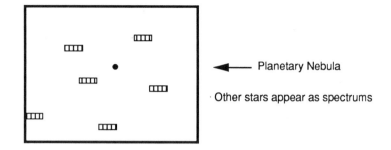

View of a planetary nebula and star field as
seen through a 200 power eyepiece and prism

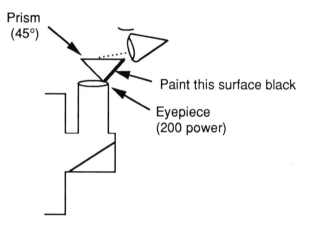

Figure 2.14
Observing a
planetary nebula
through a prism.

more accepted data. The study of the heavens is an on-going process
with rapid technological advances and changes in the body of our
astronomical knowledge.

Tips for observing variable stars, planetary nebulae, and nebulae

With our first star-hop we observed regular stars, double stars, and a
galaxy. When observing a variable star, you will be looking for the
changes in brightness of the star. On different occasions, compare the
star to how bright it looked the last time you observed it and how it
compares to the non-variables in the same field of view. Recording
your impressions of the star and multiple observations are important
if you want to be a successful variable star observer.

The observation of planetary nebulae is generally an easy task as
many of the planetaries appear as fuzzy stars, but locating those tiny
nebulae that are very faint and appear almost stellar can be more of a
challenge. Use a 45° prism, as shown in Figure 2.14, or a diffraction
grating to detect the spectrum of the stars near the planetary. The
spectrum of a planetary's gas shell shows its strongest emission lines

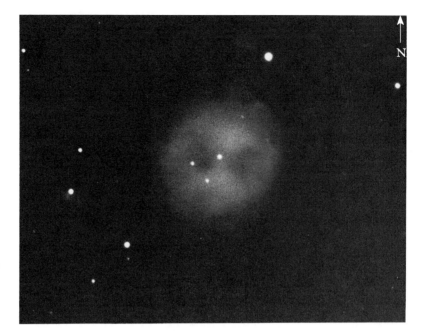

Photo 2.2 Planetary nebula NGC 3587 (PK 148+57.1); the Owl Nebula in Ursa Major.

in the neighboring areas of doubly-ionized oxygen (O-III) at 495.9 nm and 500.7 nm. When the spectrum is dispersed by the prism, these emission lines are the only ones you can visually see. However, the two lines are so close together the spectrum will appear as a point of light. Any stars in the field display a dispersed and easily seen spectrum. A sophisticated spectroscope is needed to detect the planetary's entire spectrum. Hold your eye as close to the edge of the prism as you can and look across the surface. The light is very faint as it passes through the thick prism, so cover your head with a hood to block out as much extraneous light as possible.

Many of the planetaries we will hop to have NGC numbers, but astronomers now use a more complete and specialized planetary nebula catalogue known as the P&K. The actual name for this catalogue is the *Catalogue of Galactic Planetary Nebulae* published, in 1967, by Czechoslovakian astronomers Lubos Perek (1919–) and Lubos Kohoutek. This system numbers planetaries by their galactic system coordinates rather than by celestial coordinates. They divided the sky into one-degree wide longitude and latitude strips. Within each 1° by 1° box, the coordinate for an object is truncated into its integer degrees. For example, the Owl Nebula, NGC 3587 (M97) in Ursa Major is also PK 148+57.1, since its galactic coordinates are truncated to 148° latitude and 57° north longitude. M97 is shown in Photo 2.2.

With diffuse bright nebulae you are looking for the wispy glowing patches of light in the sky. When observing these objects, take your time to study their details. Look for the filmy texture, bright and dark areas, shapes, and field stars.

The great black patches of the dark nebulae can easily be viewed by observing the star fields around them and looking for the few

Photo 2.3 Light-pollution reduction eyepiece filters.

stars peeking through thin spots in the dark nebulae. The American astronomer and pioneer in astrophotography Edward Emerson Bernard (1857–1923) catalogued many of these dark patches in his 1927 *Catalogue of 349 Dark Objects in the Sky*. This catalogue was a part his book *A Photographic Atlas of Selected Regions of the Milky Way*.

Eyepiece filters

The use of different kinds of light-pollution reduction, band-pass, or colored filters will in aid your observation of many objects. These specially coated glass filters come in two basic mechanical configurations: those that you install on the telescope in front of your eyepiece holder and those that you thread into the base of an eyepiece, like the filters shown in Photo 2.3. I have used both and prefer the eyepiece type for ease of changing from one filter to another or to no filter at all. The larger, and thereby more expensive, telescope-mounted filters are good if you are going to do astrophotography through the telescope and need a light-pollution reduction (LPR) filter. Otherwise use the eyepiece filters. The eyepiece filters are easier to change and you do not move the telescope when changing one. My telescope-mounted LPR threads onto the end of the telescope and the star diagonal threads onto the filter. I have yet to be able to install or remove the telescope-mounted filter without having to relocate the object I was looking at. Once it is on, I usually leave it on for the night. (A bad habit to get into!)

Use an LPR-type filter when you are observing from a heavily light-polluted site. You will notice that stars generally turn green, turquoise, or red with these filters as the filter blocks the red, yellow, and orange wavelengths generated by modern street lights, as shown in Figure 2.15. Wide band-pass LPR filters are best for observing objects consisting mainly of stars, such as galaxies and clusters. They show only very slight improvement when looking at faint galaxies or reflection nebulae. The wide band-pass LPR filters (such as the LUMICON Deep-Sky Filter*) are the best LPR filters for astrophotography.

High contrast filters have a very narrow band-pass which gives them much better rejection of unwanted light pollution. This makes the background sky seem even darker and makes the nebula appear to be several times brighter and therefore easier to see. Although the stars will appear two or three magnitudes dimmer, the transmission of nebula light is often over 90 percent with the light pollution being reduced by a factor of three to ten times. This makes the contrast for the nebula compared to the background sky many times higher. At a star party, a friend handed me his LUMICON O-III Filter to try on the Veil Nebula (NGC 6960/6979/6992-95), in Cygnus, the eastern portion of which is shown in Photo 2.4. Without the filter, this bright

* LUMICON Deep-Sky Filter, LUMICON O-III Filter, and LUMICON H-Beta Filter are trademarks of LUMICON, Livermore, CA, USA.

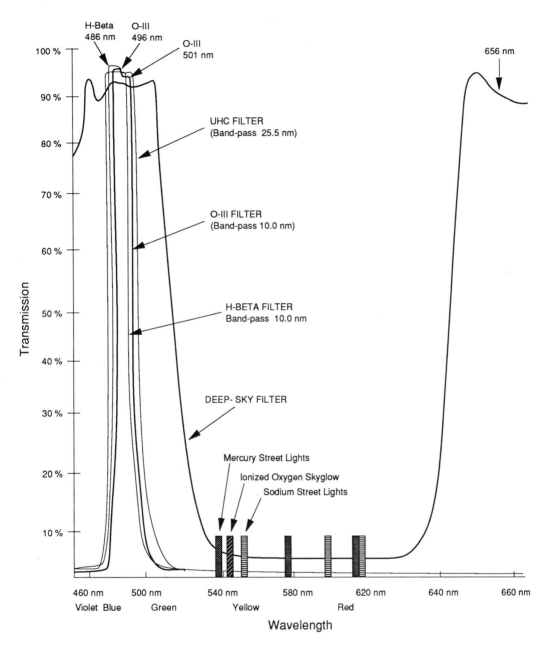

Figure 2.15 Typical characteristics for band-pass and light-pollution reduction filters. Data courtesy of LUMICON.

nebula was lost in the hazy background of the Milky Way. With the filter in place, the nebula jumped out from the hazy background as if I were looking at a three-dimensional picture. I could see the wispy, twisting filaments set against a pitch-black background. The Veil looked like white smoke from a smoldering fire. These filters are best for viewing faint emission nebulae and planetary nebulae.

The LUMICON H-Beta Filter is designed to enhance your views of certain very faint diffuse nebulae, such as the California Nebula (NGC 1499) [4:00;+36d37'] in Perseus or IC 434 located beyond the famous Horsehead Nebula in Orion. These very faint diffuse nebulae

57

Photo 2.4 NGC 6992-95; eastern edge of the Veil Nebula in Cygnus.

are strongest around the H-beta line (486.1 nm) of the spectrum. In the case of the Horsehead Nebula, by bringing out the visibility of the diffuse nebula around a dark nebula (the Horsehead), the contrast between them makes the dark nebula easier to see.

Colored filters work best on the planets and are little if no help in observing deep-sky objects. If you are planning to do any planetary observing, you should buy a set of colored filters.

Most experienced amateurs have several filters in their accessory kits. I suggest you try different filters at a star party and acquire the ones that you feel give you the best views. Try different magnifications, since each LPR filter seems to have an optimum magnification with a given telescope and observing site.

Now that we have our equipment ready, know how the sky works, and know what star-hopping is, let us venture out on our first star-hops in the next chapters. Select the chapters for the time of year in which you will be viewing. You do not have to wait until January to start using this book.

3 *January*

Taurus and Orion: The Bull and the Hunter

You that so wisely studious are
To measure and to trace each Starr,
How swift they travaile, and how farr,
   Now number your celestiall store,
Planets, or lesser lights, and trie
If in the face of all the skie
   You count so many as before!

        Sir William Davenant (1606–68),
        Salmacida Spolia
        (IIII. Song; stanza I),
        1639

The winter sky is dominated by the constellations of Taurus the Bull and Orion the Hunter. Next to Ursa Major and Scorpius, the three belt stars for the Hunter and the V-shaped Hyades cluster in the area of the Bull's eye are the easiest to recognize asterisms. The inhabitants of the winter sky watch and await the outcome of the impending, yet never to be, battle between Orion and Taurus. Accompanying their master are Orion's fierce dog, Canis Major, and timid dog, Canis Minor. Watching from over Orion's shoulder are the Gemini twins (Castor and Pollux), and Monoceros the Unicorn. At the Hunter's feet sit Lepus the Hare and Eridanus the River. West of the Bull's head are the beautiful young Pleiades, being chased by Orion, but never to be caught by him.

We will start this month's first star-hop in Taurus then take a hop around Orion. For the rest of the book, we will continue to hop our way celestially eastward until we go full circle around the celestial sphere and end the year west of Taurus in the constellation of Aries the Ram. The farther west an object is, the earlier in the year it rises in the east and sets in the west. In the early evening hours during January, Taurus is slightly west of your southern meridian, but is still well placed to see the objects we are going to star-hop to.

Star-hop in Taurus

As a constellation, Taurus the Bull is incomplete, as only his head, horns, and the front half of his body are in the sky, but a complete and multi-taled mythology surrounds him. His story ranges from the gentle snow-white bull that Jupiter disguised himself as when he seduced Europa to the cruel bull-headed Minotaur who was the offspring of the Cretan sorceress Pasiphae and a bull sent by Poseidon to

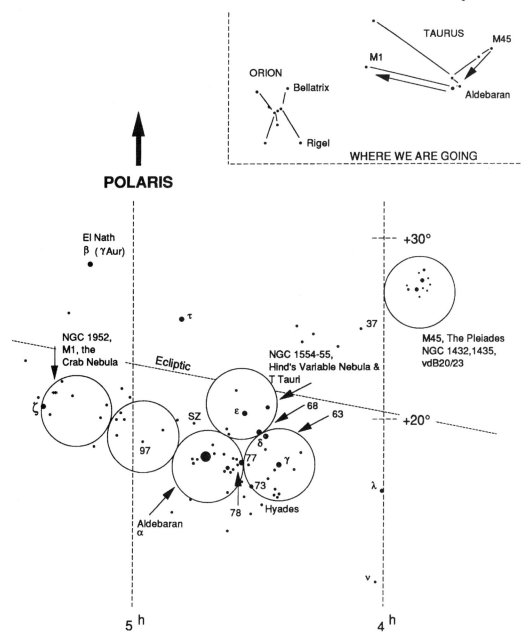

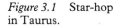

Figure 3.1 Star-hop in Taurus.

King Minos. The tale of Jupiter and Europa left the asterism with the name of "Portitor Europae." The second tale gave us another name for the constellation: "Amasius Pasipaes," the Lover of Pasipae. The use of a bull as a symbol for fertility seems to have been a worldwide rite. Many ancient cultures worshipped a bull and saw the asterism we now call Taurus as a red-eyed bull.

Taurus covers 797.249 square degrees of the sky and ranks 17th in size. Two easy to locate open clusters dominate this constellation: the Pleiades and the Hyades. As shown in Figure 3.1, we will start the star-hop at the Pleiades.

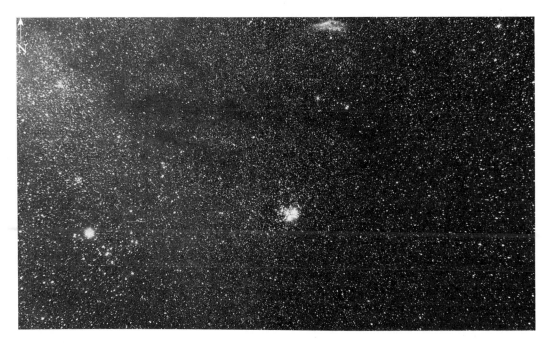

Photo 3.1
Aldebaran, the
Hyades, and the
Pleiades (M45).

The Pleiades (M45) [3:47;+24d07'] is about the easiest open cluster to find, since nine of its members are brighter than magnitude 5.65. To locate the cluster, face south and locate the three bright stars that make up Orion's belt. They point upwards (from the horizon) toward the bright red star Aldebaran. Draw an imaginary line from Orion's belt, through Aldebaran, to the hazy glow of the Pleiades; conspicuous by the lack of bright stars near it.

The cluster was well known to ancient sky watchers around the world and it was known by many names, such as the Seven Sisters (by the Greeks), the Seven Dovelets (in Sicily), a Hen and Her Six Chicks (in the Bible), the Seven Sisters of Industry (in China), the Seven Maidens (by the American Indians), the Little Eyes (in Tonga), and a Heavenly Flame (by Hindus). For some unknown reason, in 1769, Charles Messier added it as the 45th object on his list of nebulous objects, even though by then the cluster was well known.

The name we know the cluster by comes down to us from the misty world of Greek mythology. The legends about the creation of the Pleiades are numerous. About the only part of their mythology that seems constant is that the seven sisters, who became the blue-hued cluster, were the children of the Titan Atlas and the Oceanid Pleione. Being fair and beautiful, the women were chased, and all but one of them were seduced by various gods, including Zeus and Poseidon.

In one version of the creation of the asterism, the sisters were being chased by Orion. To end their torment, they were turned into doves and flew up into the heavens. This story has been discounted, because it refers to the sisters as being virgins, and when Orion began to chase them, none of them were pure anymore. Another story has the seven sisters being the seven doves that gave ambrosia to Zeus

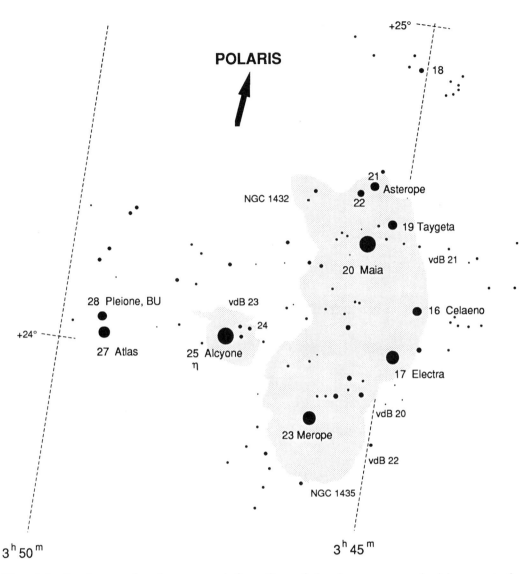

+25°

POLARIS

21
NGC 1432
22
Asterope
19 Taygeta
vdB 21
20 Maia
28 Pleione, BU
vdB 23
16 Celaeno
+24°
24
27 Atlas
25 Alcyone
η
17 Electra
vdB 20
23 Merope
vdB 22
NGC 1435

$3^h 50^m$

$3^h 45^m$

Figure 3.2 Star-hop around the Pleiades.

when he was an infant. One of the doves was crushed in a part of Homer's epic tale *The Odyssey*.

The name of the Pleiades appears to be derived from the Greek *pleo* which means "sail." This seems to be related to their setting in the west at the start of the sailing season on the Mediterranean Sea and their reappearance in autumn at the end of good sailing weather. Their name may also have come from the word *plios* meaning "many" or "full". The cluster has also been associated with the blooming of flowers in spring since the Sun reaches conjunction with the cluster in late May. In this regards, the cluster was known as the "Virgins of Spring," the "Stars of the Season of Blossoms," and the "Stars of Abundance."

The Pleiades, as shown in Figure 3.2, and listed in Table 3.1, is known to contain several hundred stars, but the seven sisters and their parents, being easy naked-eye objects, dominated the lore

63

Table 3.1. *The Pleiades, M45.*

Designation	Name	Magnitude	Spectrum
16 Tau	Celaeno	5.45	B7 IV
17 Tau	Electra	3.70	B6 IIIe
18 Tau		5.64	B8 V
19 Tau	Taygeta	4.30	B6 IV
20 Tau	Maia	3.87	B8 III
21 Tau	Asterope	5.76	B8 V
22 Tau		6.43	A0 Vn
23 Tau	Merope	4.18	B6 IVe
25 Tau (Eta)	Alcyone	2.87	B7 IIIe
27 Tau	Atlas	3.63	B8 III
28 Tau	Pleione	5.24	A2

surrounding the cluster. In order of brightness, the members of the Pleiad family are: magnitude 2.87 Alcyone (25 Eta Tau), the protector of sailors from rocks and rough weather; magnitude 3.63 Atlas (27 Tau), the father and bearer of the Heavens; magnitude 3.70 Electra (17 Tau), lover of Zeus and mother of Dardanus; magnitude 3.87 Maia (20 Tau), the first and most beautiful Pleiad, mother of Hermes, and for whom the month of May is named; magnitude 4.18 Merope (23 Tau), the only Pleiad to marry a mortal; magnitude 4.30 Taygeta (19 Tau), seduced by Zeus while she was asleep and gave birth to Lacedaemon; magnitude 5.24 Pleione (28 Tau), the mother of the Pleiades and the Hyades; magnitude 5.45 Celaeno (16 Tau), a wife of Poseidon; and magnitudes 5.76 and 6.43 Asterope (21 and 22 Tau), (also known as Sterope I and II), the lost Pleiad. Most of the other stars in the cluster cannot be seen without an optical instrument.

The cluster is sometimes mistakenly referred to as the "Little Dipper" which is generally associated with Ursa Minor, but the cluster does resemble a dipper with a very short handle. We will start at the spectroscopic binary marking the "Milk Dipper's" handle; Atlas (27 Tau) [3:49;+24d3']. Atlas is a blue giant and is the primary in a very close double system known as $\Sigma 453$ (ADS 2787).

North of Atlas shines his wife Pleione (28 Tau) [3:49; +24d8']. Pleione is an irregular shell star type variable and as such is also known as BU Tau. This type of variable is also known as a GCAS-type after 27 Gamma Cassiopeia, the first of this type of variable to be detected. The magnitude for this star ranges from 4.77 to 5.5 during periods when the star ejects a shell of gas. The star's variability may also be due to its rapid rotation. The star spins about 100 times faster than the Sun. The equator of the Sun makes one revolution about its axis in 27 days. The rapid rotation of BU Tau is believed to cause the star to occasionally eject a shell of hydrogen gas. Pleione may also be a spectroscopic binary, but the evidence is not conclusive.

Going clockwise around the dipper we first come to Alcyone (25 Eta Tau) [3:47;+24d6'], which is the primary star of a multiple system. Alcyone (pronounced al-SIGH-oh-nee) is a young hot blue-white giant star located 122 pc (400 light-years) from us. The "B" star is an 8.0-magnitude, spectrum A0, star located 117.2 arcsec from the primary at position angle 289°. The "C" star is also an 8.0-magnitude A0 star, but it is located 180.8 arcsec from the primary at position angle 312°. Look for the 8.6-magnitude yellow "D" star 190.5 arcsec from Alcyone at position angle 344°. Eta Tau is embedded in the soft glow of the reflection nebula van den Bergh 23 (vdB 23).

Maia (20 Tau) [3:45;+24d22'] is also a hot young blue-white giant, embedded in the haze of the reflection nebula NGC 1432. This nebula is part of a series of separate gas clouds that extend over much of the western side of the cluster.

Due north of Maia is her sister Asterope (21 and 22 Tau) [3:45.54+24d33' and 3:46.2;+24d31']. Asterope (pronounced as-TER-oh-pee) is a wide optical pair. Blue-white giant 21 Tau (Sterope I) shines at magnitude 5.76. 22 Tau (Sterope II) is southwest of 21 Tau and is slightly dimmer at magnitude 6.43. Both of these stars are also surrounded by the glowing gas of NGC 1435. The spectrum for these two stars shows no radial velocity, unlike the rest of the cluster which is receding from us.

West of 22 Tau is her blue-white giant sister Taygeta (19 Tau) [3:45;+24d8']. Taygeta (pronounced TAY-get-a) is the primary in a wide double system. The 8.1-magnitude comes is located 69 arcsec from the primary at position angle 239°.

Celaeno (16 Tau) [3:44;+24d17'] is a blue giant, like most of the stars in this 78-million-year old cluster, and is moving away from us at a rate of 3 km/s. The cluster is also moving south-southeast at a rate of 40.22 km/s.

At the southwestern corner of the dipper is Electra (17 Tau) [3.44;+24d6']. Electra is a blue-white giant and spectroscopic binary also bathed in the gaseous glow of NGC 1435. Of the named members of the Pleiades, this star shows the greatest radial velocity at 12 km/s in recession. The orbital period of the spectroscopic companion is 100.46 days. Electra is embedded in the glow of reflection nebula vdB 20.

The last star in the Pleiad dipper is Merope (23 Tau). She is obscured in wispy reflection nebulosity NGC 1435. The cloud around Merope (pronounced MER-oh-pee) appears as if someone smeared white shoe polish on a piece of glass in front of it. The gas cloud NGC 1435 is also called the Merope Nebula or Tempel's Nebula. Since Merope was the only Pleiad to marry a mortal, the legend is that she hides her shame behind this grayish veil.

Our next stop is the Hyades cluster. Mythologically, the Mount Nysa nymphs are sisters of the Pleiades. There is some controversy over the names of these daughters and to which star an individual name belongs. The nymphs were charged with nursing and hiding Zeus's young son Dionysus, the inventor of wine. At the completion of their task, the nymphs were placed in the sky by Zeus. Some sources consider the Hyades "The Rain Makers" and in other tales they are "The Piglets." Some Arabs called the cluster *Al Mijda* "The Triangular Spoon," whereas others labeled the V-shaped cluster *Al Kallās*, "The Rainy Hyades." The compact portion of the cluster forms Taurus's face and the extended arms of the V are his horns. Aldebaran (87 Alpha Tauri) [4:35;+16d30'] is his inflamed eye.

The cluster is 39 pc (130 light-years) from us. The Hyades is the small nucleus of a group of several hundred stars called the Taurus Moving Group. This group is moving away from us at 41 km/s. The Hyades is a very complex cluster, as it contains old and young main-sequence stars, many doubles, white dwarfs, and stars in spectral classes A, F, G, K, and M, but no B-type giants. A brown dwarf was detected in 1992. The Hyades, unlike the Pleiades, is not bathed in nebulosity, and the absence of young blue giants leads astronomers to believe the Hyades is 582 million years older than its celestial sister.

At the apex of the V is 3.63-magnitude 54 Gamma Tau [4:19;+15d37']. Gamma is a spectrum K0 IIIab: CN1, yellow giant and is the third brightest member of the cluster.

Following the northwestern leg of the V from Gamma, the first bright star is a spectroscopic double 63 Tau [4:23;+16d46']. This star is about three-quarters of the way between Gamma and 61 Delta Tau. 63 Tau is near the edge of naked-eye visibility at magnitude 5.64 and is classified as a combined spectral-type A1m F3 III. This star is the prototype for this type of star in the MK luminosity system.

Close together, but not known to be a pair, are 61 Delta[1] Tau and 64 Delta[2] Tau. Delta[1] [4:22;+17d32'] is a 3.76-magnitude, spectrum K0 III: CN 0.5, yellow giant. Delta[2] [4:24;+17d26'] is a 4.80-magnitude, A7 V spectral-type, and appears blue-white. Both stars have a radial velocity of 38 km/s in recession. Delta[1] is located 51 pc from us and Delta[2] is 30 pc. These two stars are each double star systems. Delta[1] has a 12.5-magnitude comes located 106.6 arcsec from it in position angle 341°. A 13.5-magnitude comes is located 137.3 arcsec from Delta[2] in position angle 246°.

The second brightest member of the cluster is 74 Epsilon Tau [4:28;+19d10']. This spectrum G9.5 III: CN 0.5, yellow giant sits at the end of the concentrated portion of the V and shines at magnitude 3.54. Epsilon is an open double with a spectrum K0 comes 181.6 arcsec from the primary at position angle 268° as measured in 1901.

About 1.8 degrees west of Epsilon is the irregular variable T Tauri. This star is associated with Hind's Variable Nebula, NGC 1554/1555 [4:21;+19d32']. In 1852, British astronomer John Russell Hind (1823–95) spotted a tiny nebula and a tenth-magnitude star, neither one then on any celestial charts. Over the next 50 years, the nebula faded beyond the visible range of any telescope, then reappeared on an irregular basis. Since the early 1900's, the nebula has moved from the southwest to the west of T Tauri. The nebula has steadily brightened and changed shape during the 20th century.

Dwarf star T Tauri displays erratic brightness that changes from magnitude 9 to 13 during a cycle that can last for a few weeks or can take months to go from minimum to maximum and return to minimum. T Tauri is the prototype star for eruptive variables that have an irregular light variation of at least three magnitudes, low luminosity, are F and G spectral types with emission lines similar to the solar chromosphere, and are near dark or bright nebulae. T Tauri is 138 pc from us. The evidence is not conclusive yet, but radio and infrared equipment may have detected the first planet beyond the solar system; located about 80 AU from T Tauri.

Midway between Gamma Tauri and Aldebaran is the brightest true member in the cluster, 78 Theta^2 Tau [4:28;+15d52']. This blue-white star shines at magnitude 3.42 and is the MK standard for spectrum A7 III giants. Theta^2 is a member of an open double star system. Its secondary is a yellowish F0 star located 337 arcsec away at position angle 346°. North of Theta^2 is the yellow giant 77 Theta^1 Tau [4:28;+15d57']. This 3.85-magnitude star is in the spectral/luminosity class K0 IIIb. These stars are a fine double to view in binoculars. Both stars have radial velocities of 40 km/s in recession.

At magnitude 0.85, Aldebaran (87 Alpha Tauri) is the 13th brightest star in the sky and the brightest in the area, but it is not a part of the Hyades cluster, being only 21 pc away from us. Aldebaran (pronounced al-DEB-ah-ran) is a spectrum K5+ III, reddish-orange giant that serves as the MK luminosity standard for this type of star. Its name comes to us from the Arabic *Al Dabarān* "The Follower," which seems to represent the fact the star follows the Pleiades. Since ancient times, the star has been depicted as the Eye of the Bull. The Bull's eye is receding at 54 km/s. Aldebaran is 40 times larger than the Sun, yet its density is only 0.00005 of the Sun. As the primary for multiple system ADS 3321 (ß 550), Aldebaran also has five close and faint companions, all of which are difficult to separate from Aldebaran due to Aldebaran's intense glare.

At the end of the Bull's eastern horn is the shell star and close binary 123 Zeta Tauri [5:37;+21d8']. Zeta is not a member of the Hyades, being at a greater distance from us and having a different proper motion than the rest of the members of the cluster. Zeta's location 288 pc away has made detailed study of it difficult, but astronomers have concluded the star has thrown off a shell of gas. Its variability period of 133 days and its spectra reveal its close smaller companion. This MK standard spectrum B2 IV:((e)) (shell) star has shell layers that pulsate inward and away from the star at speeds of 61 km/s.

If you place Zeta near the northeastern edge of your finder field, you should have NGC 1952 (M1; the "Crab Nebula") [5:34;+22d01'] on the opposite side. This object was first discovered, in 1731, by British physician and amateur astronomer John Bevis (1695–1771) and rediscovered by Charles Messier, on September 12, 1758. Lord Rosse is credited with giving the nebula its name after viewing its wispy filaments, in 1844.

M1 is the remnant of a supernova event seen on Earth on July 4, 1054. The "Guest Star" was near the crescent Moon that evening. The light of the explosion was visible here for 23 days in daylight and

Photo 3.2 The A.D. 1054 supernova as depicted on Mimbres pottery.

at night for about another 24 months. Until 1990, the Japanese and Chinese were thought to be the only people to have recorded the event. However, in 1990, University of Texas astronomy professor Ralph Robert Robbins (1938–) announced his discovery of what appears to be the supernova event portrayed on a piece of American Indian pottery, as shown in Photo 3.2. The New Mexico Mimbres Indian plate, first unearthed in the 1930's, shows a star-like object, with 23 rays around it, near the hind feet of a rabbit. Like many Indians, the Mimbres depicted the Moon as a rabbit in their stylized art. Carbon-14 dating shows the plate was made no earlier than A.D. 1050 and the location of its discovery was abandoned in 1070. The style of art on the plate was not used by the Mimbres after 1100.

The Crab Nebula is 1,934 pc away and has received intense study. Photographs taken over a number of years have revealed the nebula is expanding at a rate of 965.4 km/s. Astronomers suspected the nebula was the remnant of a supernova, but were not sure. By calculating back at the rate of 80.45 million kilometers per day, the age of the nebula was determined to be about 900 years, which corresponded to the Chinese record of the "Guest Star" near Zeta Tauri. The connection was made between scientific theory and historical record.

At the heart of the nebula, a 16th-magnitude hot bluish dwarf has been detected. Early theories predicted this star's surface temperature was about 500,000 K, so it therefore probably radiated most of its energy in the ultraviolet portion of the spectrum and this energy is what excited the nebula to glow. This early belief was shelved with the first studies of the nebula by radio telescopes, in 1948. The nebula was seen to be a tremendously high emitter of radio energy. In this band of the total spectrum, M1 is one of the brightest radio stars. With the advent of space exploration in the early 1960's, X-ray studies above the Earth's atmosphere could be made. Our atmosphere blocks out X-ray energy from reaching the ground. These X-ray studies of M1 detected a high level of X-ray energy output from the nebula. Current theory holds that high-speed electron movement, caused by the magnetic fields in the nebula, are the mechanism by which the nebula glows. The nebula produces about 100 times more X-ray energy than visible light energy.

Like all of the constellations we will explore in this book, many more objects within the constellation will await your discovery at the conclusion of each star-hop. I suggest you consult your star charts and see what else in Taurus interests you and to hop to these objects before going on to our next star-hop in Orion the Hunter.

Star-hop in Orion

Several versions of the mythological life of Orion abound, but the most common tales claim that he was the handsomest man alive since he was the offspring of Poseidon and Euryale. The rising of Orion early in the mornings in late autumn signified to ancient people that the rains were soon to come. The relationship between Orion and the rains was explained in a tale that portrays Orion as being the son of Mother Earth. In this strange tale, Orion (also known as Uroin) was born to Hyrieus, a poor old farmer and bee-keeper. After the death of his wife, Hyrieus was visited by Zeus and Hermes. They inquired what the man would like most in life. He replied that he would like a son. Hyrieus was instructed by the gods to sacrifice a bull, make water on its hide, and bury it with his deceased wife. Nine months later, Uroin (which means "he who makes water") was born. Urinating on a bull's hide as a means of inducing a rainy season was known by the Greeks as a ritual performed in Africa.

As a young man, Orion traveled to Hyria on the Aegean Sea island of Chios. There he fell in love with Merope (not to be confused with the Pleiad of the same name). Her father, Oenopion, decreed that, in order for Orion to have his daughter's hand, he must first rid the island of all its wild beasts. Orion accomplished the task, but Oenopion, in love himself with Merope, tried to dispatch Orion to kill more bears, lions, and wolves rumored to still be in the hills. One night, Orion got drunk and forced himself on Merope. The next day, on Oenopion's instructions, satyrs got Orion drunk again. While in a drunken stupor, Orion's eyes were put out by Oenopion.

An oracle told Orion to travel to the east and gaze toward the place where Helius rises from the ocean; his sight would then be restored. With Cedalion on his shoulder to guide him, Orion sailed for the holy island of Delos. There Helius restored Orion's eyesight. Orion fell in love with Helius's sister Eos, the personification of the dawn. Eos was also the mother of the winds and the stars. Her rosy fingers daily opened the gates of heaven for Apollo's fiery chariot.

With his sight restored, Orion returned to Chios to exact revenge. Unable to locate Oenopion, he was persuaded by Artemis (known as Diana, the Moon-goddess, to the Romans) to go on a hunt with her. Apollo, fearing his twin sister would also sleep with Orion, went to Mother Earth and claimed that Orion had boasted that he would rid the Earth of all of her wild beasts. Enraged at this boast, Mother Earth sent an armored scorpion to kill Orion.

Orion's death comes in two different versions. In one tale the scorpion stings him and he dies. In the other version, Orion fails to kill the scorpion and dives into the sea to escape. Tricked by Apollo, Artemis shoots an arrow into the sea and kills Orion. In grief over the wrongful death, Artemis places Orion in the sky in pursuit of the Pleiades, but never to catch up with them. Orion is likewise being chased from afar by his nemesis Scorpio the Scorpion.

We will start our star-hop in Orion at the multiple star Rigel (19 Beta Orionis; ADS 3832; Σ 668) [5:14;–8d12'], as shown in Figure 3.3. Orion ranks 26th in constellation size and covers 594.120 square degrees of the celestial sphere. The three belt stars are an easy asterism to locate. Rigel (pronounced RYE-jel) is the 0.12-magnitude blue-white star 9 degrees southwest of the belt stars. This young, MK standard spectrum B8 Iae, supergiant is one of the most luminous stars in the Galaxy and is about 57,000 times brighter than the Sun. The name Rigel is derived from its early Arabic name *Rijl Jauzah al Yusrā*, which means the "Left Leg of the Jauzah" (the herdsman). The name was changed to Rigel when in 1252, the *Almagest* was translated into Latin and its astronomical tables were excerpted to form the basis of the Alfonsine Tables.

Orion resides in a region where star birth is in progress. With an estimated surface temperature of 12,000 K, Rigel is thought to be a very young star, as are many of the stars condensing in the numerous dust and gas clouds in the area.

Rigel is accompanied at position angle 203.7° by a 6.8-magnitude blue star, Rigel B. You may be able to separate these stars as Rigel B is 9.5 arcsec from Rigel. Rigel B is also thought to be a visual binary system, but the evidence for Rigel C's existence is not conclusive. The magnitude 15 "D" star is located 44 arcsec from Rigel A in position angle 2°.

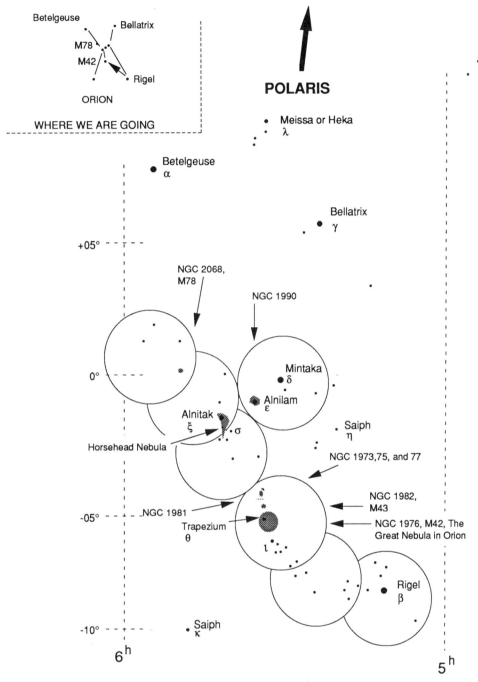

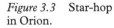

Figure 3.3 Star-hop in Orion.

Rigel sits at the southern apex of a "V" of stars. Taking the southeast leg of the "V" leads us to NGC 1976 (M42; the "Great Nebula in Orion") [5:35;–5d27']. You can spend hours gazing at all of this diffuse emission nebula's glowing gas and dark dust regions and stars, and still not truly see it all. The nebula is about 491 to 583 pc from us and has an apparent diameter of 9.2 pc (30 light-years). The mass of the nebula is estimated to contain enough material to give birth to 10,000 Sun-size stars.

71

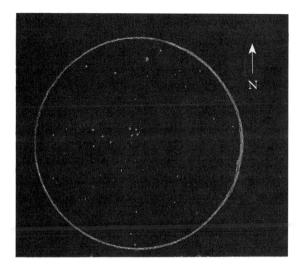

Eyepiece Impression 3.1 NGC 1976 (M42); the Great Nebula in Orion.

The red appearance of the nebula in photographs is due to its abundance of recombining hydrogen atoms. These atoms emit radiation near 6563 Å at the red end of the spectrum. Visually though, the nebula appears to be very pale green, because the nebula is also rich in doubly ionized oxygen, which glows near the green portion of the spectrum, and our eyes are more receptive to green than red. Due to the low surface brightness of the outer reaches of the nebula, you can only see a small portion of the whole gas cloud. Photographs reveal an object that appears to be about four times the size of the Moon. The use of high-contrast and O-III filters along with a low-power eyepiece will extend the amount of the nebula that you can observe.

The actual discoverer of the nebula is in doubt, since its presence was known long before Messier added it as the 42nd item on his list on March 4, 1769. The glow of the nebula was noted in ancient times and it was shown on star maps and in catalogues of many prominent early astronomers. Bayer designated it as Theta Orionis in his *Uranometria*. This designation now refers only to the star cluster 41 Theta[1] Orionis (ADS 4186; Σ748; and commonly called the "Trapezium") [5:35;–5d23'].

Italian amateur astronomer Nicholas-Claude Fabri de Peiresc (also known as Peirescus) (1580–1637) is generally regarded as the first to have observed the nebula with a telescope, having done so in 1610. Peiresc was an attorney and patron of the sciences. One of his friends was Galileo from whom he acquired his telescope. In 1618, Swiss astronomer Cysatus (Latin name of Jean-Baptiste Cysat) (1588–1657) made an independent discovery of the nebula. Beginning in 1656, Dutch astronomer Christiaan Huyghens (1629–95) made the first detailed studies and drawings of the nebula.

Though not the first to observe the quadrangle of stars embedded in the nebulosity, Huyghens discovered, in 1684, the fourth ("B") star in this famous cluster. With the advent of better telescopes, we now know of at least 11 stars in the cluster, as shown in Figure 3.4.

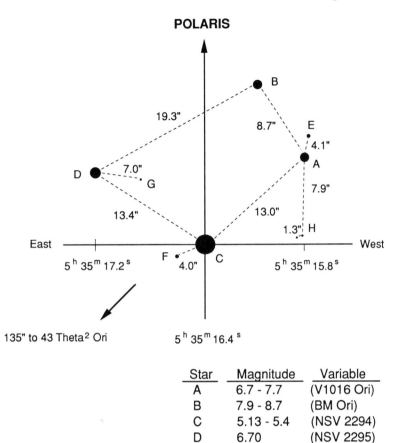

POLARIS

B

19.3"   8.7"   E

4.1"

D   7.0"   A

·G

13.4"   7.9"

13.0"

1.3" H

East   West

F ·

5ʰ 35ᵐ 17.2ˢ   4.0"   C   5ʰ 35ᵐ 15.8ˢ

135" to 43 Theta² Ori   5ʰ 35ᵐ 16.4ˢ

Star	Magnitude	Variable
A	6.7 - 7.7	(V1016 Ori)
B	7.9 - 8.7	(BM Ori)
C	5.13 - 5.4	(NSV 2294)
D	6.70	(NSV 2295)
E	11	(NSV 2291)
F	11	(NSV 2296)
G	16	
H	16	(NSV 2292)

Figure 3.4 41 Theta¹ Orionis; (the Trapezium).

The Trapezium appears to be the core of a cluster of more than 300 very young stars. The age of the cluster is estimated to be a youthful 300,000 years. Most of these stars are faint and visually lost in the glow of M42.

The stars that make up the Trapezium are designated A through H. The original designations for A through D were given to the stars according to their R.A. position. The other stars were designated in the order they were discovered. F. G. W. von Struve discovered the "E" star, in 1826, and John Herschel found the "F" star, in 1830. Using the 36-inch refractor at Lick Observatory on Mount Hamilton, CA, in 1888, Alvan Graham Clark (1832–97) discovered the "G" star and E. E. Barnard discovered the double "HH" stars.

The "A" and "B" stars are each eclipsing binary variables. The "A" star has a period of 65.432 days and the "B" star has a period of only 6.471 days. The "A" star is listed as an EA-type variable; V1016 Ori. Its eclipse lasts 19 hours and its magnitude ranges from 6.7 to 7.7. An EA-type variable exhibits an almost constant magnitude until the time of the eclipse. At that point, the magnitude suddenly decreases

73

for the duration of the eclipse followed by a rapid return to its original brightness. For "B" (BM Ori), the magnitude ranges from 7.9 to 8.7 and its eclipse lasts 16 hours. Within the area of the Great Nebula we know of over 800 variable stars. Most of these are young, irregular or flare stars too faint to be studied visually.

The "C" star is an O6 spectral type and is the brightest star in the group at visual magnitude 5.4. This star is also the primary for the cluster. The "D" star is the second brightest at magnitude 6.3 and is a spectrum B0 type star. The "E" star is harder to detect in small scopes since it is 11th-magnitude. The "G" and "H" stars are 16th-magnitude. All of these stars, except the "G" star, are variables. The "E" and "F" stars are suspected of being FU Orionis types of eruptive variables. FU Orionis types show a gradual increase in brightness that can be as much as six magnitudes. The stars then remain at maximum brightness for years, or even decades. Their return to a dimmer state is very slow. Over a long period, their spectrum begins to show emission lines as the stars slip into later spectral stages.

At 135 arcsec southeast of the Trapezium is the wide double star 43 Theta^2 Ori (ADS 4188A; DA 4) [5:35;–5d24']. This pair has a 37-year-long period and glows at magnitudes 5.2 and 6.5. The stars are embedded in the nebulosity and are believed to contribute to the illumination of the gas cloud.

About 10 arcmin northeast of M42 and seen visually as if almost attached to it is the emission and reflection nebula NGC 1982 (M43) [5:35;–5d16']. Messier added both nebulae to his list on March 4, 1769. In 1731, the French scientist Jean-Jacques Dortous de Mairan (1678–1771) first recorded the existence of a patch of nebulosity surrounding a star north of the much larger nebula. M43 is an emission nebula surrounding an eighth-magnitude star. The nebula is crossed by dark streaks and intricate flows of gas and dust.

Continuing north along Orion's sword is the triple emission and reflection nebulae NGC 1973, NGC 1975, and NGC 1977 [5:35;–4d52'] and a small open cluster NGC 1981 (Cr 73) (III2P) [5:35;–4d06']. The nebulae are next to each other, and together with NGC 1981, they resemble a fuzzy triangle. A dark nebula obscures the southern side of the nebulae. Each of the nebulae has a bright star embedded in its gassy glow. Among the 20 stars in NGC 1981 is the blue double star Σ 750 (ADS 4192). The cluster is bright and stands out from the nearby hazy glow of nebulosity and is classified as III2P in the Trumpler classification of open clusters, as shown in Table A.6 in Appendix A.

Robert Trumpler developed his open cluster classification system at the Lick Observatory, while he was studying galactic clusters and making estimates of their distances from Earth, their size, and the amount of interstellar dust. He came up with the system to make it easier to distinguish one type of open cluster from another.

Trumpler's system builds upon the system devised by Harlow Shapley. Working at Mount Wilson for a number of years and ending his career in 1952 as the Director of the Harvard College Observatory, Shapley worked on measuring the size of our Galaxy. As part of this work, he came up with his classification system for the numerous clusters he used in making his measurements. Trumpler used Shapley's data, added his own information, and introduced his system in 1930.

Trumpler is probably most noted for his work in proving the theory of Albert Einstein (1879–1955) that the Sun's gravity would bend the light from distant stars. For the total solar eclipse in 1922, Trumpler went to Australia to photograph the star positions during the eclipse. Comparing photographs taken during the eclipse and those taken six months later, the stars near the Sun did indeed appear to be slightly shifted, thereby proving Einstein's theory.

In most of the clusters, the individual stars are very, very faint, but when grouped together their glow is intensified. Though an individual star may have a visual magnitude of 10, it will appear brighter than a cluster with the same magnitude since the cluster's light is spread out and the star's light is concentrated at one point.

Imagine you are in a gigantic ballroom with only a single one-watt bulb glowing at the other end of the hall. The room is still dark. Now add 1,000 one-watt bulbs and the room becomes brighter as each additional light is turned on. If you turn on a single 1,000-watt bulb instead, its light would be so intense and concentrated it would nearly blind you. The same effect happens when many faint stars are grouped into a cluster. The cluster seems brighter than its component parts, but not as bright as a single point of light of equal apparent luminosity.

Multiple star Alnitak (50 Zeta Ori; ADS 4263A and B) [5:40; –1d56'] at the eastern end of Orion's belt is bathed in the bright nebulosity of IC 434. This emission and reflection nebula is shaped like a thin map of Africa. Unfortunately, the 1.79-magnitude blue-white, spectrum B0 III, supergiant and spectroscopic binary Alnitak (pronounced al-nih-TAK) illuminates IC 434 and the surrounding area so that you cannot see (without the aid of an H-Beta filter) the famous Horsehead Nebula (B 33) intruding into the eastern side of IC 434. Alnitak is a multiple binary with at least two comes. As the primary star in the system, Alnitak has an absolute magnitude of -6.4 and is about 30,000 times more luminous than the Sun. The companion stars are magnitude 4.2 and 10.2. Its name is derived from the Arabic *Al Nitāk*, "The Girdle." The dark Horsehead Nebula is 368.3 pc from us whereas IC 434 is 491 pc. The Horsehead was discovered photographically, in 1889, by Edward Pickering.

The middle star in Orion's belt is 46 Epsilon Ori [5:36;–1d12'] and is known as Alnilam. Alnilam (pronounced al-NIGH-lam) is derived from the Arabic *Al Nitham* meaning "The String of Pearls." Being a B0 Iae spectral type, but slightly brighter, it is otherwise similar to Zeta Ori in type, luminosity, and distance, and is embedded in an emission and reflection nebula (NGC 1990). Epsilon Ori is the MK standard for this spectrum class.

West of Epsilon Ori is the B1 IV spectral-type eclipsing binary variable VV Ori [5:33;–1d09']. The period for this EB-type star to go from magnitude 5.7 to 6.1 is a rapid 1.485 days.

The last major star in Orion's belt is the triple system of Mintaka (34 Delta Orionis; ADS 4134A; β 558) [5:32;–0d17']. Mintaka (pro-

Photo 3.3 IC 434 and dark nebula Bernard 33; the Horsehead Nebula.

nounced min-TAK-ah) comes from *Al Mintakah* meaning "The Belt." Spectrum B0 III Delta Ori shines at apparent magnitude 2.23, but is actually 20,000 times brighter than the Sun. The star is 460 pc distant and speeding away from us at a radial velocity of 16 km/s. This star is an EA-type eclipsing variable with a magnitude fluctuation of 1.94 to magnitude 2.13 during a period of 5.732476 days. The changes in magnitude are due to the orbital motion of a spectroscopic binary with an orbital period of 5.7324 days. This was the first spectroscopic binary for which stationary lines were observed on its spectrogram.

Almost directly north of, and 52.8 arcsec away from, Delta is a bluish 6.85-magnitude, spectrum B2 V, binary companion 34 Delta (ADS 4134C; Σ 14). No orbital movement of this comes around the primary has been detected, so the belief is that these are simply a physical pair about 27,000 AU apart.

At the Potsdam Observatory in 1904, German astronomer Johannes Franz Hartmann (1865-1936) discovered the existence of interstellar gas as he studied the spectra of Delta Ori. He noticed the stationary line of calcium in Mintaka's spectrum and the lack of this line in the spectra of the other belt stars, but could not explain the phenomenon. American astronomer Vesto Melvin Slipher (1875–1969) studied Hartmann's findings and deduced that the cause was absorption by calcium atoms in very low pressure gas in the line of sight between us and Delta. Until that time, space was thought to be a perfect vacuum.

Armed with the concept of interstellar gas and dust, Trumpler, in 1930, was able to revise the estimated size of the Galaxy by determining that this dust and gas absorbed part of the light passing through it, thereby dimming the light at a ratio of one magnitude per kpc. This effect is known as "interstellar extinction" and causes a slight reddening in the spectra of objects more distant from us than the gas and dust clouds.

Almost due east of Delta is the bright reflection nebula NGC 2068 (M78) [5:46;+0d03']. This gas cloud, shaped somewhat like the head of a comet, was discovered by Pierre Méchain, in 1780. Two stars are very noticeable in the middle of the nebula.

Near M78 are three other patches of gas and dust clouds glowing in the reflected light of stars embedded in them. In the same field as M78 are NGC 2071 (northeast of M78), NGC 2067 and NGC 2064 (both southwest of M78). These patches of glowing gas are, like M42 and M43, part of a gas cloud that extends over much of Orion, but most of it is too faint to be seen visually or on astrophotographs. The presence of the gas cloud is known by the radiation energy of hydrogen and to a much lesser extent other chemicals detected by radio telescopes. Only those parts of the gas and dust cloud illuminated by the glow of nearby stars or glowing itself in reaction to stellar radiation are visually detectable.

Both Taurus and Orion contain many more targets for your optical equipment than I have covered here. Each of these constellations is littered with clusters, easy doubles, and faint gas clouds. December and January are the best months in which to observe the Bull and the Hunter for during these two months they are in your southern sky before midnight. If you are observing the sky early in the mornings during September and October you can see them rise over your eastern horizon and begin to study them then. They will still be in your evening western twilight sky into March and April.

4 *February*

Canis Minor, Canis Major, and Puppis: Dog days and the stern of Jason's *Argo*

The Pythagoreans bid us in the morning look to the
heavens that we may be reminded of those bodies which
continually do the same things and in the same manner
perform their work, and also be reminded of their purity
and nudity. For there is no veil over a star.

> Emperor Marcus Aurelius Antoninus (A.D.
> 121–80),
> *The Meditations*,
> (Book XI, paragraph 27)

Usually you think of the "dog days" as those hot lazy days of summer, but you have to look skyward in the cold frosty nights of February for the best views of the wonders to be found within the boundaries of Orion's two faithful canine companions as well as the poop deck of the most famous sailing craft of all time – Jason's mythological ship the *Argo*.

Star-hop in Canis Minor

Though Canis Minor (CMi) is usually referred to as a companion of Orion, this is not the role this dog played in Greek mythology. Canis Minor was originally known as Procyon and is associated with the tales of the hunter Actaeon. The hapless Actaeon was turned into a stag by Artemis, the Moon-goddess, and devoured by his own 50 dogs as punishment for seeing Artemis naked while she was bathing in a stream.

The star-hop through Canis Minor, as shown in Figure 4.1, contains some interesting variable and double stars which we will explore this month. The Small Dog (also known as the Lesser Dog) is one of the smallest constellations, ranking 71st. It covers 183.367 square degrees of sky and has just two bright stars, Procyon and Gomeisa.

We will start this month's short star-hop at the eighth-brightest star in the sky: Procyon (10 Alpha Canis Minoris; ADS 6251) [7:39;+5d13']. Procyon (pronounced PRO-see-on) is the MK standard for spectrum F5 IV-V white subgiants and shines in the winter sky at magnitude 0.38. The Little Dog Star is the primary for a multiple star system located close to us at 3.46 pc and getting closer at a rate of 3 km/s.

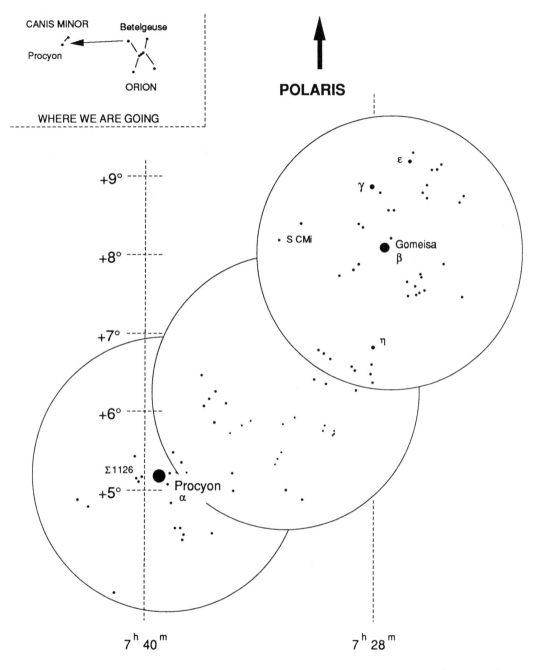

Figure 4.1 Star-hop in Canis Minor.

To find Procyon, first locate Orion. Procyon is the bright star about 27 degrees (three fists) to the east of Orion's eastern shoulder star, Betelgeuse (58 Alpha Orionis) [5:55; +7d24'].

In 1718, Edmond Halley noticed the coordinates for Procyon, along with Aldebaran, Arcturus, and Sirius, had shifted when compared to old Greek and Arab star catalogues. Most of the other stars were still at their coordinates as listed in the catalogues. Procyon had moved over $\frac{1}{2}$ degree; the apparent angular diameter of the Moon. Halley deduced the movement of these stars was due to their individual movement and

79

not the movement of the Earth. Precession affects all of the stars equally so individual actual stellar movement (proper motion) had to be the cause of Procyon's 1.25 arcsec annual shift. Halley's work ended the belief that all of the stars were "fixed" in space.

Procyon is accompanied in space by a degenerate white dwarf star, Procyon B. Located about 15 AU from Procyon A, this 11th-magnitude comes encircles Procyon in an orbit that takes 40.65 years to complete one revolution. Procyon B is about twice the size of the Earth, but contains about 0.65 of the mass of the Sun. German astronomer Georg Friedrich Julius Arthur von Auwers (1838–1915) noticed the slight wobble in Procyon's proper motion. In 1861, he calculated that Procyon probably had a companion. Due to the brilliance of Procyon, its companion is only visible in very large telescopes and was not seen visually until German-American astronomer James Martin Schaeberle (1853–1910) discovered it with the 36-inch refractor at Lick Observatory, in November 1896. Three other faint stars make up this multiple star system.

Studies show the degenerate white dwarf Procyon B fits into a rare spectral class designated DF. The D stands for dwarf. With a surface temperature estimated at 6,000 K, stars like this one are slowly cooling off and will die out as a spectral-type DC (continuum without emission lines) black dwarf. The term "degenerate" when used with a white dwarf refers to the compression of the space within an atom which allows more mass (up to 1,000 kilograms per cubic centimeter; 1,000 kg/cm^3) to be gravitationally packed into a given space than would happen under normal atomic activity. The mass of a white dwarf is tightly compressed at a density of about 300 kg/cm^3. The highest mean density observed in a solar system body is for the Earth at 5.52 g/cm^3. The mean density of the Sun is 1.410 g/cm^3.

East of Procyon are two double stars. The closest to Procyon is an eighth-magnitude pair. The second pair is Σ1126 (ADS 6263) [7:40;+5d14']. These stars shine at magnitudes 6.6 and 6.9 and are about 1 arcsec apart. The primary is a spectral-type A0.

Head slightly northwest of the string of stars connecting Procyon and Gomeisa (3 Beta CMi) until you come to the spectrum A5 double star 5 Eta CMi (ADS 6101; β 21) [7:28;+6d57']. Eta CMi shines at magnitude 5.3 whereas its secondary is located at position angle 25° and is faint at magnitude 11.1. At a distance of 4 arcsec, you may be able to separate these stars.

Moving northeast across the star string connecting Procyon and Gomeisa is the variable S CMi [7:32;+8d19']. This long-period M-type pulsating class variable has a spectral range of M6e–M8e. Over 90 percent of all the known variables are of the M-type (more commonly known as Mira stars). The majority of M-type variables show intense emission lines of hydrogen in their spectra and therefore fall into the Me spectra class. These stars also belong to the MK luminosity classes II and III. Mira stars have periods ranging from 80 to over 700 days and amplitudes of 2.5 to 11 magnitudes. The period for S CMi is 332.94 days and its magnitude fluctuates between 6.3 and 13.2.

The first star to become known as a periodic variable was discovered on August 13, 1596, when Dutch astronomer David Fabricius (1564–1617) saw a "new star" in the constellation Cetus. Soon after its discovery, the pulsating red supergiant faded beyond the naked-eye range. Bayer saw it a few years later and added it to his 1603 atlas as Omicron Ceti. Again it faded, only to be rediscovered in 1638. Polish astronomer, Danzig (Gdansk today) city councillor, and brewer Johannes Hevelius (also known as Johann Hewel, Johann Hewelcke, and Johann Howelcke) (1611–87) called the star "Mira" for wonderful. Omicron Ceti was the first star known to exhibit a regular period of magnitude fluctuation. Over 6,000 catalogued Mira-type variables are located throughout the sky.

Gomeisa (Beta CMi) [7:27;+8d17'] has an apparent magnitude of 2.90, but at 42 pc its absolute luminosity is about 105 times as bright at the Sun. Gomeisa (pronounced GO-me-sa) was known as one of two stars named *Al Murzim* by the Arabs. The name refers to the fact the rising of this star let the observers know that Procyon would soon rise. This spectral-type B8 Ve shell star also exhibits a radial velocity of 22 km/s in recession. Gomeisa is the primary in a multiple system that has four faint very open secondaries. This star is also a Gamma Cassiopeiae type of eruptive variable. When a massive flare erupts from the equatorial zone on this rapidly rotating star, its magnitude fluctuates between 2.84 and 2.92 in a period lasting 0.09 day. The 4.3 angular degree distance between Procyon and Gomeisa was used by the Arabs as one of their sky distance measuring tools.

Northeast of Gomeisa is the multiple stars of 4 Gamma CMi (ADS 6100) [7:28;+8d56']. This open multiple shines at magnitudes 4.3, 12.0, and 13. The 13th-magnitude "B" star is at position angle 240° and has an orbital period of 11 years. The "C" star is located 110 arcsec from the primary at position angle 262°. The primary is a yellow K3 III: Fe-1 spectral type. Gamma is accompanied by a spectroscopic secondary that has an orbital period of 389 days.

Star-hop in Canis Major and Puppis

The portions of Canis Major (CMa) and Puppis (Pup) that we will tour on our second star-hop this month, as shown in Figure 4.2, lie along the fringe of the Milky Way. Covering 380.118 square sky degrees, Canis Major ranks 43rd in size. Puppis is 20th in size as the aft portion of the Jason's ship covers 673.434 square sky degrees. These constellations are rich in bright stars, doubles, clusters, nebulae, and lore.

Puppis is one of those Milky Way constellations you do not hear talked about much at star parties. Certainly it is not as well observed by amateurs as its summer Milky Way counterpart, Sagittarius. I am not sure why since it is loaded with over 50 clusters and nebulae, 60-plus variables, and 125-plus multiple star systems. Most of these objects are located in the narrow portion of the constellation located between Canis Major and Pyxis the Compass.

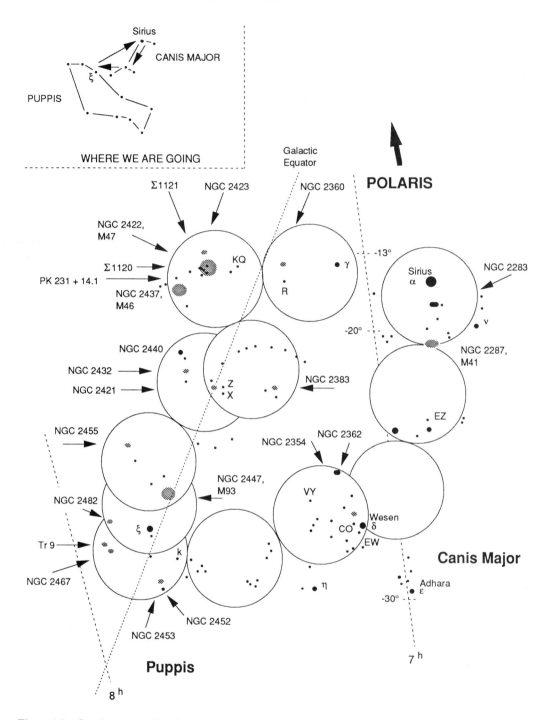

Figure 4.2 Star-hop in Canis Major and Puppis.

To the ancient Greeks, Canis Major represented the dog Laelaps. Poor Laelaps was passed from one god to another several times. He seems to have entered the mythological stage as a hunting dog owned by Artemis. She gave the great hunting hound to Minos on Crete. As a bribe of adulterine seduction, Minos gave the dog to Procris, the jilted wife of Cephalus. Leaving Minos, Procris disguised herself as a boy, changed her name to Pterelas, and joined Cephalus in a hunting expedition. Not realizing who she was, Cephalus offered to buy the dog and Procris's darts that never failed to hit their target. She accepted, but only if they would go to bed. There she revealed herself to him and they reconciled their marriage.

Artemis was so upset because her gifts were being traded so freely for sex that she plotted to get even with Cephalus and Procris. She convinced Procris that Cephalus was still visiting Eos, who had borne him a son. Suspicious, Procris followed him on a hunt one night. Cephalus heard a noise behind him. Thinking it prey, he shot a dart at it and killed his wife. Charged with murder, he fled with Laelaps to Thebes.

In Thebes, King Amphitryon borrowed Laelaps to hunt the vixen Teumessian. Under divine guidance, Laelaps always caught whatever he was after and for the same reason Teumessian never could be caught. A dilemma arose. Zeus settled the matter by turning Laelaps and Teumessian into stones.

At one time Puppis was part of a much larger constellation called Argo Naivs. Its name came from the famous ship on which Jason and his crew of 50 had sailed in search of the Golden Fleece. The year after French astronomer Nicolas Louis de Lacaille (1713–62) died, his *Coelum Australe Stelliferum* ("Star Catalogue of the Southern Sky") was published. In this work, he broke up Argo Naivs into Carina, Vela, Pyxis, and Puppis. He retained Bayer's Greek letter designation for each of Argo Naivs's stars, so the brightest star in Puppis is designated zeta instead of alpha.

In 1801, German astronomer Johann Elert Bode (1747–1826) published his *Uranographia*. In this star atlas, he depicted the northern area of Puppis as a new constellation called Officina Typographica for the Printing Press. This constellation along with several others he proposed are no longer used. For a time, Argo was also called Noah's Ark. The Arab world knew it as *Al Safinah*, "The Ship."

The goddess Athena honored the great and successful voyage of Jason and his Argonauts by placing their ship in the sky. Puppis now depicts the stern or poop deck of the *Argo*. Five crew members were also honored by the gods who placed them in the sky as the constellations Hercules (Heracles), Cepheus, Lyra (the Lyre of Orpheus), and the Gemini twins (Castor and Polydeuces [Pollux in Latin]).

The Milky Way has been the subject of numerous tales and myths devised to explain its existence. To some tribes of American Indians, the hazy band of light stretching across the sky was caused by the Breathmaker when he blew a frosty breath across the sky. This became the pathway to the City of the West where the souls of honorable Indians went. Dishonorable Indian souls stayed in the ground. To most of the Mediterranean cultures it was the River of Heaven or a Milky Pathway. In Greek mythology, Heracles was the mortal son of Amphitryon and Alcmene. Hera despised the baby Heracles and wanted to keep him from becoming the ruler of Argos. She coveted that position for Eurystheus instead. As Hera slept, Hermes placed

Heracles at her breasts to suckle and become immortal. When the goddess awoke, she pushed the baby away. The milk that squirted from her breast became the Milky Way.

The easiest place to start this star-hop, as shown in Figure 4.2, is at Sirius (pronounced SEAR-ee-us). Except for the Sun, this is the brightest star in our sky. The spectrum A0m Vm: CN1; Ca1, Dog Star (9 Alpha Canis Majoris; ADS 5423) [6:45;–16d42'] is 2.7 pc from us and has an absolute luminosity 26 times as bright as the Sun. Several centuries ago, the term "dog days of summer" arose with the widely held belief that the close proximity of the Sun and Sirius during July and August caused the heat of summer.

During the period 1834 to 1844, German astronomer Friedrich Wilhelm Bessel (1784–1846) studied the proper motion of this -1.46-magnitude blue-white star. He detected a slight wavelike proper motion and deduced that Sirius was probably accompanied by an invisible secondary, which affected the star's straight line motion. The 8.65-magnitude white dwarf Sirius B was discovered, in 1862, by Alvan Graham Clark, the son of the famous American telescope maker Alvan Clark (1804–87). Sirius B the "Pup" was the first white dwarf to be seen. The stars in this spectral DA-type class of stars have small diameters, small mass, low luminosity, great density, and high temperatures. This class was not developed as a separate class until about 50 years after Clark first saw the Pup.

The orbital period for Sirius B is 50.090 years. On a night of great seeing with a large telescope, you might be able to separate Sirius A and B. Try placing Sirius A outside your field of view to reduce its glare.

With Sirius on the northern edge of your finder, you will have the faint Sc-type galaxy NGC 2283 [6:46;–18d11'] near the center of the field and the open cluster NGC 2287 (M41) (II3m) [6:47;–20d44'] on the southern fringe of your field of view. NGC 2283 shows as a faint diffuse patch near a trio of 12th- and 13th-magnitude stars. M41 is one of the few naked-eye open clusters. It contains ten stars of magnitude 6.93 to 8.5. The brighter of its approximately 100 stars form a twisted X with a red giant central star. This cluster extends over an area the size of the full Moon and is located 721.3 pc from us. This cluster was first seen telescopically sometime before 1654 by Italian court astronomer Giovanni Batista Hodierna (1597–1660) in his 20 power Galilean refractor. Flamsteed rediscovered it in 1702, and 47 years later French astronomer Guillaume-Joseph-Hyacinthe-Jean-Baptiste Le Gentil de la Galazière (1752–92) wrote about it. Messier added it to his list, on January 16, 1765.

In 1760, Le Gentil sailed from France to observe the transit of Venus that would be visible from Pondicherry, India. The war between France and England prevented him from landing at Pondicherry, so he prolonged his voyage to see the June 3, 1769, transit from Manila. Just before the transit, he was ordered back to Pondicherry. In India, the transit was clouded out, but not in Manila. The Royal Society and the Admiralty sent Naval Captain James Cook (1728–79) to the South Pacific to observe this same Venusian transit.

On this voyage, Cook also discovered many Pacific islands, including New Zealand, from the decks of his ship *HMS Endeavour*. The Command Module for *Apollo 15* and a US space shuttle have been named for Cook's ship.

Head southeast to 1.8-magnitude Wesen (25 Delta CMa) [7:08;–26d23']. Wesen (pronounced WE-zen) forms the apex of the triangle marking the dog's hips and legs. At an estimated distance of 940 pc, this spectrum F8 Ia, white supergiant is the most distant naked-eye individual star. Its absolute luminosity of -8.0 means it may be 60,000 to 70,000 times brighter than the Sun. Wesen is the MK luminosity standard for this spectral type. East of Wesen is a reversed J-shaped line of stars. In the middle of these stars is the BE-type eruptive variable EW (27 CMa) [7:14;–26d21']. This star has a magnitude amplitude range of 4.3 to 4.6 with an irregular period. Just north of 27 CMa is the magnitude 8.7 to magnitude 9.5 variable CO CMa. These two variables point you in the direction of the open cluster NGC 2354 (III3p). NGC 2354 [7:14;–25d44'] is a very loose cluster of about 100 stars; being so loose you can easily miss it in the multitude of background stars.

On the way to NGC 2447 (M93) is another interesting variable, VY CMa [7:23;–25d46'], which is also a double star (ADS 6033; van den Bos 46). VY CMa has at least three very faint companions which also exhibit some variability. The period for this spectrum M5, red supergiant lasts between 200 to 1,900 days and its magnitude ranges between 6.5 and 9.6. The star is shrouded by a reflection nebula, but is easy to locate.

Crossing the boundary line, we enter Puppis and come to the third brightest star in this constellation, 7 Xi Pup (ADS 6393, ß1063) [7:49;–24d51']. Xi Pup is a spectrum G3 Ib supergiant with an apparent magnitude of 3.34 and an absolute magnitude of -4.5. The star is also racing away from us at 3 km/s. Xi has a 12.8-magnitude comes located 4.8 arcsec from it.

In your finder along with Xi Pup is the open cluster NGC 2447 (M93) (IV1p) [7:44;–23d52']. Messier discovered this easy to spot cluster of about 45 to 50 stars on March 20, 1781. It has an integrated magnitude of 6.2, yet its brightest star is a faint magnitude 9.7. M93 straddles the galactic equator at the 240° mark. This cluster is right on the border of being a naked-eye object. M93 is 1,100 pc from us.

Seven clusters (NGC 2453, NGC 2467, Trumpler 9 [Harvard 2], NGC 2482, NGC 2455, NGC 2421, and NGC 2383) and a planetary nebula (NGC 2452; PK 243-01.1) form a wide semicircle south, east and north of Xi Pup and M93. All of these are faint targets, but worth the effort to locate them.

Open cluster NGC 2421 (I2m) [7:36;–20d37'] has an integrated magnitude of 9, yet the brightest of its 70 members is a faint magnitude 11 star. NGC 2421 is located on the galactic equator at a distance of 1.9 kpc from us.

Slightly west of NGC 2421 are the variables X Pup [7:32;–20d55'] and Z Pup [7:32;–20d40']. The magnitude amplitude for X Pup is from 8 to 9.2. The spectral class changes from F6 to G0 for this

Photo 4.1 Planetary nebula NGC 2440 (PK 234+02.1) in Puppis.

long-period Cepheid as its magnitude fluctuates. The M- type variable supergiant Z Pup has a greater magnitude change, 7.3 to 14.5, but its emission spectral class remains fairly constant at M6e to M9e.

Follow the thin line of stars northeast from Z Pup to the 9.1-magnitude planetary nebula NGC 2440 (PK 234+02.1) [7:41;–18d13']. In 1992, the Hubble Space Telescope, orbiting above the blurring effects of our atmosphere, took the first clear images of the central star. Preliminary studies of the data indicate that this dense white dwarf may have a surface temperature of 200,000 K, making this star the hottest known to date. The nebula appears as a slightly out of round bluish halo around its 16th-magnitude central star.

North of NGC 2440 is the open cluster NGC 2437 (M46) (II2r) [7:41;–14d49']. This very rich 500-member star cluster was discovered by Messier, on February 19, 1771. About 150 of the stars shine at magnitudes of between 10 and 13. M46 contains several deep blue giant spectrum A0 stars which exhibit rapid rotational speeds under spectroscopic study. The cluster is estimated to be 1,660 pc from us and getting farther away at a rate of 42 km/s. Use a low-power eyepiece in dark skies.

In the field with M46 is the planetary nebula NGC 2438 (PK 231+04.2) [7:41;–14d44'], discovered by Sir William Herschel. The

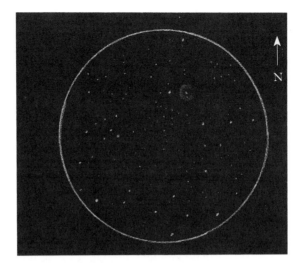

Eyepiece Impression 4.1 Open cluster NGC 2437 (M46) in Puppis.

Photo 4.2 Planetary nebula NGC 2438 (PK 231+04.2) in Puppis.

planetary nebula is believed to be much closer at 900 pc and moving faster at 76 km/s in recession than M46. These facts tend to confirm that NGC 2438 is not a member of the M46 cluster. The 16th-magnitude central star is a deep-blue A0 class object. Its surface temperature is calculated to be a very hot 75,000 K. The planetary nebula appears as a fuzzy star among the other members of the cluster when you have them in sharp focus.

About $1\frac{1}{2}$ degrees northwest of M46 and in your same finder field as M46 is the naked-eye open cluster NGC 2422 (M47) (I3m) [7:36;–14d30']. Giovanni Hodierna is credited as being the first to telescopically see this cluster. This was sometime before 1654. Messier first saw this cluster on the same evening as M46, but he wrote down the wrong coordinates. For many years, M47 was one of his missing objects. The cluster contains about half a dozen bright stars, which give the cluster an apparent magnitude of 4.3. In the middle of the cluster is a group of stars that form a miniature dipper. The brightest star is on the eastern edge of the group and shines at magnitude 5. About 45 additional stars are considered to be members of the cluster. The estimated distance to M47 is 480 pc. The cluster is estimated to be 78 million years old. I have easily located and observed M47, even in the light-polluted skies of the San Francisco Bay Area, so you should have little problem locating it in almost any unclouded sky.

Mixed in with the 50 members of M47 are two multiple systems, Σ 1120 and Σ 1121. Σ 1121 (ADS 6216) is located near the center of the quadrangle of bright stars in the middle of the cluster. Its primary and "B" stars are 7.9-magnitude spectrum B9 stars located about 4,100 AU apart. Eight other stars make up this system. Σ 1120 (ADS 6208) is an easy double to separate. The primary is a pale-blue, spectrum B9, sixth-magnitude star located on the northwestern side of the quadrangle. The "C" star in this system is not visible in small telescopes.

North of M47 is the faint, but easy to locate, open cluster NGC 2423 (IV2m) [7:37;–13d52']. The 40 members of the 350-million-year old cluster shine with an integrated magnitude of 6.7. A close double is visible near the center of the cluster.

West of, and in the same field as, M47 is the strikingly intense orange variable KQ Pup [7:33;–14d31']. This spectral-type M2 Iabpe star is suspected of being an eclipsing binary system. Its magnitude fluctuates between 4.88 and 5.17.

Back in Canis Major, we come to the open cluster NGC 2360 and variable R CMa. NGC 2360 (I3r) [7:17;–13d58'] contains about 50 compact stars and shines with an apparent combined magnitude of 9.1. The cluster is sparse and not easy to locate in a small telescope. NGC 2360 is in an area with few stars, so the cluster stands out from the blackness around it. Four bright blue-white stars form a quadrangle in the middle of the cluster.

R CMa [7:19;–16d24'] is an EA/SD eclipsing binary variable. Its magnitude ranges from 5.7 to 6.3 in a period of 1.135 days. The eclipse phase lasts four hours. EA/SD-type variables are semidetached (SD) systems where the beginning and the end of the eclipse phase can easily be identified by plotting its light curve. At other times in the stars' period, its magnitude remains constant.

The volume around all celestial bodies, out to about $2\frac{1}{2}$ times its radius, is called a *Roche lobe*, as shown in Figure 4.3. This lobe, first calculated to exist by French astronomer Édouard Roche (1820–83), encompasses the object's gravitational field. Where the Roche lobes for two close objects (such as binary stars or planets and their moons) meet, there is a point of equal gravitational forces between them known as the Lagrangian point (L_1). Each body also has an L_2 point

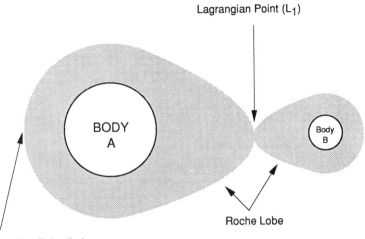

Lagrangian Point (L$_1$)

BODY
A

Body
B

Roche Lobe

Lagrangian Point (L$_2$)

Figure 4.3
Gravitational fields
between celestial
bodies showing
Roche lobes and
Lagrangian points.

located on the edge of the gravitational field opposite from its companion body. The mathematical theory behind this phenomenon was developed, in 1772, by the French-Italian mathematician and astronomer Joseph-Louis Comete de Lagrange (1736-1813). In a semidetached binary system, the stars can expand to the point where one star's material expands beyond its L$_1$ point and loses matter to its companion's stronger gravitational field.

We make our last stop on this star-hop at the "secular variation" 23 Gamma CMa [7:03;−15d38']. This star probably was much brighter 400 years ago when Bayer designated it Gamma. He may have observed it as the third brightest star in Canis Major. But since Bayer did not always designate the lucida star for a constellation as alpha, we cannot be sure if Gamma was the third brightest in his day. For some constellations, Bayer designated his stars according to the position of the star in the pattern of the constellation. Today, this red giant is the 14th brightest star in the constellation. Currently, Gamma shines at magnitude 4.11 and with a B8 II spectra and luminosity rating. This star has been known by several names over the centuries. At one time it was called Muliphen, but by mistake as that is one of the names Delta CMa is known by. Gamma CMa has also been called Mirza and Isis.

Though small, these three constellations contain a rich variety of objects to view. I suggest you also take a look south along the Milky Way in Puppis. The southeastern corner of Jason's ship is adorned with clusters, doubles, and variables.

5 *March*

Cancer, Leo, and Corvus:
The Crab, the King of Beasts, and the Crow

I stood by the open casement
  And looked upon the night,
And saw the westward-going stars
  Pass slowly out of sight.

Till the great celestial army,
  Stretching far beyond the poles,
Became the eternal symbol
  Of the mighty march of souls.

And long let me remember,
  That the palest, fainting one
May to diviner vision be
  A bright and blazing sun.

Thomas Buchanan Read (1822–72),
"The Celestial Army"
(lines 1–4, 9–12, and 41–4),
1860

This month we will continue to head east, keeping ahead of the Sun. The two zodiac constellations, Cancer (Cnc) and Leo as well as Corvus (Crv) have generally been known by these designations since ancient times. These three animals inhabit an area of the sky surrounded by Gemini the Twins to the west, Canis Minor to the southwest, tiny Sextans the Sextant and elongated Hydra the Water Snake stretched along their southern borders, Virgo the Virgin to the east of them, and the Coma Berenices to the northeast. Leo Minor resides north of Leo.

Star-hop in Cancer

The idea of a crab as a constellation has its origins deep in antiquity. For this faintest of the ecliptic constellations, most of the early civilizations had names that meant crab or some other related clawed creature; a lobster or crayfish. A beetle was a sacred animal of immortality to the Egyptians. They honored this creature by naming this asterism *Scarabaeus*.

To some of the Greeks, this area of the sky was the "Gate of Men" where souls passed through as they descended to Earth to enter human bodies. The name we know this constellation by comes from their mythological crab that was placed in the sky by Hera after Heracles had crushed it. The tale surrounding the crab begins when Zeus declared the next grandson of Perseus would rule Mycenae. Hera persuaded Eilithyia, the goddess of childbirth, to hold back the impending birth of Heracles, a grandson of Perseus and Andromeda, and to speed up the birth of Eurystheus instead. Eurystheus thereby inherited Mycenae.

As grown men, King Eurystheus sent his rival Heracles on the famous Twelve Labors. Some sources on antiquity claim that Eurystheus was afraid of Heracles and gave him the tasks hoping that Heracles would be killed. Other sources claim the two men were lovers and that Heracles completed the Labors for his lover. In the second of these labors, Heracles was to kill the Lernaean Hydra. This monster had been raised by Hera near the source of the River Amymone in Lerna, Greece.

The multiheaded snake Hydra was empowered with the ability to regrow a head if it was cut off. Heracles used a sword to cut off the heads and flaming arrows or branches to cauterize the neck-stumps in an effort to keep the heads from regenerating. He finally succeeded in killing the monster. Hera sent a giant crab to kill Heracles to avenge the death of her pet monster. The crab nipped Heracles on the heel, but the mighty hero was able to crush it. Hera then rewarded the crab with a place in the heavens. The head of Hydra and Cancer are next to each other in the sky. The protagonist of this tale has his own place in the summer sky – Hercules.

Cancer contains few bright stars in the 505.872 square degrees of sky assigned to it. Cancer ranks 31st in size, has two Messier objects and several other interesting objects which we will stop at. We will hop from Procyon to Acubens and end in the area of the Beehive Cluster.

Since Cancer's lucida star (65 Alpha Cancri) is a faint 4.25 magnitude, we will start this month's first star-hop, as shown in Figure 5.1, at Procyon (10 Alpha Canis Majoris) [7:39;+5d13'] then hop to Cancer. Being the eighth brightest star, Procyon serves as one of the guide stars in this area of the sky. From Procyon, follow the north-easterly path of faint stars to orange, spectrum K4 III: Ba 0.5, magnitude 3.52, double star 17 Beta Cnc (ADS 6704) [8:16;+9d11']. This suspected variable star was called *Al Tarf* by the Arabs and signifies the end or the southern foot of the Crab.

Hop along the string of stars leading east toward 65 Alpha Cnc. Forming an elongated triangle with 37 Cnc [8:38;+9d34'] and 36 Cnc [8:37;+9d39'] is the Delta Scuti-type of pulsating variables, VZ Cnc [8:40;+9d49']. This type of variable contains stars in spectral classes AO to F5 with luminosities in the range of III to V. Their magnitudes fluctuate in fractions of a magnitude over a period of an hour to two days. VZ Cnc's photovisual magnitude ranges from 7.18 to 7.9 in a period of about two hours. During its pulsations, VZ's spectra shows a greater amount of spectral change than of magnitude as it goes from being an A7 III spectral class star to a F2 III type.

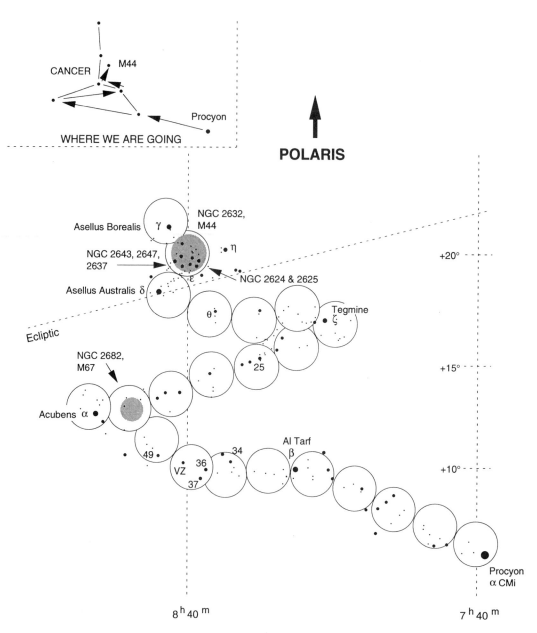

Figure 5.1 Star-hop in Cancer.

Continue to follow the trail of bright stars to the wide double Acubens (65 Alpha Cancri; ADS 7115) [8:58;+11d51']. This spectroscopic binary star's name is derived from the Arabic *Al Zubānah* and means "The Claws." Acubens (pronounced ACK-u-bens) shines at magnitude 4.25 and is an A3m spectral-type star. Its companion is an 11th-magnitude star located 11.3 arcsec from Alpha at position angle 325°.

West of, and in the same finder field as, Alpha is the open cluster NGC 2682 (M67) (II2r) [8:50;+11d49']. This object was discovered by German astronomer Johann Gottfried Koehler (or Köhler) (1745–1801) sometime between 1772 and 1779. Messier added it to his list on April 6, 1780. M67 is believed to be one of the oldest open clusters, since its core stars are closer together than the core stars of other open clusters, it contains few young blue stars, and has about 80 white dwarfs among its 500-plus members. Estimates for its age range from 4 to 10 billion years. Unlike most open clusters, M67 is located far above the plane of the Galaxy. On a very dark night with keen eyesight, you might be able to spot this object as it is just beyond the normal naked-eye range at magnitude 7. Its brighter stars range in magnitude from tenth- to 16th-magnitude. M67 is 800 pc from us. Use a low-power eyepiece or binoculars for the best views of M67. It takes a larger than 6-inch scope to distinguish as individuals many of the members of this cluster.

Heading northwest across Cancer we come to the visual binary Tegmine (16 Zeta Cancri; ADS 6650A, B, and C; Σ 1196) [8:12;+17d38'] marking the edge of the Crab's shell. In 1756, German astronomer and self-taught mathematician Johann Tobias Mayer (1723–62) discovered that Tegmine (pronounced teg-MY-nee) is a close double. Sir William Herschel discovered a third star in the system, in 1781. The inner stars, Zeta[1] A and Zeta[2] B, are both main-sequence stars. Spectral-type G5 V, Zeta[2] B orbits spectrum F8 V, Zeta[1] A in a period of 59.7 years and is too close to Zeta[1] A to be separated in most small telescopes. The stars appear as a single yellowish star, because their magnitudes are about equal at 5.44 for "A" and 6.01 for "B". The third member of the system, Zeta[1] C, is easy to locate being almost 6 arcsec away from the primary stars. Zeta[1] C consists of a yellow F9 V star and maybe two low-luminosity white dwarfs. Zeta[1] C takes about 1,150 years to make a complete orbit of Zeta A and B. Spectral analysis and its irregular orbit reveal the presence of Zeta[1] C's companions. The Zeta Cancri system is located 16 pc away and is getting closer to the Sun at a rate of 8 km/s.

Use the string of faint stars heading east of Zeta Cnc to reach the open double 31 Theta Cnc [8:31;+18d05']. Then sweep due east to the magnitude 3.94, open double Asellus Australis (47 Delta Cancri; ADS 6967) [8:44;+18d09']. To Ptolemy, Asellus Australis and Asellus Borealis (43 Gamma Cancri) were the southern and northern Ass Colts. Delta Cnc sits almost on the ecliptic near the 128.75° position. The Sun is at this ecliptic location on August 1st. Delta is a KO III-IIIb spectral-type object and is accompanied by a 12th-magnitude optical star at position angle 121°.

Northwest of Delta Cnc is the naked-eye visible open cluster, known since ancient times as the Praesepe; NGC 2632 (also known as M44 and the Beehive Cluster) (II3m) [8:40;+19d59']. The Praesepe (pronounced pray-SEE-pee) which means the "Manger" was one of the few deep-sky objects considered to be a nebula (cloud) before the invention of the telescope. The cluster was designated as Epsilon by Bayer, but that name now refers only to the brightest star in the cluster. Ptolemy listed the cluster as "The nebula called the Crab" in his *Almagest*, since it appeared to him as a cloudy spot in the sky. For several of the stars in Cancer, he noted the star's relationship

Photo 5.1 NGC 2632 (M44); the Praesepe in Cancer.

to the nebula. In 1610, this nebula was one of the first objects Galileo studied through his telescope. In his *Sidereus Nuncius* ("The Starry Messenger" [Galileo considered the title to be "The Starry Message," but his humble title has been corrupted over time]), he reported the nebula was "a mass of more than forty starlets." Messier added the cluster as the 44th object to his list, on March 4, 1769.

Since no nebulosity has been detected within the cluster, and M44 is composed mostly of yellow and orange main-sequence stars similar to the Sun, scientists believe that M44 is an old cluster. A few white dwarfs are included in the 350-plus member cluster, which is located 160 pc away. The brightest member of the cluster is 41 Epsilon Cnc [8:40;+19d32']. This spectrum A6 III star is 70 times brighter than the Sun and lies near the southeastern edge of the cluster. Most of the stars in the cluster are fainter than tenth magnitude. Use low-power and a wide-field eyepiece or binoculars to view the entire cluster at one time.

Near the center of M44 is the seventh-magnitude triple star system β 584 [8:39;+19d33']. These stars form a small triangle near the center of the cluster. The three AO stars are only 45 and 93 arcsec apart and are easy to separate, even in binoculars. The "C" star is a double-line spectroscopic binary.

M44 also contains several variable stars. S Cancri [8:43;+19d02'] is an EA/DS-type eclipsing binary with a period of 9.5 days and magnitude amplitude of 8.3 to 10.3. The dwarf eclipsing binary TX Cancri has a period of only 0.38 day. Its magnitude change is a mere 0.3 of a magnitude. Mira-type U Cancri changes from magnitude 8.5 to a very faint 15.5 in a period of 304.78 days. On the northern fringe of the cluster is the pulsating slow irregular (LB-type) variable UV Cancri [8:38;+21d10']. This variable takes 299 days to cycle from magnitude 8 to 9.5 and return to magnitude 8. LB-type variables fit into spectral classes K, M, C, and S, and are generally very red in appearance. UV Cancri is located about midway between 33 Eta Cancri [8:32;+20d26'] and 43 Gamma Cancri.

In our line of sight with M44 are five very faint galaxies. In a large scope, NGC 2624, NGC 2625, NGC 2637, NGC 2643, and NGC 2647, appear as out-of-focus stars in the southern half of the cluster.

Our last stop in Cancer is northeast of M44 at the magnitude 4.67 multiple system Asellus Borealis (43 Gamma Cancri) [8:44;+18d09']. The Northern Ass Colt is an A1 Va spectral-type star and is the primary for a spectroscopic binary system. This system is a member of the Hyades Moving Group. To some ancient cultures, the southern and northern Ass Colts portended disaster for those born under their astrological sign if bad weather obscured these stars in an otherwise clear sky.

Star-hop in Leo

For our second star-hop, we will sweep eastward into the constellation of Leo, as shown in Figure 5.2. On this hop we will observe Regulus, Denebola, Zosma, and several Messier galaxies. Leo sits on the western fringe of the massive cluster of galaxies stretching north and south through Virgo and Coma Berenices. Many of these galaxies, along with dozens of multiple star systems, are visible in Leo.

Most of the ancient western and Middle Eastern cultures depicted the asterism as a lion. The Chinese, at various times in their history, considered it as a "Horse," a "Red Bird," and the "Quail's Fire." The sickle-shaped portion of Leo was their "Yellow Dragon."

The constellation's name comes to us from the mythological world in which Leo, the lion monster, was the offspring of Orthrus and Echidna. The lion was raised by Hera and set loose to ravage the region of Nemea. The lion lived in a cave and his coat was said to be impervious to fire and metal weapons. In the first of his Twelve

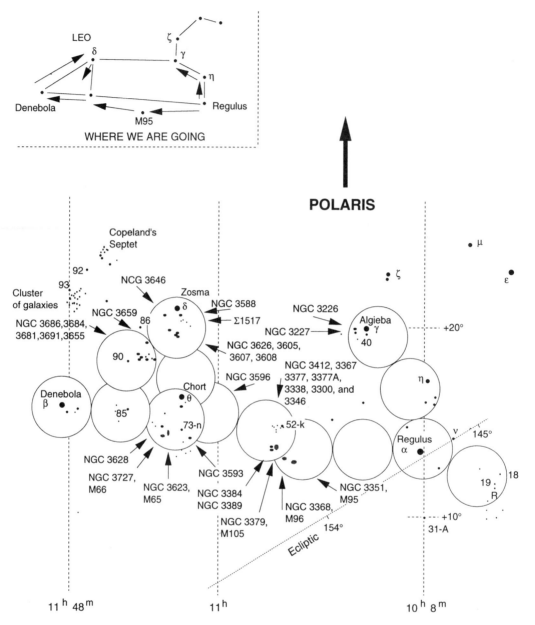

Figure 5.2 Star-hop in Leo.

Labors, Heracles was sent to slay the Nemean Lion. He trapped the lion in its cave and clubbed it to death. Heracles then skinned the lion with its own claw. He wore its coat as a cape with the lion's head serving as his new helmet. In honor of Heracles, Zeus placed the lion in the sky.

Leo rests on his haunches facing the setting Sun at this time of the year. He consists of two easy to locate asterisms. The head of the 12th largest constellation appears as a backwards question mark or a farmer's sickle, and his hind quarters are marked by a right triangle of bright stars. Leo covers 946.964 square degrees of the night sky.

The starting point for this hop is the 21st brightest star in the sky, and Leo's heart; Regulus (32 Alpha Leonis; ADS 7654; Σ II 6) [10:08;+11d58']. Blue-white, spectrum B7 Vn, main-sequence Regulus (pronounced REG-you-luss) sits at the base of the question mark asterism. Regulus shines at magnitude 1.3 and is located 26 pc away. Its absolute magnitude is calculated to be -0.6, which means the star is 150 times brighter and three times larger in diameter than the Sun. The star was named by Polish astronomer Nicolas Copernicus and means "Little King."

Regulus is the primary star in a triple system. Its companion, Regulus B, shines at magnitude 7.7 and is about 177 arcsec northeast of Regulus in position angle 307°. Regulus B has a very close dwarf companion, Regulus C, but since it is faint at magnitude 13.2 and in the glare of Regulus, you probably will not be able to separate these stars. This star is also a spectroscopic binary.

Regulus is about $\frac{1}{2}$ degree north of the ecliptic and is often occulted by the Moon. If you are in the right place when Regulus is hidden by the northern or southern limb of the Moon, you can watch the star blink off as its light is blocked by the mountains and the light flashes back on as it passes through the lunar valleys. Watching this "grazing" also gives you a feeling for how fast the Moon is moving eastward relative to the westward-moving background stars.

On your star charts you may have the dwarf elliptical galaxy Leo I (UGC 4570) [10:08;+12d18'] plotted just north of Regulus. This galaxy is close to our Galaxy at 230 Mpc (750,000 light- years) and is therefore a member of the Local Group of Galaxies. It has extremely low surface brightness, which makes it too faint to be seen visually in any telescope.

Continue north to 30 Eta Leonis [10:07;+16d45'] then move one hop to the yellow giant Algieba (41 Gamma[1] and Gamma[2] Leonis; ADS 7724; Σ 1424) [10:19;+19d50']. The name Algieba (pronounced al-GEE-bah) may have derived from the Arabic word *Al Jeb'hah* meaning "The Brow" or "The Forehead," but some scholars believe it comes from either the Latin word Algieba or the Arabic *Al Jabbah* which represented Leo's forehead and the 8th *manzil*. A manzil was the daily resting place or "Mansion" of the Moon (*Al Manāzil al Ḳamr*). This resting place idea was common among many ancient civilizations. This close four star system marks part of Leo's mane. The actual magnitude for Algieba is 2.2, but combined with the magnitude of 3.5 for its yellow giant comes, the pair shine at magnitude 1.9. Gamma[1] is a spectrum KO IIIb: Fe-0.5 star. Sir William Herschel discovered that Algieba was a double in 1782. Beginning in 1831, F. G. W. von Struve recorded the changing positions of Gamma[2], which is located 4.3 arcsec from the primary at position angle 123° (as measured in 1980). Gamma B is a G7 III: Fe-1 type star. The orbital period of Gamma[2] around Gamma[1] is between 407 and 618 years. The other two visual stars in this system are both ninth-magnitude stars located 260 and 333 arcsec from Gamma[1]. This system also includes spectroscopic binary stars. Algieba is located 33.7 pc distant and is moving toward us with a radial velocity of 37 km/s.

At 22 arcmin south of Algieba is the white giant, spectrum F6 IV, 40 Leonis [10:19;+19d28']. Do not confuse this 4.79-magnitude star as being one of Algieba's companions. This star is located 29 pc from us and receding at a radial velocity of 7 km/s.

About $\frac{1}{2}$ degree east of Algieba are NGC 3226 [10:23;+19d54'] and NGC 3227 [10:23;+19d52']. NGC 3226 is a round 12.3-magnitude E2 galaxy located at the northern end of the elongation of NGC 3227. This pair of galaxies looks almost like an exclamation point. NGC 3227 is an 11.55-magnitude Sb-type Seyfert galaxy seen face on. These interacting galaxies are listed as number 209 in Vorontsov-Vel'iaminov's catalogue.

About one hop west from Regulus you will find two stars, reddish 5.63-magnitude 18 Leo [9:46;+11d48'] and yellowish-white 6.45-magnitude 19 Leo [9:47;+11d34']. These stars point southeastward to the pulsating red giant variable R Leonis. To the west of R Leonis are two ninth-magnitude stars, C Leonis and D Leonis. These three stars form a tight triangle with R Leonis at its eastern apex. This long period Mira-type variable was discovered, in 1782, by Polish astronomer J. A. Koch. In a period lasting 312.43 days, the star's spectrum cycles from M6.5e to M9e as its visual magnitude goes from 4.4 to 11th or fainter levels. During its cycle, R Leonis remains a very distinct red and sometimes appears almost purple. Watch it over a period of months and you can easily see it dim and brighten when compared to its neighbors. R Leonis is located about 184 pc from us and is getting farther away with a radial velocity of 13 km/s.

From R Leonis, hop east (passing through Regulus) following the trail of fifth-, sixth-, and seventh-magnitude stars until you are about halfway between Regulus and 70 Theta Leonis. This area contains about ten easy (depending on your instrument) to locate galaxies, and includes three Messier objects; M95, M96, and M105. This cluster of galaxies is a part of the Leo Group of Galaxies.

The SBb-barred spiral galaxy NGC 3351 (M95) [10:44;+11d42'] is about $2\frac{1}{4}$ degrees south-southwest of 5.48-magnitude, yellowish spectrum gG4, 52-k Leonis [10:46;+14d11'] and was discovered by Méchain, in late March 1781. This face-on galaxy has a visual magnitude of 9.71 (-19.26 absolute) with a fairly bright core surrounded by wispy arms. The bars, which are hard to see in small instruments, project easterly and westerly from the core and sweep around it to form a near circle. Thin spots of mottled whiteness appear above and below the very faint bars. M95 is 9 Mpc from us with a receding radial velocity of 673 km/s.

At 42 arcmin east of M95 is the Sa-type spiral galaxy NGC 3368 (M96) [10:46;+11d49']. This galaxy was discovered by Méchain, on March 20, 1781. M96 appears slightly smaller and brighter overall than M95, yet its core is smaller and fainter. This is because its denser arms give it a brighter visual magnitude of 9.24 (-19.62 absolute). M96 is moving away from us at a speed of 773 km/s. The mass of M96 is calculated to be equal to about 160 billion Suns. Both M95 and M96 can be seen in the same low-power or binocular field.

About a degree north of M95 and M96 are the E1 elliptical galaxy

Photo 5.2 Spiral galaxy NGC 3351 (M95) in Leo.

NGC 3379 (M105), E7 galaxy NGC 3384, and Sc galaxy NGC 3389. Méchain discovered M105 [10:47;+12d35'] on March 24, 1781, but it was not included by Messier in the 1781 edition of the catalogue. Like all E1 ellipticals, M105 appears like a bright star which you cannot focus into a sharp point of light. Its visual magnitude is listed at 9.26 (–19.39 absolute). M105 is moving away from us with a radial velocity of 756 km/s.

At magnitude 9.96, NGC 3384 [10:48;+12d38'] is dimmer than M105 and appears slightly ovoid. NGC 3389 [10:48;+12d32'] is much fainter at magnitude 11.80 (-19.82 absolute) and appears as a thin streak of weak light southeast of NGC 3384.

About a degree north of M105 is a scattered clustering of faint galaxies NGC 3412 (E5; magnitude 11.45), NGC 3367 (Sc; magni-

Photo 5.3 Galaxies NGC 3623 (M65) and NGC 3727 (M66) in Leo.

tude 11.48), NGC 3377 (E5; magnitude 11.05), NGC 3377A (SBm; magnitude 14.15), NGC 3338 (Sb+; magnitude 11.30), NGC 3300 (SBa; magnitude 13.19), and NGC 3346 (Sc; magnitude 12.49). In the midst of these galaxies is the star 52-k Leonis. Use this yellowish star to locate the galaxies, as all of them, except NGC 3346 and NGC 3300, are south of 52-k. All of these galaxies appear very small and

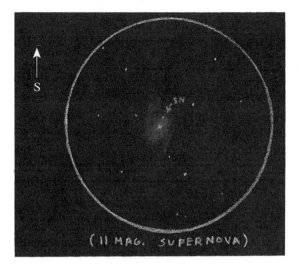

(II MAG. SUPERNOVA)

*Eyepiece Impression
5.1   NGC 3727
(M66) in Leo.*

are difficult if not impossible to see in scopes smaller than eight inches. You will also need a good dark site on a moonless night to see them. Search carefully, looking for smudges of light that you are unable to focus into sharp points. These will be the galaxies. Challenge your eyesight and use averted vision.

Sweep east to 3.34-magnitude Chort (70 Theta Leonis) (also known as Chertan or Coxa) [11:14;+15d25']. Chort (pronounced CH- ourt) depicts the hips of the Lion and is the western apex of the easy to locate triangle of stars marking the rear haunches of Leo. The Arabs also called Theta, along with Delta, *Al H·arātān*, which means "The Two Little Ribs." This name was sometimes written as *Chortan* which means the "Two Holes" through which the viewer could see into the Lion's innards. Over time, the name was used to refer only to Theta and was changed to the singular Chort. Theta is a blue-white, spectrum A2 V, star located 24 pc from us and has a radial velocity of 8 km/s in recession.

The faint smudge of light about $\frac{3}{4}$ degree south of Chort is the Sc-type galaxy NGC 3596 [11:15;+14d47']. This nearly round object glows at magnitude 11.55. NGC 3596 has a bright small core.

About $2\frac{1}{2}$ degrees south-southeast of Chort is magnitude 5.32 red 73-n Leonis [11:15;+13d18']. Four Sb galaxies, M65, M66, NGC 3593, and NGC 3628, will be in the same finder field of view as 73-n Leonis. Not only do these galaxies appear to us to be visually close, but they are actual neighbors in space. NGC 3593 [11:14;+12d49'] is 5.5 Mpc from us, M66 is 6.6 Mpc, M65 is 7.3 Mpc; and NGC 3628 [11:20;+13d36'] is 7.7 Mpc. They are all moving away from us with similar radial velocities.

About a degree southeast of 73-n Leonis are the galaxies NGC 3623 (M65) [11:18;+13d05'] and NGC 3727 (M66) [11:20;+12d59']. Both of these Messier objects were discovered by Méchain prior to March 1, 1780. Compared to the other galaxies we have seen on this tour, M65 and M66 are larger and slightly brighter. Both can be seen in the same low-power eyepiece and are best viewed from a dark site.

When comparing these two objects, M65 is dimmer by one-half magnitude and is slightly longer and narrower than M66. These galaxies, along with NGC 3628, are interacting with each other and are number 308 in Vorontsov-Vel'iaminov's system of interacting galaxies.

Forming a triangle with M65 and M66 is NGC 3628 [11:20;+13d36']. This elongated Sb-type galaxy is at the northern apex of the triangle. Midway between M65 and NGC 3628 is a 7.1-magnitude, spectrum F8 V, star. This star is located 42 pc from us and getting farther away at a rate of 6 km/s. At magnitude 10.31 (-19.96 absolute) and appearing large to us, NGC 3628 can easily be viewed in small scopes. A dust lane bisects the galaxy. Its radial velocity is 728 km/s in recession.

Another easy to spot galaxy is NGC 3593 [11:14;+12d49']. Look for its elongated glow about $\frac{1}{2}$ degree southwest of 73-n Leonis. We view this 11.7-magnitude (-17.01 absolute) galaxy edge on. Like the other three galaxies we have just looked at, NGC 3593 has a radial velocity in recession of 543 km/s.

The hind quarters of Leo are marked by the triangle composed of Chort, Zosma (68 Delta Leonis) [11:14;+20d31'] to the north, and Denebola (94 Beta Leonis) [11:49;+14d34'] to the east. In the center of the triangle is another cluster of galaxies with magnitudes ranging from 10.0 to 13.5.

Zosma (pronounced ZOSE-mah) is the very open multiple system β 1282. Located 95 arcsec from Delta A at position angle 43° is a magnitude 12.1 star designated Delta a. At position angle 344° and 191.4 arcsec from Delta A is a 8.6-magnitude star designated Delta B. Delta A Leonis is a 2.6-magnitude, spectrum A4 IV, blue-white main-sequence star located 24.5 pc away and moving closer to us at a rate of 21 km/s. With a proper motion of 0.20 arcsec in position angle 133°, Zosma appears to be gravitationally attached to and moving with the Ursa Major star cluster, along with Sirius and several other bright stars.

The name Zosma seems to have been made by mistake as it derives from a Greek word meaning the "girdle" yet this star is located in Leo's rump area. Two other anatomically misplaced names often used in the past have Arabic origins and are Duhr, which is derived from *Al Thaḥr al Asad* meaning "The Lion's Back," and Zubra meaning "Mane." John Flamsteed was studying the area around Delta Leonis on the night of December 13, 1690. He noted in his logbook the position of an object, but gave it no further attention. Years later, this object was determined to have been Uranus, officially discovered by Sir William Herschel 91 years later.

Located 23 arcmin southwest of Delta Leonis is the visual binary star Σ 1517 (ADS 8094) [11:13;+20d08']. The two stars in this system are each 7.7-magnitude, spectrum G0, stars. Their combined magnitude gives them a visual magnitude of 6.9. When first discovered by F. G. W. von Struve in 1829, the stars were 1.0 arcsec apart with the comes at position angle 288°. The comes, as measured in 1940, was at position angle 212°, 0.2 arcsec from the primary. In 1958, the star had moved to be 0.3 arcsec from the primary at position angle 349°. Calculations based on this recorded movement give the pair an orbital period of 4,050 years.

About 2 degrees east of Σ 1517 is the Sc-type spiral galaxy NGC 3646 [11:21;+20d10']. Being small and diffuse with a magnitude of 11.21, this galaxy is best viewed in large scopes. In my 10-inch SCT, it is difficult to separate NGC 3646 from the background glow of hazy stars.

Midway between Zosma and Denebola is the close triple star 90 Leonis (ADS 8220; Σ 1552) [11:34;+16d47']. This system consists of a magnitude 6.0, spectrum B4 V, primary, a 7.3-magnitude comes 3.3 arcsec away at position angle 209°, and an 8.7-magnitude star located 63.1 arcsec from the primary at position angle 234°.

Almost centered within the triangle formed by Beta, Delta, and Theta is a clustering of ten spiral and elliptical galaxies in two groupings. With 90 Leonis on the east side of your finder, magnitude 5.52, spectrum K0 III, 86 Leonis [11:30;+18d24'] along the northeastern edge, and magnitude 5.74, 85 Leonis [11:29;+15d24'] on the southeastern edge, you should have the eastern of the two groups near the center of your view. This group consists of NGC 3691, NGC 3686, NGC 3684, and NGC 3681. Southwest of this group is NGC 3655, and to the northwest is NGC 3659. The western group consists of NGC 3626, NGC 3607, NGC 3608, and NGC 3605.

The E3 galaxy NGC 3605 [11:16;+18d01'] and the S-type spiral NGC 3691 [11:28;+16d55'] will be hard to see as they are very small and faint at magnitudes 13.15 and 13.5, respectively. Also difficult to see is the magnitude 12.78, S-type, NGC 3659 [11:23;+17d49']. These three galaxies are visible in 8-inch or larger scopes. The seven other galaxies are 10th- and 11th-magnitude objects.

The eastern group contains all spiral galaxies. Forming a north-south line are NGC 3686, NGC 3684, and NGC 3681. East of, and in the same field of view as, these galaxies is NGC 3691. The largest and brightest, and therefore easiest to locate is NGC 3686 [11:27;+17d13']. This round Sc-type spiral glows at magnitude 11.4 with a bright core. NGC 3686 is located 23.5 Mpc from us. South of NGC 3686 is another Sc-type spiral, NGC 3684 [11:27;+17d02']. This galaxy appears slightly oval with only the hint visually of any arms. We see this galaxy more face on than the other members of this group. At the southern end of the line is the S(B)b+ spiral NGC 3681 [11:26;+16d52']. This is the smallest of the three galaxies, yet its bright core makes it easy to locate. Use NGC 3684 as your guide as NGC 3691 [11:28;+16d55'] is at a right angle from NGC 3684 and the other two galaxies. Averted vision will help you to locate the very small, and at magnitude 13.5, very faint, NGC 3691.

In the western group, the ellipticals NGC 3607 [11:16;+18d03'] and NGC 3608 [11:17;+18d09'] look like enlarged out-of- focus stars. Though about the same size as seen from our vantage point, NGC 3607 appears much larger since it is a full magnitude brighter at 10.95, whereas the magnitude for NGC 3608 is 11.90. NGC 3607 is 19.9 Mpc and NGC 3608 is 32.4 Mpc from us. Both galaxies are speeding away from us.

Midway between Delta and Beta is the 5.74-magnitude, gK4 III spectral-type star, 85 Leonis [11:29;+15d24']. About $\frac{1}{2}$ degree southwest of 85 Leonis is the pulsating SRB-type, spectrum M5, variable AF Leonis [11:27;+15d09']. SRB-type variables are semiregular giants which fall into spectral classes M, Me, C, Ce, S, or Se. These variables have periods ranging from 20 to over 2,300 days. Some of these stars have overlapping periods of change and some show long periods when their brightness does not change. AF Leonis can be as bright as magnitude 9.4 or as dim as 11 during a 107-day-long period.

Multiple star Denebola (94 Beta Leonis; ADS 8314; β 604) [11:49;+14d34'] marks the rump of Leo. Denebola (pronounced de-NEB-oh-la) is a 2.1-magnitude, blue-white, spectrum A3 V, main-sequence star with three very faint companions. Beta is located about 12 pc from us and shows almost no radial velocity. Denebola's luminosity is estimated to be 20 times brighter than the Sun and its diameter is about twice as large.

In 1870, American amateur astronomer Sherburne Wesley Burnham (1838–1921) began his systematic study of double stars with Denebola. The magnitude 15 "B" star is located 39.7 arcsec from the primary at position angle 346°. He discovered the 13.1-magnitude "C" star 77 arcsec from the "A" star, in 1878. The 8.5-magnitude "D" star is located 264 arcsec from the primary at position angle 203°. The "D" star is a whitish F8 spectral object. Burnham published his first list of visual double stars in an 1873 edition of the *Monthly Notices* of the Royal Astronomical Society. This paper, entitled "Catalogue of Eighty-one Double Stars, Discovered with a Six-Inch Alvan Clark Refractor" formed the basis for his 1906 catalogue that contained the listings for 13,655 double stars.

About $\frac{1}{8}$ degree southwest of Denebola is the very close visual binary star (ADS 8311; β 603) [11:48;+14d17']. The 8.3-magnitude "B" star is located 1 arcsec from the 6.0-magnitude white, spectrum A5 V primary. This is too close to separate, except in large telescopes. The orbital period is calculated to be 122 years. These stars are located 46 pc from us and show a radial velocity of 9 km/s in recession.

If you have access to a large (greater than 15 inches) scope, then look for the large cluster of very faint galaxies southwest of 93 Leonis. Most of these galaxies are fainter than magnitude 14. Another cluster to look for in this area of the sky is Copeland's Septet [11:30;+22d01']. These 14th- to 15th-photographic magnitude spirals are located about 1 degree northwest of 92 Leonis.

Denebola will be one of our jumping-off points for next month's tour of Virgo and Coma Berenices. Before we leave for the realm of the galaxies, we will first take a look at some easy to find objects in Corvus the Crow.

Star-hop in Corvus

The 183.801 square degrees of the sky known as Corvus have since ancient times been referred to as being the depiction of a bird, in particular a crow or raven. The ancient Greeks had two main tales of how the trapezoidal asterism of third-magnitude stars became known as Corvus. Both tales recount events in the life and loves of the Sun-god Apollo.

In the more renowned story, Apollo sent Corvus, his pet white raven or crow, to fetch him a cup of water. The bird took off on his mission with Crater "the Cup" in his talons. He came across a fig tree and waited for its fruit to ripen. After filling himself on ripe figs, he flew back to his master with the water and Hydra "the Water Snake" in his beak. He tried to convince Apollo the snake had delayed him. Apollo did not believe the bird, and in a rage, he turned the bird black and dispatched him to the sky to be near Hydra and Crater. This fable was used to explain why ravens and crows are black.

In another tale of Apollo and his bird, Apollo was the lover of Coronis, the beautiful daughter of Phlegyas, the king of Lapiths. Apollo left his white crow to guard her when he went to attend to matters in Delphi. Coronis had a secret passion for Ischys, the son of Elatus and, though pregnant with Apollo's child, she slept with Ischys. The bird saw this and sped to tell Apollo of his lover's unfaithfulness, but Apollo, being a god, already knew of the affair. He cursed the crow for not doing his duty to guard Coronis from other men. Mad at the bird for allowing Ischys to sleep with his lover, Apollo turned the bird black. He told his sister, Artemis, of the infidelity. Artemis avenged the act by killing Coronis with her arrows.

Saddened at the sight of his dead lover, and realizing she was still with child, Apollo had Hermes remove the baby from the dead woman's womb. Apollo's new son, Asclepius (Aesculapius to the Romans), was taken to live with Cheiron the Centaur. The boy learned how to heal people and became the father of medicine. Astronomers know him today as the summer constellation Ophiuchus the Serpent Bearer. Another interesting outcome from these tales of the Sun-god Apollo is that the name for the bright outer layer of gas from the sun is called the corona, which is derived from the name of his unfaithful lover.

To the Arabs, the constellation represented a camel, *Al Ajmāl*, and a tent, *Al Hiba` al Yamāniyyah*. I see the four main stars of the constellation looking more like a fez than a camel or a tent, or a bird for that matter.

We will begin this tour of the 70th constellation in size at 9 Beta Corvi [12:34;–23d23'], as shown in Figure 5.3. This magnitude 2.65, spectrum G5 IIb, yellow-white giant marks the southeastern corner of the trapezoid. Beta Corvi is located 89 pc from us and has a radial velocity of 8 km/s in approach. Studies of its proper motion and radial velocity indicate the star is moving through space at the speed of 11 km/s.

About $2\frac{1}{2}$ degrees southeast of Beta Corvi is the class 10 galactic globular cluster NGC 4590 (M68) [12:39;–26d45'] in the constellation of Hydra. This cluster is about 0.6 degrees northeast of the 5.5-magnitude close double star B 230 (ADS 8612) [12:37;–27d08']. Both the star and cluster should easily fit into your eyepiece. M68 is a

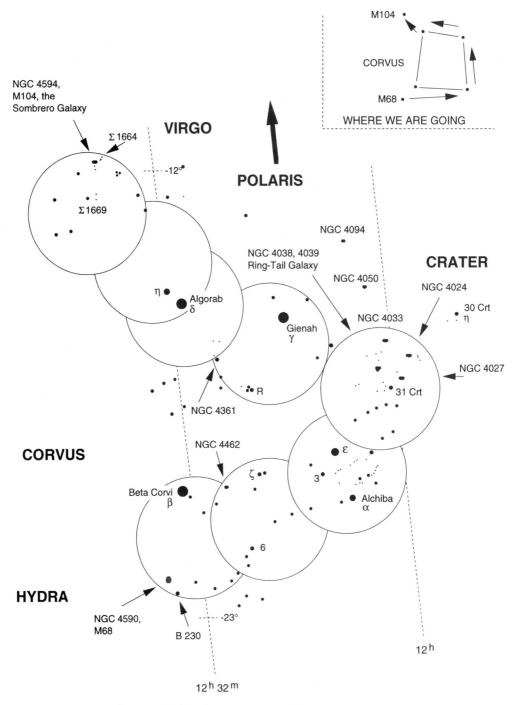

Figure 5.3 Star-hop in Corvus.

large and rich cluster that was discovered by Méchain, in April 1780, and contains an estimated 100,000 stars. The composite spectral type for this object is F2. Its visual magnitude is 8.2, and if we were 10 pc from M68 instead of 9.6 kpc, it would shine as brightly as a -6.81-magnitude object. This cluster is racing toward us at a radial velocity of 117 km/s. The stars in the center of the cluster are not as tightly packed as in most other globulars. This loose core gives you the impression you can see through it.

The close double B 230 consists of an F0 spectral-type primary and its 11.4-magnitude comes 1.3 arcsec away at position angle 170° (as measured in 1959).

To the northwest of Beta Corvi is the Sb III spiral galaxy NGC 4462 [12:29;–23d10']. This galaxy is fairly elongated with a bright core even with its dim magnitude of 12.49.

An interesting note about Corvus is that its α star is not the lucida for the constellation, at least not today. At magnitude 3.00, 2 Epsilon Corvi [2:10;–23d37'] is brighter than 4.02-magnitude 1 Alpha Corvi [10:08;–24d43'], but orange Alpha was given an Arabic name, *Al Chiba*, and Epsilon was not. The current theory is that Alpha has dimmed over the years and changed color as the star was listed as being red by the Persian astronomer 'Abd ar-Rahmān Ibn 'Umar as-Sūfī (more commonly known as al-Sūfī (the Mystic)) (A.D. 903–86). Ptolemy listed both stars at magnitude 3 in his *Almagest*. The two stars are visible together in your finder.

Alchiba (pronounced al-CHE-ba) was originally the name for the whole constellation, and this star was known as *Al Minhar al Ghurāb* meaning "The Raven's Beak." The star is an F2 IV spectral-type star located 21 pc from us and getting farther away at a rate of 4 km/s.

Epsilon Crv is a spectrum K2 IIIa: Ba 0.2 red giant located 32 pc from us. Its luminosity is about 100 times brighter than the Sun. Epsilon is racing toward us at a rate of 5 km/s.

With Epsilon Crv on the southeastern edge of your finder, the 5.2-magnitude, spectrum B1.5 V, star 31 Crateris [12:00;–19d39'] should be near the opposite side. This is one of the few stars numbered in one constellation, but now officially in another. The star used to be in Crater (Crt) the Cup.

To the north of 31 Crt and in the same low-power field, you may be able to locate the interacting galaxies NGC 4038 and NGC 4039 (VV 245) [12:01;–18d52']; these galaxies are known as the "Ring-Tail Galaxy" or the "Antennae." Some astronomers consider this object to be two gravitationally interacting galaxies, whereas others think they may be colliding. A third possibility states the galaxies are undergoing some kind of fissioning process. They show differing radial velocities of 1447 km/s for NGC 4038 and 1430 km/s for NGC 4039. These galaxies are 2,762 Mpc from us. Some people see this object as a question mark. It looks more like a grayish embryo to me. NGC 4038, the larger Sc-type and brighter (magnitude 11.48) galaxy is to the north. NGC 4039 is much dimmer at magnitude 13. You can see its mottled image in 6-inch or larger instruments. Bright and dark patches are visible.

A little more than $\frac{1}{2}$ degree to the southwest of the Ring-Tail is a galaxy some astronomers think might be a companion of the Ring-Tail. Peculiar galaxy NGC 4027 (VV 66) [11:59;–19d16'] has a single arm which extends in the direction of the Ring-Tail. This 11.65-magnitude Sc-type galaxy is only about a million light-years from the other two galaxies. The thought is that NGC 4027 is moving

Photo 5.4 Planetary nebula NGC 4361 (PK 294+43.1) in Corvus.

away from NGC 4038, but the gravity of the larger object appears to be pulling mass away from NGC 4027.

North of NGC 4027 are the E2 galaxy NGC 4024 [11:58;–18d21'] and the E5 galaxy NGC 4033 [12:00;–17d51']. These two galaxies appear as slightly elongated blobs of faint light as both are dimmer than magnitude 12.60.

With the brightest star in Corvus, magnitude 2.59, Gienah (4 Gamma Corvi) [12:15;–17d32'], slightly off center to the northwest in your finder, you may have the planetary nebula NGC 4361 and the Mira-type variable R Corvi near the southeastern edge of the field of view. Gienah (pronounced GEE-nah) is a spectrum B8 IIIp: Hg; Mn, blue-white giant located about 300 light-years from us and getting closer at a rate of 4 km/s. The star's name is derived from the Arabic *Al Janāḥ al Ghurāb al Aiman* meaning "The Right Wing of the Raven." As depicted on modern mythological atlases of the heavens, this star marks the left wing rather than the right, which is Delta Corvi. Gienah may be a very long-period or an irregular variable as it was listed in Ptolemy's *Almagest* as being magnitude 3. It has faded slightly since the 1890's, when it was listed as being magnitude 2.3. This star is about 500 times brighter than the Sun with its absolute magnitude estimated to be -2.0.

The planetary nebula NGC 4361 (PK 294+43.1) [12:24;–18d48'] appears as a misshapen circle of light with its northeastern edge missing. The low surface brightness of this magnitude 10.3 planetary makes it difficult to see, except in dark skies. In 8-inch scopes, the magnitude 13.24, spectrum O6, central star becomes visible. Use a high-power eyepiece for the best views of it. The gas shell around the star is expanding at a rate of 38 km/s. This rapid movement of the gas is quite high. The majority of planetary nebulae with known expansion velocities have rates of less than 20 km/s.

The long-period variable R Corvi [12:19;–19d15'] is almost in the middle of the trapezoid formed by the Raven's bright stars. This M4.5e–M9e star fluctuates from magnitude 6.7 to 14.4 during a period of 317.03 days.

Algorab (7 Delta Corvi; ADS 8572; SHJ 145) [12:29;–16d30'] is a wide double star first noted, in 1823, by British astronomers James South (1785–1867) and John Herschel. Algorab (pronounced al-GORE-ab) is listed as number 145 in their 1824 catalogue. The name used to apply to Gamma, but over the years it came to designate Delta instead. The primary is a white-yellow 3.0-magnitude, spectrum B9.5 IV, star with its purplish, magnitude 9.2, spectrum DK2, dwarf comes 24.2 arcsec away in position angle 214°. The separation and position angle have not changed since their discovery. These stars are located 36 pc from us and have a common radial velocity of 9.1 km/s in recession. The glare of the primary makes it hard to see the comes in small scopes.

To the northeast of Algorab is 8 Eta Corvi [12:32;–16d11']. This 4.31-magnitude, spectrum FO IV, spectroscopic binary star points you in the direction of NGC 4594 (also known as M104, the "Sombrero Galaxy"). Head northeast to the triple star system Σ 1669 (ADS 8627) [12:41;–13d01']. This system consists of a 6.0-magnitude, spectrum F5 primary, a 6.1-magnitude comes 5.4 arcsec from it, and a 10.5-magnitude "C" star 59 arcsec from the primary.

The Sombrero Galaxy is located about two degrees northwest of Σ 1669 at coordinates [12:40;–11d37']. M104 is situated on the border between Corvus and Virgo. This object was discovered by Méchain, on May 11, 1781. At magnitude 8.27, this Sb galaxy is one of the easiest galaxies to locate, even though it is 20 Mpc (40 million light-years) from us. M104 is also one of the largest and brightest galaxies we know of. Its diameter is about 130,000 light-years and its mass is calculated to be equivalent to about 1.3 trillion (10^{12}) Suns, which gives the galaxy an absolute magnitude of -22.98. Studies done on this galaxy in 1914 by Vesto Slipher led to the belief that objects like this one were probably not members of the Milky Way. Slipher discovered this object has a redshifted spectrum which indicates the nebula is moving away from us at a rate of 963 km/s and that it is rotating. No previously studied object exhibited such speed. His conclusions were validated by Hubble's work ten years later.

The Sombrero Galaxy looks somewhat like a hazy Saturn with its rings seen almost edge on. The nucleus appears as a bright bulge with a band of light extending outward to the east and west. In small scopes, the dark absorption lane of gas and dust bisecting the southern part of the nucleus is difficult to distinguish. In scopes of ten inches or larger, it becomes very noticeable as a black streak. Many sixth- and seventh-magnitude Milky Way stars surround M104.

To the west-northwest of M104 is the wide multiple star Σ 1664 [12:38;–11d31'], first measured in 1830. The system consists of six stars. The primary is an 8.1-magnitude, spectrum KO, star. The "B" star is a magnitude 9.3, G5 star located 26.3 arcsec from the primary as position angle 237°. The other four stars are magnitude

Photo 5.5 NGC
4594 (M104); the
Sombrero Galaxy in
Virgo.

11.5 and fainter. The "E" star is located 118.8 arcsec from the "A" star.

In this month, we have looked at a great variety of objects in Cancer, Leo, and Corvus. We have not seen them all, yet. Study your star charts and see what else you can star-hop to.

6 *April*

Ursa Major:
A Dipper round tripper

> To go into solitude, a man needs to retire as much from his
> chamber as from society.... But if a man would be alone,
> let him look at the stars. The rays that come from those
> heavenly worlds, will separate between him and what he
> touches.... But every night come out these envoys of
> beauty, and light the universe with their admonishing
> smile.

> Ralph Waldo Emerson (1803–82),
> *Nature*,
> 1836

For this month's star-hop adventure, we will circle the bowl of
the Big Dipper of Ursa Major (UMa) and take a look at other
deep-sky objects within this very popular constellation. These
two trips are loaded with galaxies upon which you can test your ability
to locate faint objects.

This prominent asterism, as shown in Figure 6.1, has been known
by many names throughout historical times. Seven thousand years
ago, the Egyptians saw the stars as a great hippopotamus. In the last
few centuries B.C., their hieroglyphics depicted the group as a boat
for the god Osiris. The Chinese called this constellation the *Pei-Tou*
meaning the "Northern Basket" or a "Bushel." Over the centuries in
India, it was referred to as the Seven Sages, the Seven Wise Men, the
Seven Bears, the Seven Antelopes, the Seven Bulls, or the Great
Spotted Bull. To the British, the bear looks more like a plough. The
British have also used several different designations for the constellation, including calling it a wagon (wain).

The Greeks began to use the name Ursa Major sometime prior to
900 B.C. The most common tale for Ursa Major revolves around
Zeus; his nymph lover Callisto (also known as Kallisto); their son
Arcas; the eponymous hero of his native Arcadia; and Hera, the wife
of Zeus. This version of the story has Hera enraged with jealousy
when she found out that her husband had sired Arcas. To get back at
Zeus, she turned Callisto into a bear and caused her to wander the
forests for eternity. Meanwhile, Arcas grew up to be a great hunter.
Hera conspired to have Arcas hunt down his mother and slay her.
Seeing the plot unfolding, Zeus saved his former lover from certain
death. As he whipped Callisto over his head and threw her up into the
northern heavens, her tail stretched out to an unnatural length. Arcas

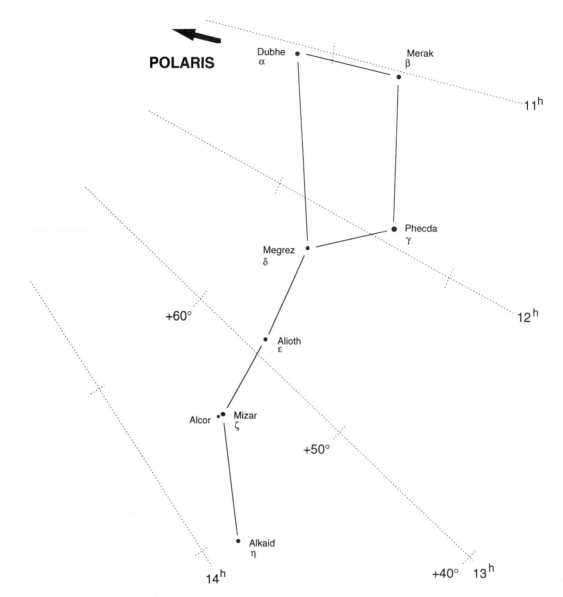

POLARIS

Dubhe
α

Merak
β

11^h

Phecda
γ

Megrez
δ

+60°

12^h

Alioth
ε

Alcor Mizar
ζ

+50°

Alkaid
η

14^h

+40° 13^h

Figure 6.1 Ursa Major in the spring evening sky.

later joined his mother as the constellation Ursa Minor the "Lesser Bear."

The asterism which makes up this circumpolar constellation is actually scattered in space. Alioth (77 Epsilon Ursae Majoris) at coordinates [12:54;+55d57'] is 19 pc from us. Dubhe (50 Alpha UMa) [11:03;+64d45'] is 23 pc, and Alkaid (85 Eta UMa) [13:47;+49d18'] is at 33 pc. The rest of the naked-eye stars of the Dipper form a cluster at 24 pc. Further proof that these stars are not a single group is that Alkaid and Dubhe (pronounced DUBB-ee) have proper motions taking them toward the southwest whereas the others are headed northeast. It will take thousands of years before this movement is noticeable on Earth and the Dipper loses the shape we are familiar with.

Though the main asterism of the Dipper is bright and easy to see, the whole of the third largest constellation spreads out over 1,279 square degrees of sky, and contains a portion of the galaxies which make up the Virgo supercluster. About 25 of these galaxies are within the box formed by the four Dipper stars. Many more galaxies surround the Dipper.

Star-hop from Alkaid to M101

Our first star-hop this month will be an easy trip to a cluster of 12 galaxies southeast of Mizar, as shown in Figure 6.2. This cluster surrounds the spiral galaxy NGC 5457 (M101). In some Messier sources, NGC 5457 is mistakenly listed as M102. The galaxies in this cluster have magnitudes ranging from 8.2 down to 14.2. Many of these will be too faint to observe, unless you use a large scope at a dark site on a night of excellent seeing, but do search for them. M102 is generally thought to be NGC 5866 in Draco. Pierre Méchain is credited with discovering M101, in early 1781. For unknown reasons, Messier and Méchain gave incorrect coordinates for some of the objects they discovered. This leaves some ambiguity about whether M101 and M102 are really the same object or two objects.

You can take an easy path to M101 from Mizar by following the trail of fourth- and fifth-magnitude stars (Flamsteed numbers 81, 83, 85, and 86 UMa), but we will begin this hop at Alkaid instead and take a more interesting route. Alkaid is the last bright star in the handle of the Big Dipper and glows blue-white at magnitude 1.9. Eta was one of the stars British astronomer James Bradley (1693–1762) studied, in 1725, in his unsuccessful attempt to prove the Earth revolves around the Sun. By using the effect of parallax when the Earth is on opposite sides of its orbit, he had hoped to find a star which exhibited a shift in position relative to other stars. This shift would then prove that we in fact had moved. Due to inaccurate equipment, he was unable to measure any parallax shifts, but his work led to his discovery of the Earth's wobbling motion which causes the stars to precess across the sky. In 1742, Bradley succeeded Edmond Halley to become the third Astronomer Royal at the Greenwich Observatory.

About halfway between Alkaid and 17 Kappa Boötis (ADS 9173; Σ1821) [14:13;+54d47'] is a visual double galaxy consisting of NGC 5480 [14:06;+50d43'] and NGC 5481. These galaxies are hard to see, because they are almost magnitude 13. NGC 5480, an Sc/SBc-type galaxy, is 32 Mpc from us. NGC 5481, an SO/E2-type galaxy, is 36 Mpc away and only 3 arcsec east of NGC 5480.

Shift your scope northeast to place the yellow and bluish wide double stars of Kappa Boötis in your view. These 4.6- and 6.6-magnitude stars are 13.5 arcsec apart. The comes is at position angle 236°.

Just before you get to M101, stop and observe the magnitude 10.85, I/SC-type galaxy, NGC 5474 [14:05;+53d40']. This galaxy is relatively close to us at 6 Mpc. It appears slightly distorted with almost no distinctive features in a small scope.

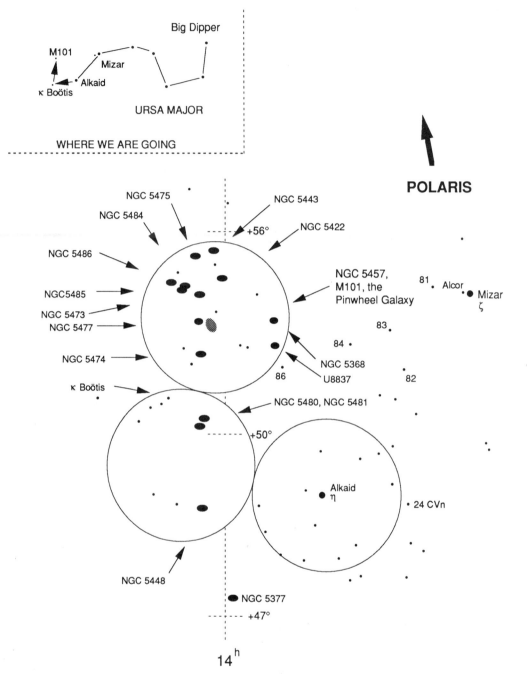

Figure 6.2 Star-hop from Alkaid to M101 (the Pinwheel Galaxy).

One of the three Messier objects to be called the "Pinwheel Galaxy" is NGC 5457 (M101) [14:03;+54d21'], which is a magnificent sight even in small telescopes. The other two Pinwheel Galaxies are NGC 598 (M33) in Triangulum and NGC 4254 (M99) in Coma Berenices. Compared to the other galaxies in the area, M101 is very large. Calculations indicate that M101 has an absolute magnitude of -20.45, but we see its apparent magnitude at 9.6. Eight dark nebulae partly obscure and break up the spiral arms into sections. The central core of the galaxy is quite noticeable even though M101 is 5.4 Mpc (15 million light-years) from us. The arms for this face-on Scd-type galaxy extend for a great distance from the central core. Use averted vision as you try to follow the arms to the knots of stars at their tips.

With M101 centered in your finder, you may have a cluster of four galaxies near the northern edge of your field of view. Not only are these galaxies a line of sight cluster, but they are fairly close to each other in space. A fifth galaxy, NGC 5477 [14:05;+54d28'], is very close to M101, but, with a magnitude of 14 and being very small, you may have difficulty seeing it. The cluster consists of NGC 5473, NGC 5484, NGC 5485, and NGC 5486. All of them are dimmer than magnitude 11. NGC 5485 [14:07;+55d00'] is the brightest of these space islands. Though NGC 5485 is very small and its arms are difficult to discern, it is a good example of an SO/Sa-type lenticular galaxy. NGC 5485 has an absolute magnitude of -20.8 and is located 32.8 Mpc from us. At a distance of 33 Mpc, NGC 5473 [14:04;+54d54'] is an E2/SO-type galaxy with a round outer haze of stars surrounding a small dense core.

Also in your finder, you may have the galaxies NGC 5368, NGC 5422, NGC 5443, NGC 5475, and UGC 8837. NGC 5368 [13:54;+54d20'] is very faint and extremely difficult to see. NGC 5422 [14:00;+55d10'] is an easy to locate SO lenticular/spiral patch glowing at magnitude 13.01; it is 31 Mpc from us. Being small and faint at 13.2 magnitude, NGC 5443 [14:02;+55d49'] is a hard to locate SBb-type galaxy. NGC 5475 [14:05;+53d40'], an Sa-type galaxy, is slightly fainter and a little smaller than NGC 5443. Irregular galaxy U8837 [13:54;+53d54'] is a little blob of grayish light glowing at magnitude 13.3 northeast of 86 UMa. This Im-type galaxy is 3.9 Mpc from us and has an absolute magnitude of -14.82.

Star-hop around the bowl of the Big Dipper

Our second star-hop this month is a trip around the bowl of the Big Dipper, as shown in Figure 6.3, followed by a look at a sampling of the deep-sky objects contained within the bowl. We will begin at yellow Dubhe (50 Alpha UMa; ADS 8035; β 1077), which is the most northerly of the seven naked-eye stars in the Big Dipper. At magnitude 1.79, Dubhe is the 34th brightest star in our night sky and fits into spectral class KO IIIa. About 1 arcsec south of Dubhe is its magnitude 4.8, blue visual binary companion. The orbital period for the companion star is 44.66 years. Unless you are using a very large telescope, you will not be able to resolve them into separate points of light. This star is also a spectroscopic binary system.

Five degrees south of Dubhe is Merak (pronounced ME-rak). These stars are your pointers to Polaris. Merak (48 Beta UMa) [11:01;+56d22'] is a blue-white, spectrum AOm A1 IV-V, star shining at magnitude 2.37. Merak's absolute luminosity of 1.0 and distance of 19 pc means this star is actually 33 times as bright as the Sun. Merak is moving toward the Sun at a rate of 12 km/s. Place Merak at

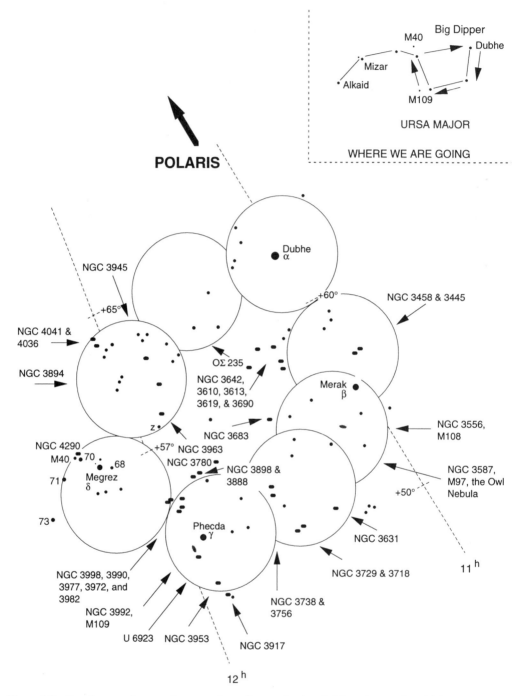

Figure 6.3 Star-hop around the bowl of the Big Dipper.

the northwest edge of your finder field and you should have both the galaxy NGC 3556 (M108) [11:11;+55d40'] and the planetary nebula commonly called the Owl Nebula (NGC 3587; M97; PK 148+57.1), [11:14;+55d01'] in your field of view.

At apparent magnitude 10.2 (absolute magnitude of -20.77), M108 is a faint Scd-type barred-spiral galaxy seen edge on. It looks

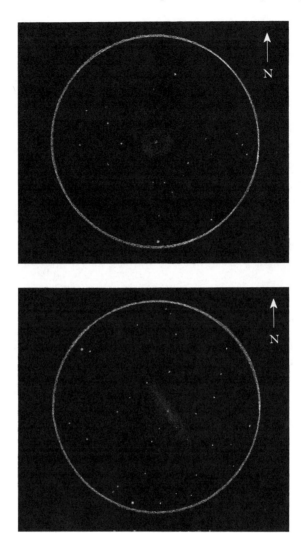

*Eyepiece Impression
6.1* NGC 3587
(M97); the Owl
Nebula.

*Eyepiece Impression
6.2* Galaxy NGC
3556 (M108) in Ursa
Major.

like a hazy streak with a slight brightening and widening near its center. Méchain discovered this galaxy in February, 1781 or 1782. This galaxy is 14.1 Mpc from us.

The type IIIa planetary nebula M97 was first seen by Méchain, on February 16, 1781. In large telescopes, the magnitude 9.9 nebula gives the appearance of an owl's face, but only if you are able to see the two dark openings in the gas shell, which look like its eyes. The gas shell is expanding at a rate of 41 km/s. Three bright stars form a triangle near the center of the nebulosity. The brightest of these three stars is a white dwarf. The nebula has a radial velocity of 6 km/s in recession and is 0.4 kpc from us. Averted vision helps to some degree with this object. Also try a high-contrast filter and different magnifications. The nebula has an apparent diameter about the size of Jupiter.

Five galaxies, NGC 3718, NGC 3729, NGC 3738, NGC 3756, and NGC 3631, are visible on the path between M97 and 2.4-magnitude, spectrum A0 Van, blue-white Phecda (64 Gamma UMa) [11:53;+53d41']. These galaxies are visible in 6-inch and larger

Photo 6.1 Galaxies NGC 3992 (M109) and UGC 6923.

scopes. NGC 3718 [11:32;+53d04'] and NGC 3729 [11:33;+53d08'] are Sa-type galaxies 17 Mpc from us. NGC 3729 is bright at magnitude 10.9 and small whereas NGC 3718 is a magnitude fainter at 11.9, is twice as large as NGC 3729, and shows no nucleus in small scopes.

Irregular Im-type galaxy NGC 3738 [11:35;+54d31'] also appears to be a peculiar-type galaxy. This object has an absolute magnitude of -16.11, but we see its light at magnitude 12. In the same field as NGC 3738 is NGC 3756 [11:36;+54d18'], an Sc-type spiral galaxy located 23.5 Mpc from us. It appears slightly elongated.

Phecda (pronounced FECK-dah) shines with an absolute magnitude of 0.6. Its luminosity equals that of about 75 Suns. The name Phecda derives from the Arabic word for thigh: *Al Faḥdh*. The name designates its position in the thigh of the Bear. Phecda is located 23 pc from us and has a radial velocity in approach of 13 km/s.

Located southeast of Phecda is the SBc-type oval-shaped galaxy NGC 3992 (M109) [11:57;+53d23']. NGC 3992 has a bright nucleus and is located 17 Mpc from us. Do not be confused by the bright star near M109's core. This star is closer to us than the galaxy and is not a part of M109. Méchain discovered this object on February 16, 1781. To the southwest is 13.98-magnitude Ir+ -type galaxy UGC 6923 [11:56;+53d10'].

On the way northeast to Megrez, stop and explore the U-shaped cluster of five galaxies. NGC 3982, NGC 3972, and NGC 3998 are 11th-magnitude galaxies. NGC 3977 [11:56;+55d24'] and NGC 3990 [11:57;+55d28'] are difficult to see in small telescopes. Use a high-power eyepiece to view NGC 3982 [11:56;+55d08'], which is a small Sb-type galaxy, but has a bright center. Its red-shifted spectrum indicates that NGC 3982 has a radial velocity of 1,208 km/s.

The last of the bright stars in the bowl is magnitude 3.31 blue-white, Megrez (69 Delta UMa) [12:15;+57d01']. Megrez (pronounced MEE-grez) is an A2 Van spectral-type star located 20 pc from us. Its absolute magnitude of 1.7 indicates that it shines about 20 times brighter than the Sun. This star's name derives from the Arabic *Al Meghrez* meaning "The Root of the Tail."

One and a half degrees northeast of Megrez is the double star Winnecke 4 (M40) [12:22;+58d05'] and the SBb-type galaxy NGC 4290 [12:20;+58d06']. M40 is slightly northeast of the bright orange 70 Uma. This wide double star is a strange addition to Messier's catalogue, because it is the only double star in the listing and is much fainter than so many other doubles he could have added to his list. Messier was looking for a *Supra tergum neboulosa* (a "nebula above the back" of Ursa Major) first seen by Johann Hevelius in 1660 and listed as object number 1496 in his 1690 catalogue *Prodromus Astronomiae* and atlas *Firmamentum Sobiescianum*. Both works were published three years after his death by his widow and collaborator Catherina Elizabetha Hevelius (also known as Elizabeth Margarethe and Elizabeth Koopman) (1646–93). She is considered to have been the first female astronomer. Messier could only find a pair of ninth-magnitude stars where Hevelius had listed a nebula. He concluded that Hevelius had made a mistake, but included the stars on his own list anyway. This pair was first measured by German astronomer Friedrich August Theodor Winnecke (1835–97) and was added to his list of double stars, in 1863. The 9.3-magnitude comes is at position angle 83° and about 50 arcsec from the 9.0-magnitude primary. Line up Delta and magnitude 5.5, spectrum K2 III, 70 UMa [12:20;+57d51'] to point you in the direction of M40.

The two stars of M40 point you toward NGC 4290. Being 42.1 Mpc from us and faint at magnitude 12.4, this galaxy is difficult to locate, but you can use the right triangle of 70 UMa, M40, and NGC 4290 to find it. NGC 4290 is at the 90° angle.

Heading back to Dubhe, you will pass by six more galaxies, the variable Z UMa [1-1:56;+57d52'], and the double star OΣ 235 (ADS 8197) [11:32;+61d05']. Z UMa is a semiregular, spectrum M5e III, giant with a magnitude range of 6.2 to 9.4 and periods of 98 and 196 days. This is an easy variable to observe, even in binoculars. With the southeastern edge of your finder on Delta UMa, this variable will be near the southwestern edge and located slightly southeast of NGC 3963. The visual binary OΣ 235 consists of a magnitude 5.75, spectrum F5, primary and its magnitude 7.1 companion at position angle 325°. The comes has an orbital period of 5.75 years. In 1943, the position angle of the comes was 39°.

NGC 3963 [11:55;+58d30'] is an Sb/Sc-type spiral. Its spectrum reveals that it has a radial velocity of 3,316 km/s in recession.

Now that we have gone around the outer edge of the Dipper, let us take a look at some of the objects within the Dipper's bowl. This area is crowded with galaxies. You can almost assume that any object you cannot focus into a sharp star-like image is a galaxy. From Dubhe to Phecda, make a slightly southwesterly sweeping curve following these hazy patches of light. With a moderate-size scope and good seeing, you may be able to observe NGC 3642, NGC 3610, NGC 3613, NGC 3619, NGC 3690, NGC 3683, NGC 3780, NGC 3888, and NGC 3898. Near the end of this path we come again to NGC 3998, NGC 3972, and NGC 3982. These galaxies range from 11th to 13th magnitude and are good examples of a variety of galaxy types. You may be able to see about a dozen more galaxies if you have a large telescope.

On these hops we have only scratched the surface of the abundant objects you can see in Ursa Major. Study your star charts and make up your own star-hops to visit the galaxies, double stars, and variables populating this most prominent of constellations. These additional objects surround the areas we explored this month.

7 *May*

Coma Berenices and Virgo:
The sparkling hair of Berenice and the Wheat
Maiden with her bushel of galaxies

salimmo sù, el primo e io secondo,
tanto ch'i' vidi de le cose belle
che porta 'l ciel, per un pertuigo tondo.
E quindi uscimmo a riveder le stelle.

"He first, I second, without thought of rest
we climbed the dark until we reached the point
where a round opening brought in sight the blest
and beauteous shining of the Heavenly cars.
And we walked out once more beneath the stars."

Dante Alighieri (1265–1321),
La Commedia: L'Inferno
("The Comedy: Hell");
Canto XXXIV;
(lines 136–9),
1307?

In this month's star-hops we will tour the realm of galaxies. Spread throughout the constellations of Virgo (Vir) and Coma Berenices (Com) is the Virgo Cluster of Galaxies. This area of the spring sky is a deep-sky observer's paradise. The Virgo Cluster consists of about 3,000 galaxies. The cluster, which is located 12.8 Mpc (42 million light-years) from us, is thought to be near the center of the Local Supercluster. We are not a member of the Virgo Cluster, but our Milky Way Galaxy is a member of the Local Supercluster. A separate cluster of galaxies in Coma Berenices is the Coma Cluster of Galaxies. This 1,000-plus member cluster is located between 92 and 123 Mpc from us and is too faint to be seen except in photographs and the largest telescopes. Nineteen Messier objects, along with hundreds of other interesting targets, are located within the boundaries of these two constellations.

Coma Berenices and Virgo are surrounded by Ursa Major to the northwest, Canes Venatici to the north, and Boötes (Boo) to the northeast. Leo is west of them. Crater, Corvus, and Hydra reside to the southwest and south. Libra and Serpens Caput (SerCp) are Virgo's eastern neighbors.

Compared to the dense and beautiful star fields near the plain of the Milky Way, this region of the sky is sparsely populated by bright stars. We are looking up into the halo of our Galaxy in the vicinity of the north galactic pole (NGP), which is located in Coma Berenices. This lack of star fields and obscuring gas and dust clouds allow us to see far beyond the boundaries of our Galaxy. We are afforded magnificent views of an area crammed with faint extragalactic objects.

On photographic plates, about 3,000 Virgo Cluster galaxies can be detected, but visually in amateur equipment (6-inch reflector or larger) we can see about 250 to 500 of these island universes. Most of the galaxies are spirals and bright ellipticals. All of the galaxies in Coma Berenices and Virgo are faint and you should hunt for them from a dark site on a moonless night. The closer the galaxies are to your meridian, the better are your chances of locating them. You should have a good atlas in hand to keep from getting lost among these faint fuzzies. Use a low-power eyepiece (30x to 50x) when sweeping and switch to higher power to get the best view of an individual galaxy.

Most of the Messier objects in this area are surrounded by other galaxies, and unless you carefully locate the Messier object it is easy to think you are looking at one when you are actually looking at something else. The Messier objects in these two constellations are: M49, M53, M58, M59, M60, M61, M64, M84, M85, M86, M87, M88, M89, M90, M91, M98, M99, M100, and M104. We have already been to M104, the "Sombrero Galaxy," in the southern reaches of Virgo. While in the area, we will also hop to the magnificent globular cluster M3 in Canes Venatici.

Star-hop in Coma Berenices

We will start this month's celestial tour in Coma Berenices and work our way into Virgo on the second star-hop. As a group of fourth- and fifth-magnitude stars, Coma Berenices was known since ancient times as a part of Leo and was commonly referred to as the Lion's tail or as a separate asterism. To the Arabs, the asterism was a part of Leo and was known as *Al Halbah* or *Al Ḍafīrah* meaning "The Course Hair" or "The Tuff."

The Alexandrian-Greek astronomer, poet, and scientific writer, Eratosthenes of Cyrene (*c*. 276 to *c*. 194 B.C.) gave the asterism two names. He named the starry cloud Ariadne's Hair in honor of the daughter of King Minos of Crete. In the pages of mythology, Ariadne led Theseus through the labyrinth after Theseus had slain Ariadne's half-brother, the feared Minotaur. Eratosthenes also considered these stars to be a part of Leo, yet named them for the hair of Queen Berenice of Egypt.

The Dutch mathematician and cartographer Gerhardus Mercator (Latin name for Gerhard Kremer) (1512–94) depicted the constellation as a separate entity on a celestial globe that he made, in 1551, for the Holy Roman Emperor, Charles V (1500–58).

According to astronomical tradition, Danish astronomer Tycho Brahe (1546–1601) made the constellation's name official in 1602, but he had died from a stroke in October the year before. Upon his death, Tycho's meticulous observational records were entrusted to his German assistant Johannes Kepler (1571–1630). Having lost his beloved observatory Uraniborg (Castle of the Heavens) in Denmark, in 1597, Tycho took some of his instruments and moved to Germany. The next year, he circulated among influential men in Germany manuscript copies of a 1,000-star catalogue in an attempt to raise funds for a new observatory near Hamburg. Kepler developed his three laws of orbital motions based largely on Tycho's observations.

Tycho followed Eratosthenes' path and named the 42nd largest constellation after the amber tresses of Queen Berenice II of ancient Egypt. She was the sister and wife of King Ptolemy III (Euergets I) (reigned 246–221 B.C.). Ptolemy went to war against the Assyrians and their king, Seleucus II (also known as Callinicus) (reigned 246–226 B.C.), in the Third Syrian War (246–241 B.C.). To assure his safe return, Berenice pledged to sacrifice her long hair to the goddess Arsinoe Aphrodite. Upon her husband's victorious return, Berenice had her hair cut off and placed in a locked sanctuary in a temple at Zephyrion (Zephyrium). During the victory celebrations, the lock of hair disappeared. The court astronomer Conon of Samos (*fl. c.* 245 B.C.) declared that Zeus had taken the hair and honored its donor by placing it among the stars. This calmed everyone and the festivities continued.

For the first part of this hop, we will work our way across the eastern galaxies in Coma then go to Denebola (94 Beta Leonis) and look at the galaxies on the western side of Coma. The constellation of Coma Berenices has few bright guide stars, so we will start this month's first star-hop at -0.04-magnitude Arcturus (16 Alpha Boötis), as shown in Figure 7.1. Golden Arcturus (pronounced arc-TO-rus) in the constellation of Boötes is at coordinates [14:15;+19d10'], and shines as the fourth brightest star in the sky. The name for this spectrum K1.5 IIIb star has been carried down from ancient Greece and means the Bear Guard or Keeper of the Bear as this star is said to be watching over Ursa Major.

Follow the westerly trail of stars to reach 4.32-magnitude yellow main-sequence, spectrum F5 V, and binary double star 42 Alpha Comae Berenicis (ADS 8804; Σ 728) [13:09;+17d31']. This star is slightly east of the midpoint between Arcturus and Denebola. Alpha Comae is a very close double, first observed by F. G. W. von Struve, in 1827. The components of this system are seen almost edge on, so at times the stars appear as one and at times they are separated by as much as 9 arcsec. The orbital period for the spectrum F6 V secondary star is 25.87 years.

Alpha Comae is at the southeastern end of the U-shaped open cluster known as the Coma Berenices Star Cluster (Melotte 111) (II3p) [12:25;+26d00']. The distance to this widely scattered cluster is listed at 80 pc, yet many of its brighter members are between 18 and 31 pc distant. The average radial velocity for most of its 80 members is zero to 1 km/s in recession. Most of the cluster members are blue-white A to yellow G spectral-type main-sequence stars. No members are

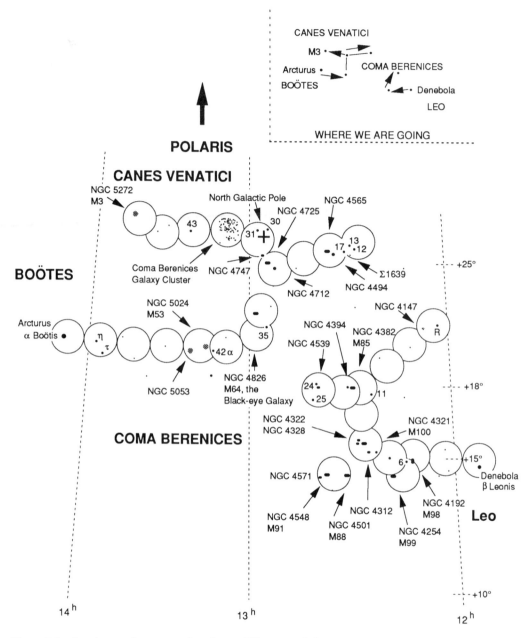

CANES VENATICI

M3 •

Arcturus •

BOÖTES

COMA BERENICES

• Denebola

LEO

WHERE WE ARE GOING

POLARIS

CANES VENATICI

NGC 5272
M3

North Galactic Pole

NGC 4725
30

NGC 4565

43

31

NGC 4747

Coma Berenices
Galaxy Cluster

NGC 4712

17 13
12

Σ1639

NGC 4494

+25°

BOÖTES

NGC 5024
M53

NGC 4147

35

NGC 4394

NGC 4382
M85

Ṙ

Arcturus
α Boötis •

η
τ

42 α

NGC 4539

NGC 4826
M64, the
Black-eye Galaxy

24
25

11

+18°

NGC 5053

COMA BERENICES

NGC 4322
NGC 4328

NGC 4321
M100

6

+15°

NGC 4571

Denebola
β Leonis

NGC 4548
M91

NGC 4312

NGC 4501
M88

NGC 4192
M98

NGC 4254
M99

Leo

+10°

14^h

13^h

12^h

Figure 7.1 Star-hop
in Coma Berenices.

known to be giants. The age of the cluster is estimated at 400 million
years. The brighter members of the cluster include the stars 12, 13-
GN, 14, 15 Gamma, 16, 17-AI, 21-UU, 22, and 31 Comae.
Magnitude 4.94, 31 Comae is the MK standard for spectral-type G0
IIIp stars. This cluster is best viewed in binoculars or finder scope due
to its wide apparent size (about ten Moon diameters).

About $\frac{1}{2}$ degree northeast of 42 Alpha Comae is the globular cluster
NGC 5024 (M53) [13:12;+18d10']. This class 5, magnitude 7.72,
globular was discovered by Johann Bode, on February 3, 1775.
Messier, while rechecking earlier listings of nebulous objects, con-

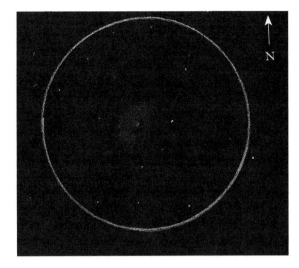

*Eyepiece Impression
7.1* NGC 4826
(M64) in Coma
Berenices.

firmed Bode's sighting on February 26, 1777, and noted it as being a "Nébuleuse sans étoiles" (nebula without star). Over 45 members of the cluster are known to be variable. The center of the cluster appears as a tightly compacted ball of stars surrounded by a glowing halo of stars. In a scope larger than six inches, you can resolve a few stars into individual points of light on the fringe of the central core. Estimates for the distance to M53 range from 17 to 21 kpc. Its radial velocity is 79 km/s in approach.

About 1 degree southeast of M53 is the loose class 11 globular cluster NGC 5053 [13:16;+17d42']. You can see individual stars in the core of this 9.8-magnitude cluster. NGC 5053 was first seen by Sir William Herschel, in 1784. The cluster is 15.2 kpc from us. It shines with the equivalence of 16,000 Suns, which is about the lowest luminosity known for a globular cluster.

Hop about two finder fields northwest from Alpha Comae to the 4.90-magnitude, yellow, spectrum G8 III, very close double star 35 Comae (ADS 8695; Σ 1687) [12:53;+21d14']. The magnitude 7.23, spectrum F6, comes is 1 arcsec from the magnitude 5.06 primary at position angle 182° (equinox 2000). Its orbital period is estimated to be 510 years. Being so close, this is a difficult pair to separate in small scopes.

About $\frac{3}{4}$ degree east-northeast of 35 Comae is NGC 4826 (M64; the Black-Eye Galaxy) [12:56;+21d41']. Johann Bode discovered this Sb-type galaxy in 1779. Messier first saw it on March 1, 1780. The core of this oval-shaped galaxy is very bright, but a thick dust lane obscures the north and eastern sides of the core. M64 is located somewhere between 4.1 Mpc and 13.5 Mpc from us and has a radial velocity in recession of 414 km/s. Its absolute magnitude is calculated to be -19.15, but we see it as a 8.51-magnitude object. Since the distance to M64 is in dispute, its actual luminosity is also uncertain. Luminosity estimates range from a few hundred million to ten billion suns. If it is the latter, then M64 may be the most luminous object we know of.

Take about two hops north-northeast from M64 to 43 Beta Comae [13:11;+27d51']. This magnitude 4.26, spectrum F5.9 V, main-sequence visual double and spectroscopic binary star is the only star in Coma Berenices to have received an Arabic name, but the name *Al Dafīrah* has fallen out of use. This star is coming toward us at a rate of 6 km/s.

The globular cluster NGC 5272 (M3) [13:42;+28d23'] is located about 6 degrees north-northeast of 43 Comae. It is ironic that Messier noted this beautiful class 6, magnitude 6.35, cluster appeared as a "nebula without star" but with a "center brilliant" when he observed it through a very low-power instrument, on May 3, 1764. M3 may contain as many as 500,000 stars, most of which are old red giants and subgiants. The absence of young hot O and B stars leads astronomers to the conclusion that this is a very ancient object, but how old remains a big question. Age estimates for M3 range from 5 to 26 billion years.

This globular also contains more variable stars than have been detected in any other cluster. Almost 200 have been catalogued. One is known to have a period lasting only ten minutes from minimum to maximum. The true diameter for the cluster is about 67 pc (220 light-years). Observation of the streamers of stars, visible in 4-inch or larger scopes, gives you the feeling they are being pulled into the bright core.

About one hop west of 43 Comae is the north galactic pole. The NGP is about $\frac{1}{2}$ degree south of 31 Comae. When looking at the NGP, you are looking at the point in the sky directly over the center of our Galaxy. An observer situated at the NGP would see our Galaxy face on just like we view M100 in Coma Berenices.

Located between 43 Comae and 31 Comae is a mass of about 40 faint galaxies known as the Coma Berenices Galaxy Cluster. Almost all of the galaxies in this distant galaxy cluster are very faint and almost impossible to see, except in a large scope, under ideal skies from a very dark site. The brightest member of this galaxy cluster is NGC 4889 at magnitude 11.4. A 16th-magnitude planetary nebula PK 049+88.1 [12:49;+27d38'] is on the southern fringe of the cluster.

One half degree west of 31 Comae is the magnitude 5.8 open double star 30 Comae (ADS 8674) [12:49;+27d33']. The magnitude 11.5 secondary star is located in position angle 13° 42.5 arcsec from the spectral-type A0 primary. This system has a radial velocity of 1 km/s in approach from its distance of 75 pc.

Star-hop 2 degrees south of 30 Comae to locate the large S(B)b barred-spiral galaxy NGC 4725 [12:50;+25d30']. At magnitude 9.21, this is one of the brightest galaxies in the Coma area. The core is bright in this face-on galaxy. A sprinkling of 8th- to 12th-magnitude foreground stars will also be in your view. The distance to NGC 4725 is estimated to be 12.4 Mpc. To the northeast in the same eyepiece view is the faint (magnitude 12.3) P-type galaxy NGC 4747 [12:51;+25d47']. Since we see this galaxy almost edge on, it visually lacks any definitive spiral shape, but in photographs a bar can be seen.

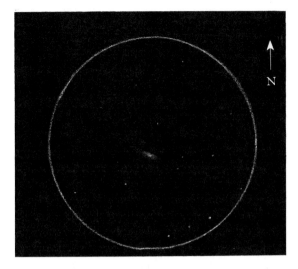

Eyepiece Impression 7.2 Galaxy NGC 4565 in Coma Berenices.

Southwest of NGC 4725 is the very faint (magnitude 13.0) S-type galaxy NGC 4712 [12:49;+25d28']. You will need dark skies and good seeing in a large scope to observe this object. NGC 4712 is speeding away from us at 44,453 km/s.

Considered to be the most beautiful edge-on galaxy and certainly the longest is NGC 4565 [12:36;+25d59']. This Sb-type galaxy is located about 3 degrees north-northwest of NGC 4725 and 2 degrees east of 17 Comae. At magnitude 9.5, this streak of light is one of the few galaxies visible in small scopes, and with excellent skies it can even be seen in binoculars. The arms of this target stretch out about 8 arcsec from both sides of its bright core. A dust lane is visible bisecting the core. NGC 4565 is 9.7 Mpc from us, which means that its apparent 16.2 arcsec length translates into an actual diameter of about 27.6 kpc (90,000 light-years). This galaxy also has a high absolute magnitude of -20.34.

About 1 degree west of NGC 4565 is the E1 elliptical galaxy NGC 4494 [12:31;+25d47']. This ball-like object glows at magnitude 9.86. Use high power to examine this galaxy.

In the heart of the Coma Berenices Star Cluster are three triple stars: 17 Comae, Σ 1639, and 12 Comae. The A0 spectral-type primary for the wide double 17 Comae (ADS 8568; and variable AI Comae) [12:28;+25d54'] shines between magnitudes 5.23 and 5.40. The magnitude 6.6 "B" star is located 145.4 arcsec from the primary in position angle 251°. In 1889, Burnham measured a magnitude 14.6 "C" star (β 1080) 1.8 arcsec from the "B" star. AI Comae is an Alpha^2 Canum Venaticorum-type (Alpha^2 CV; also ACV) of rotating variable with strong magnetic fields. As the star rotates, during a period of 5.0633 days, the intensity of its spectra changes simultaneously. This type of rotating variable is noted for its anomalously strong spectral lines of strontium, silicon, rare earths, and chromium.

About 1 degree south-southwest of 17 Comae is the triple star Σ 1639 (ADS 8539) [12:24;+25d35']. F. G. W. von Struve first measured this visual binary system in 1836. The magnitude 7.79 comes orbits its magnitude 6.76 primary at a distance of 1.7 arcsec in a period of 678 years. In 1952, a faint "C" star was located 90 arcsec from the primary in position angle 160°.

The third triple system is 12 Comae (ADS 8530) [12:22;+25d50'], which is located $1\frac{1}{2}$ degrees west of 17 Comae. James South and John Herschel first listed 12 Comae as a multiple in their 1824 catalogue. With a combined magnitude of 4.79, this star is the brightest one on the western side of the cluster. The primary shines at magnitude 4.8. The magnitude 11.8 "B" star is located 35 arcsec from the primary at position angle 54°. The A3 spectral-type, magnitude 8.3, "C" star is 65.2 arcsec from the "A" in position angle 167°. The system shows a radial velocity of 1 km/s in approach. 12 Comae is 27 pc from us.

About $\frac{1}{2}$ degree north-northeast of 12 Comae is the variable 13 Comae (GN Comae) [12:24;+26d05']. This suspected Alpha^2 CV rotating variable's magnitude fluctuates between 5.15 and 5.18.

To locate the galaxies we want to observe on the western side of Coma Berenices, first aim your scope at Denebola on the eastern edge of Leo. From Denebola, take about one and a half star-hops east to whitish, magnitude 5.10, spectrum A2 V, main-sequence star 6 Comae [12:16;+14d53'].

The S(B)c-type spiral galaxy NGC 4192 (M98) [12:13;+14d54'] can be seen to the northwest in the same low-power eyepiece as 6 Comae. This faintly glowing edge-on galaxy was first seen by Pierre Méchain, on March 15, 1781. Messier confirmed the sighting on April 13, 1781, and added it to his list. This galaxy is not a member of the Virgo Cluster as its radial velocity is 220 km/s in approach whereas most of the members of the cluster have radial velocities in the range of 2,000 km/s in recession. M98 is located at a distance (10 Mpc) about halfway between us and the cluster. The apparent magnitude for M98 is 10.9 whereas its absolute magnitude is -21.08. The galaxy looks like a thin smudge of light with a slight bulge near the center.

The brighter face-on Sc-type spiral NGC 4254 (M99) [12:18;+14d25'] is located about 1 degree east-southeast of 6 Comae. Even in small scopes, the sweeping arm structures in this roundish object are visible. Méchain found this object on the same night as M98 was located, and Messier confirmed its existence on April 13, 1781. This is another galaxy sometimes called the "Pinwheel Galaxy." M99 is a member of the Virgo Cluster and has a radial velocity of 2,324 km/s in recession. In total mass and diameter, M99 is about half the size of our Galaxy. Its absolute magnitude is -20.84.

Two fifth-magnitude stars heading east-northeast from 6 Comae point you to the large Sc-type galaxy NGC 4321 (M100) [12:22;+15d49']. Méchain discovered this, the largest face-on spiral in the Coma area, on March 15, 1781. Messier first saw this member of the Virgo Galaxy Cluster on April 13, 1781. In 1850, Lord Rosse noted its spiral shape as being similar to that of M51 in Canes Venatici. M100 glows at magnitude 9.37 and has an absolute magnitude of -21.13. This galaxy is 16.8 Mpc from us and its red-shifted spectra indicates that it is receding at a speed of between 1,543 and 1,590 km/s. The

Photo 7.1 Galaxies NGC 4321 (M100), NGC 4312, NGC 4328, and NGC 4322 in Coma Berenices.

galaxy appears circular with very faint arms. You can see the nucleus a lot easier than the diffuse arms. In a low-power, wide-field eyepiece, you may be able to include the elongated Sb-galaxy NGC 4312 [12:22;+15d32']. This object is much fainter at magnitude 11.78 and is much smaller than M100. A ninth-magnitude star is located equidis-

Photo 7.2 Galaxies NGC 4382 (M85) and NGC 4394 in Coma Berenices.

tant between M100 and NGC 4312. Two 13th-magnitude spiral galaxies, NGC 4322 [12:23;+15d54'] and NGC 4328 [12:23;+15d48'], are less than $\frac{1}{4}$ degree east and northeast of M100.

About $1\frac{3}{4}$ degrees north-northwest of M100 is the yellow G8 III spectral-type, magnitude 4.78, double star 11 Comae (ADS 8521; Ho 52) [12:20;+17d47']. This star leads us to the Ep-type (also listed as an SO) galaxy NGC 4382 (M85) [12:25;+18d11']. This small circular patch of magnitude 9.22 light is located about 1 degree northeast of 11 Comae. Méchain determined its position on March 4, 1781, and Messier confirmed the discovery two weeks later. This member of the Virgo Galaxy Cluster is estimated to be as massive as 100 to 400 billion Suns. M85 is located 16.8 Mpc from us with a redshift giving it a recessional velocity of 773 km/s. To the east in the same low-power field is the barred-spiral galaxy NGC 4394 [12:25;+18d13']. NGC 4394 glows at magnitude 10.92.

From 11 Comae, star-hop about $2\frac{1}{4}$ degrees north-northwest to the class 6 globular cluster NGC 4147 [12:10;+18d33']. This tight ball of stars glows at magnitude 10.26 and is 17.5 kpc from us. Some individual stars can be resolved around the fringes of the cluster.

About 1 degree northwest of NGC 4147 is the Mira-type variable R Comae [12:04;+18d49']. This red star fluctuates from magnitude 7.1 to 14.6 during a period of 362.82 days.

The easiest way to find our next two Messier objects, M88 and M91, is to return to 6 Comae then to M99. M88 and M91 are at the same declination as M99, so with an equatorially mounted telescope all you have to do is center M99 in your eyepiece, turn off your motor drive, and wait about 14 minutes for M88 to drift into your view. Wait approximately three more minutes and M91 will slip into view.

On the busy evening of March 18, 1781, Messier observed nine new objects (M84 to M92). One of these was the Sc-type galaxy NGC 4501 (M88) [12:32;+14d25']. He noted that it resembled his nebula number 58. A string of faint foreground stars trail southward from the southeastern edge of the galaxy. M88 appears elongated as we are looking at it slightly face on. The core shines brightly and appears almost stellar. We see M88 at an apparent magnitude of 9.5, but its absolute magnitude is -21.3. Though the majority of the Virgo Cluster galaxies are south of M88, this galaxy is believed to be near the center of the galaxy cluster. M88 has a radial velocity of 1,989 km/s in recession.

Let the SBb-type barred-spiral galaxy NGC 4548 (M91) [12:35;+14d30'] drift into your view. At magnitude 10.19, it is about $\frac{1}{2}$ magnitude fainter than M88. This galaxy has low surface brightness with a faint core. The thin bar is visible in scopes larger than 12 inches. Some controversy surrounds the discovery of this object. For years it was considered to be a "missing" Messier object as there is no object located at the coordinates originally listed by Messier on March 18, 1781. For some of the objects in the Virgo/Coma area, Messier used M58 as the starting point in plotting an object's location. When you move the number of degrees in R.A. and dec. from M58 as Messier did, you end up in empty space. Some Messier authorities therefore concluded that M91 was a resighting of M58 as that galaxy is at about the same right ascension, but is $2\frac{3}{4}$ degrees to the south of M91. By plotting M91's coordinates from M89 instead of M58, NGC 4548 comes into view. Most Messier authorities now agree that NGC 4548 is Messier's 91st object. One other factor in the confusion is that Messier listed each object's coordinates as of the date discovered instead of precessing them all to one equinox. We know he was poor at mathematics and maybe failed to unify all of the coordinates because of this handicap. M91 displays a radial velocity of 403 km/s in recession. In a low-power, wide-field eyepiece you may be able to see M88, M91, and NGC 4571 together.

To the southeast of M91 is the magnitude 11.31, Sc-type galaxy NGC 4571 (IC 3588) [12:36;+14d13']. This galaxy is 16.8 Mpc from

131

the solar system. The slight redshift in its spectrum reveals a slow radial velocity of 282 km/s in recession. It appears roundish with an even illumination from side to side.

From M91, hop 2 degrees north to magnitude 5.68, spectrum K5 III, 25 Comae [12:36;+17d05'], then hop about one degree north-northwest to the color-contrast double star 24 Comae (ADS 8600; Σ 1657) [12:35;+18d22']. The yellowish primary is a spectrum K2 III, giant that shines at magnitude 5.02. Located at position angle 271° and separated from the primary by 20.3 arcsec, is the bluish, spectrum A9 Vm, magnitude 6.56 comes. These stars are coming toward us at a speed of 5 km/s. Their rotational velocity as a system is about 25 km/s. The blue star is also a spectroscopic binary with a spectrum F1 V companion. The companion has an orbital period of 7.3366 days. This is a very fine pair to observe and is easy to separate in small scopes.

About $\frac{1}{4}$ degree southwest of 24 Comae is the magnitude 12.02, Sba-type galaxy NGC 4539 [12:34;+18d12']. This elongated galaxy has low surface brightness due mainly to our slight edge on view of it.

Star-hop in Virgo

For the second star-hop this month we will hop through the realm of galaxies in Virgo, as shown in Figure 7.2. The mythology behind who is Virgo the Virgin (or Maiden) twists through several tales and contains numerous claimants to the title. One facet about this constellation that appears in most of the tales is the widely held belief that the bright star Spica represents either an ear of wheat or corn held in a woman's hand. Virgo holding an ear of corn would most likely be shown due to the agricultural ignorance of the mapmaker since corn was unknown in Europe until after the voyages of Columbus. The plant is indigenous to the Americas, but the word corn meaning "grain" comes from Old or Middle English and the Latin *granum*.

The pretenders to the title of Virgo range from Minerva, Pantica, Medusa, Panda, Diana, to the Muse Urania. Tyche, the goddess of fortune, and Atargatis, the Syrian goddess of fertility, have also held the title. To the Egyptians, the stars represented Isis (a goddess also well-known to the Greeks and Romans) who spread ears of wheat across the sky to form the Milky Way. In the Middle Ages, this constellation was depicted as the Virgin Mary holding the Christ child.

The two most accepted tales come from the Greeks (the Romans supplanted their gods for the Greek ones) and are allegories behind the cause for the decline of mankind, the creation of the seasons, and the rising of plants in spring and the barren Earth in winter. The first tale involved the decline of mankind from the golden age to the iron age. The second tale explained the cause for the seasons.

In the first tale, the world deteriorated from the Golden Age of innocence and happiness down through the Silver Age and Brazen Age to the Iron Age. All was well in this Golden Age as spring was eternal. Food for man and beast grew on its own and man did not

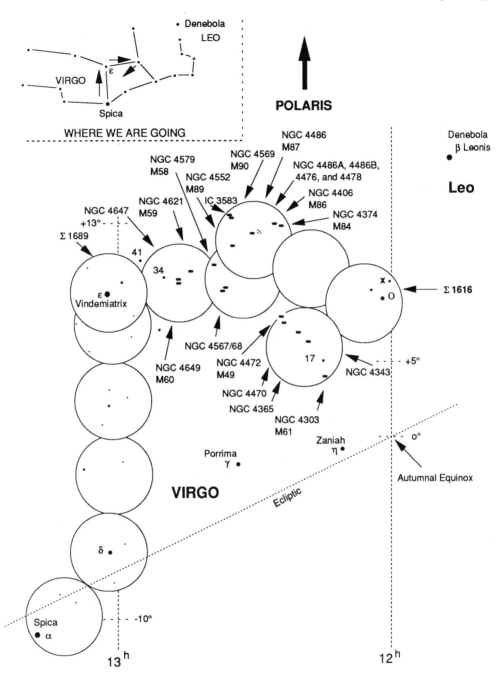

Figure 7.2 Star-hop in Virgo.

need shelter from winter's cold. War was unknown. The rivers flowed with milk and wine.

In the Silver Age, Zeus created seasons, which caused man to seek shelter from the heat of summer and the cold of winter. For sustenance, man was forced to learned how to farm, because food would no longer grow without the assistance of man.

The Brazen Age followed the Silver and was marked by strife and fighting. This age gave way to the Iron Age in which man went to war.

133

The forests were cut down to build ships and greed ruled the land. As sons slaughtered fathers to gain their inheritance, the gods began to abandon the earth and move along the Milky Way to palaces in the heavens. The last goddess to leave was Astraea. She usually represented justice, but also portrayed innocence and purity. Astraea was placed among the stars as Virgo. Her mother, Themis, is nearby holding Libra's scales of justice.

The second tale revolves around Demeter the Mother Goddess of Earth and her daughter, Persephone. The defeat of the Titans and then the Giants opens the tale. These early gods and monsters were banished to the Underworld. Their continued unrest and struggle to regain their freedom from Hades' domain was said to be the cause of earthquakes and volcanoes. With all the rumblings from these imprisoned former gods, Hades was afraid they would split open his world, Tartarus, and expose it to the light of day. Hades took his chariot and hastened his black horses to rush him on an inspection tour of the Earth to be sure all was well.

While on his travels, Hades was seen by Aphrodite and her son Eros. Aphrodite was upset that some goddesses despised her and tried to hold back her power. Demeter was one of the goddesses so despised. Aphrodite instructed Eros to hit Hades with one of his love arrows so their empire would be extended. Eros cocked his bow and let his sharpest arrow fly. The arrow pierced Hades' breast striking his heart, turning his thoughts to love.

While this encounter between Hades and the arrow was taking place, nearby Persephone was in the meadow of Enna (also known as Henna on the island of Sicily) picking lilies and violets by a tree-shaded lake. In this meadow flowers bloomed all year for spring never ended. Hades entered the meadow, saw the maiden and instantly fell in love. He reined his horses to charge across the meadow and snatched her from the bank of the lake. Persephone cried for help. At the River Cyane, Hades attempted to enter his world. The river goddess Cyane refused to part her waters. In a rage, Hades struck the river bank with his trident. The goddess obeyed his command and opened. As Hades and his prize entered the Underworld, the earth closed over them.

Demeter had heard her daughter's cries and began to search the world over. No one would help her out of fear of the wrath of Hades. In frustration, Demeter cursed the ground and no crops came forth. Having searched the earth, Demeter returned to the River Cyane and met the wood nymph Arethusa. The nymph had passed underground to escape from the river god Alpheus. While in the Underworld she saw Persephone on a throne. She passed this information on to Demeter.

Demeter rushed to Zeus and implored him to free their daughter. Zeus consented on the condition that she had not eaten any food while with Hades. Unfortunately, Persephone had eaten the seeds of a pomegranate and could not be set free. In a compromise, Persephone was allowed to rise to the surface of Earth and be with her mother on Mount Olympus for half of the year and to return during the other half to Hades.

Grateful for the return of her daughter, Demeter lifted her curse and crops again burst forth each spring as Persephone annually rises from Tartarus. As she descends from Olympus for her return to the Underworld, the leaves begin to fall. While she is with Hades in winter, the ground is barren.

Virgo is second in size of the constellations and covers 1,294.428 square degrees of sky. The Maiden is surrounded by Coma Berenices to the north; Leo and Crater on her western border; Corvus and Hydra at her feet; and Libra, Serpens Caput, and Boötes on her eastern side.

On the second star-hop this month, we will spend most of our time observing the brighter members of the Virgo Galaxy Cluster located in the northern half of Virgo. We will also hunt down several bright stars, variables, and multiple systems.

We begin this star-hop at the brightest star in the area and 16th brightest star in the sky: Spica (67 Alpha Virginis) [13:24;−11d09']. Spica (pronounced SPY-ka) is the MK standard for spectrum B1 III-IV stars and shines blue-white at magnitude 0.97. The absolute magnitude for this star is calculated to be -3.5. Being rich in helium (He), this star shines about 2,300 times brighter than the Sun. A magnitude 11.8, spectrum B2 V, secondary star is 147 arcsec away in position angle 32°. In 1890, German astronomer Herman Carl Vogel (1841–1907) discovered the spectrum B3 V spectroscopic binary orbiting Spica with a period of 4.0145 days. Spica is also an ellipsoidal-type (Ell) rotating variable. This type of variable consists of a binary pair in close orbit, but from our line of sight, the two stars do not eclipse each other. Their combined brightness changes of less than 0.1 magnitude corresponds to their orbital motions. About 80 percent of the light from Spica is from the primary star. Its name derives from the Latin *Spicum* meaning "Ear of Wheat." After Isaac Newton's death, there was a failed movement to change the name to honor him. To find Spica, follow the curve of the handle from the Big Dipper, arc south through golden Arcturus in Boötes and continue to curve to bright Spica.

About 21 degrees north of Spica is the yellow-white giant, magnitude 2.83, spectrum G8 III, Vindemiatrix (47 Epsilon Virginis) [13:02;+10d57']. The dawn rising of Vindemiatrix (pronounced vin-DEE-my-A-trix) in late August has long been associated with marking the time to begin harvesting grapes. The name derives from the Latin meaning "grape-gather." The Greek name of Protrygeter had the same meaning. The star seems to have brightened slightly since Ptolemy's time as he listed it at 3+ magnitude in his *Almagest*. In 1879, Burnham discovered a magnitude 11.7 comes 240.4 arcsec from the primary in position angle 120°. This system is 32 pc from us and has a radial velocity of 14 km/s in approach.

Before we look back at how a portion of the universe appeared 42 million years ago, hop $1\frac{1}{2}$ degrees northwest from Vindemiatrix, to the double star Σ 1689 (ADS 8704) [12:55;+11d30']. The pale yellow, spectrum M, primary shines at magnitude 7.1. Its blue, magnitude 9.4, comes is located 29 arcsec away in position angle 211°. F. G. W. von Struve first measured this pair in 1827.

Hop about $1\frac{3}{4}$ degrees northwest to the double star 34 Virginis [12:47;+11d57']. The white magnitude 6.1 primary is accompanied by a magnitude 9.3 secondary at position angle 2°. These stars have a separation of 139.4 arcsec.

A hop 1 degree southwest brings us to the first of many galaxies in Virgo – the giant E1 galaxy NGC 4649 (M60; VV 206) [12:43;+11d33']. Between April 11 and 13, 1779, Johann Koehler discovered M60 as a comet passed near it. Messier made an independent discovery on April 15, 1779. The mass of the galaxy is estimated to be equal to one trillion Suns. A halo seems to surround a bright core of this magnitude 8.8 object (-21.36 absolute magnitude). M60 is 16.8 Mpc from us and shows a red shift of 1,127 km/s in recession.

To the north of M60 is its companion Sc-type galaxy NGC 4647 (VV 206) [12:43;+11d35']. These galaxies appear to be interacting even though NGC 4647 is moving away from us at a radial velocity of 1,285 km/s compared to M60's 1,127 km/s. This galaxy is faint at magnitude 12.03 and appears roundish with a bright core and is slightly smaller than M60.

In a wide-field, low-power eyepiece, you may also be able to observe E3 galaxy NGC 4621 (M59) [12:42;+11d39'] northwest of M60. Both Koehler and Messier discovered M59 on the same evenings they each independently found M60. At magnitude 9.79, M59 appears oval with a bright core. M59 has a radial velocity of 341 km/s in recession and is about 18.6 Mpc from us.

About 1 degree northwest of M59 is the barred-spiral (Sb) galaxy NGC 4579 (M58) [12:37;+11d49']. Messier first saw this magnitude 9.78 object on April 15, 1779. M58 is considered to be almost a twin to our Galaxy in size and mass. It appears elongated with low surface brightness.

An interesting object consisting of a pair of apparently attached galaxies is located about $\frac{1}{2}$ degree southwest of M58. The "Siamese Twins" are Sc galaxies NGC 4567 [12:36.5;+11d15'] and NGC 4568 [12:36.6;+11d14']. In the catalogue of interacting galaxies these objects are listed together as VV 219. NGC 4568 is slightly brighter at magnitude 10.80 and appears larger than NGC 4567. The recessional radial velocity of NGC 4567 is 2,121 km/s, whereas NGC 4568 is cruising at 1,055 km/s.

Look for the magnitude 9.81 elliptical (E0) galaxy NGC 4552 (M89) [12:35;+12d33'] about $1\frac{1}{2}$ degree north-northwest of the Siamese Twins. Messier found this round galaxy on March 18, 1781. M89 is estimated to be 250 billion solar masses and has an absolute magnitude of -20.40. This galaxy appears as a round glowing patch with a bright center. It is receding from us with a radial velocity of 165 km/s.

About $\frac{3}{4}$ degree northeast is the beautiful magnitude 9.48 Sb+ spiral NGC 4569 (M90) [12:36;+13d10'] and its irregular companion IC 3583 (UGC 7784) [12:36;+13d15']. Messier saw M90 on the evening of March 18, 1781. At magnitude 13.6, IC 3583 was too faint for him to see. M90 appears as an elongated oval as we see it at about a 30° angle from face on. The core is bright, and in a large scope dust lanes can be seen. The galaxy is about 24.5 kpc (80,000 light-years) in diameter and contains the mass of 80 billion Suns. The absolute magnitude of M90 is -21.6.

About 1 degree southwest from M90 is the giant elliptical (E0) galaxy NGC 4486 (M87) [12:30;+12d24']. M87 is estimated to contain 790 billion solar masses of material. We see it glow at magnitude 9.56 whereas its absolute magnitude is -21.64. What you see is not all you get with this monster in the sky. M87 is a strong radio and X-ray emitter known as "Virgo A" and has been given a radio source catalogue number of 3C 274. For many years, a jet of material spewing from the core has been visible on photographs of M87. As I am writing this chapter, the Hubble Space Telescope has captured what appears to be the preliminary evidence for the existence of a black hole in the core of M87. A black hole is believed to be the collapsing concentration of matter (in M87 it is thought to be about 2.6 billion solar masses) with such tremendous gravitational pull that light cannot escape its pull. The Hubble pictures show a beam of light shining through the core of the galaxy. The light appears to be coming from a mass of stars packed 300 or more times tighter than in a normal elliptical galaxy. These stars are circling the black hole. Another interesting feature of M87 is its high number of globular clusters. Estimates run as high as 4,000 globulars. In no other elliptical galaxy have we seen anything close to that number. Messier discovered M87 on March 18, 1781, and noted it with his usual comment of "Nébuleuse sans étoile, dans la Vierge,..." (nebula without star, in Virgo,...).

In the same field of view as M87, you may be able to see the faint elliptical galaxies NGC 4486B, NGC 4476, NGC 4478, and NGC 4486A. These galaxies range from 10th to 13th magnitude and are located in a semicircle around the western side of M87.

One degree northwest of M87 is another pocket of faint galaxies surrounding the brighter pair of elliptical galaxies NGC 4406 (M86) [12:26;+12d57'] and NGC 4374 (M84) [12:25;+12d53']. These galaxies may be on the western edge of the core of the Virgo Cluster. There is a difference of opinion in the literature on the distance to M86. Some sources give a distance of 12.5 Mpc (42 million light-years) and others place it at about 6 Mpc (20 million).

At magnitude 9.82, M86 appears near circular with a slight brightening in the middle. Unlike the majority of galaxies in the Virgo Cluster, which are receding from us, this E3 galaxy has a radial velocity of 419 km/s in approach. It contains about 130 billion solar masses of matter that shine with an absolute magnitude of -19.1.

If SO-type galaxy M84 is twice the distance from us than is M86, then it must be larger and shines with a greater absolute magnitude (-21.5) as its apparent magnitude of 9.29 is just slightly brighter. The mass of M84 is calculated to equal 500 billion Suns. This galaxy has a radial velocity of 845 km/s in recession. The distance to this object is estimated to be 12.5 Mpc. M84 is slightly oval with a noticeable brightening at the core.

Head about 8 degrees south-southwest to yellowish, magnitude 4.12, MK standard spectrum G8 IIIa: Cn-1; Ba-1; Ch-1, 9 Omicron

Virginis [12:05;+8d43']. About 1 degree northwest of Omicron is the variable X Virginis [12:01;+9d04']. This spectrum F IV-Vp star fluctuates between magnitude 7.3 and 11.2. Its variability has not yet been classified.

About $2\frac{1}{4}$ degrees east of Omicron is the multiple star Σ 1616 (ADS 8473) [12:14;+8d47']. F. G. W. von Struve first measured this system in 1828. He found a reddish 9.8-magnitude secondary 23.3 arcsec from the yellowish 7.6-magnitude primary. The "B" star is in position angle 296°. The 9.7-magnitude "C" star is located at position angle 293°. This star had moved 7 arcsec away from the "A" star during the period between 1864 and 1920. In 1920, it was 156.8 arcsec from the primary star.

Sweep 3 degrees east to the yellowish magnitude 6.37, spectrum G8 III, star [12:27;+8d36']. From this star, hop about $\frac{1}{4}$ degree south-southeast to the giant elliptical galaxy NGC 4472 (M49) [12:29;+8d00']. Messier discovered this magnitude 8.37 object on February 19, 1771. In total mass, M49 contains about one trillion solar masses. M49 appears as an oval with almost even luminosity across most of it. Its edges seem to fade into space.

In the same low-power field with M49 are several 12th-magnitude and fainter galaxies. The brightest of these is the Sa-type galaxy NGC 4470 [12:29;+7d49']. This galaxy is directly south of M49.

Two large swarms of faint galaxies are situated 1 and 2 degrees southwest of M49. Each group contains over a dozen galaxies. In the first group, E2-type NGC 4365 [12:24;+7d19'] and Sa-type NGC 4343 [12:23;+6d57'] are the brightest members. The second cluster is located to the northeast of the wide double star 17 Virginis (ADS 8531; Σ 1636) [12:22;+5d18']. The brighter members of this cluster are 11th and 12th magnitude.

The yellow magnitude 6.6 primary and orange 9.4-magnitude comes of 17 Vir were first measured by F. G. W. von Struve, in 1829. Their separation is 20 arcsec with the comes at position angle 337°.

About $\frac{1}{2}$ degree south of 17 Vir is the face-on Sc-type spiral NGC 4303 (M61) [12:21;+4d28']. The Italian astronomer Barnabus Oriani (1752–1832) first saw this galaxy on May 5, 1779, while tracking a comet. Messier saw it the same night, but mistook it for a comet. He realized his mistake on May 11 and added it to his list. This galaxy is large at about 18.4 kpc (60,000 light-years) in diameter, but with only 50 billion solar masses, which is about one-quarter the mass of our Galaxy. The arms and dust lanes are visible in this magnitude 9.67 object. M61 is 12.57 Mpc (41 million light-years) from us and has a radial velocity of 1,483 km/s in recession.

Four and a half degrees to the southwest of M61 is the ecliptic and the autumnal equinox. The Sun crosses the ecliptic at this point on September 23/24. The stars Zaniah (15 Eta Vir) [12:19;–0d40'] and Porrima (29 Gamma Vir) [12:41;–1d26'] point at the equinox.

Zaniah shines at magnitude 3.89. This whitish, spectrum A2 IV-V, star is located 32 pc from us. Eta is suspected of being a Delta Scuti-type variable. This star is also a spectroscopic binary system.

The orbital period for the companion is 71.9 days. This system has a radial velocity of 2 km/s in approach.

The magnitude 3.5, yellow four-component star Gamma Vir (ADS 8630; Σ 1670) is named for the Roman goddess of prophecy, Porrima (pronounced POUR-ih-mah). In 1718, James Bradley discovered that 29 Vir was actually a double star. The primary and "B" star are both yellow, spectrum F0 V, main-sequence stars of magnitude 3.5. The "B" star has an orbital period of 171.37 years. These stars are close to us at 11 pc.

As you can see, Coma Berenices and Virgo are very interesting constellations. There are so many galaxies in this area, that I could have written an entire book just on them. Consult your sky atlas and behold what additional objects these two constellations have to offer.

8 *June*

Libra and Lupus:
The Balance Scales and the Wolf

"Arcturus" is his other name —
I'd rather call him "Star."
Its very mean of Science
To go and interfere!

What once was "Heaven"
Is "Zenith" now —
Where I proposed to go
When Time's masquerade was done
Is mapped and charted too.

Emily Dickinson (1830–86),
Poem 70,
c. 1859

The splendors in the constellations of Libra the Balance and Lupus the Wolf await us on the warm nights of late spring and early summer. Though these constellations are south of the celestial equator, they are visible low on the horizon for mid-northern observers at this time of the year. The southern edge of Lupus (Lup) is at declination -55° 30m. Neither of these constellations contain a Messier object, but they do have some interesting Milky Way star fields, multiple systems, clusters, bright and dark nebulae, and galaxies for us to search for.

Libra (Lib) and Lupus are neighbors to a mixture of celestial beasts and beauties. Virgo is to the northwest of Libra. Hydra and Centaurus are on the western sides of Libra and Lupus. Circinus the twin Compasses borders Lupus on the south along with Norma the Square. Scorpius and Ophiuchus are to the east and Serpens Caput is on the northern boundary of Libra. We will visit Libra first this month as we continue easterly ahead of the Sun.

When ancient civilizations were dividing up the sky into mythological beasts and monsters, the asterism we know today as Libra outlined the claws of Scorpius. Libra, as a separate constellation, came about during the time of Gaius Julius Caesar (100–44 B.C.) when the Romans revised their calendar and added Libra as the 12th sign of the zodiac. Introduced in 46 B.C., the Julian calendar was used in most of Europe until Pope Gregory XIII (reigned 1572–83) introduced the Gregorian calendar, in 1582.

Since Libra represents a balance scale, the name appears to be associated with the Sun crossing the celestial equator (the autumnal equinox) while in Libra 3,000 years ago. Due to precession, the Sun now crosses the celestial equator in Virgo. It reaches its highest northern declination position in the sky around June 21 (the summer solstice) while Libra is visible near the meridian during the late evening hours. On the days of the two equinoxes the hours of daylight and darkness are equal.

The length of hours during daylight and the dark of night, as calculated during ancient times, were unequal due to man's dependence on the position of the Sun to tell what time it was. The actual length of darkness changes daily, so some of the early civilizations simply divided however long each night was into 12 equal segments; the length of their hour changed daily. With the invention of mechanical clocks, during the Middle Ages, time was divided into equal hours. With modern timekeeping, we know that June 21–22 is the longest day and shortest night of the year.

Libra has also been considered as the scales of justice held by Virgo when she is depicted as Astraed, the blindfolded goddess of justice. The last of the zodiac constellations to be devised is also the only inanimate one. Libra ranks 29th in size and covers 538.052 square degrees of sky.

Star-hop in Libra and Lupus

We will start this month's star-hop, as shown in Figure 8.1, at blue-white Spica (67 Alpha Vir) in Virgo at coordinates [13:25;–11d09'] then hop through Libra and Lupus. The 16th brightest star stands out as the easiest to locate guide star in the area and leads us to the trapezoidal-shaped Libra.

First find magnitude 0.98 Spica and sweep east to the 2.6-magnitude, blue-white spectrum B8 IV-V, Zubeneschamali (27 Beta Librae) [15:17;–9d22']. Prior to the Julian calendar, this star was known by its positional name meaning the "Northern Claw of the Scorpion." Zubeneschamali (pronounced zoo-BEN-ess-sha-MAY-lee) comes from the Arabic *Al Zubān al Shamāliyyah* and was carried forth by the Arabs from Ptolemy's *Almagest*.

Beta Librae may have been much brighter in ancient times than we see it. Eratosthenes noted it as the brightest star in the area, even brighter than Antares (21 Alpha Scorpii). Beta Lib shines about 120 to 150 times brighter than the Sun. It is located 37 pc from us and is getting closer at a rate of 35 km/s.

About 11 degrees due north of Beta Librae, in Serpens Caput (SerCp), is the magnitude 6.2 globular cluster NGC 5904 (M5) [15:18;+2d05']. German astronomer Gottfried Kirch (1639–1710) first observed this object on May 5, 1702. Kirch became the Director of the Royal Observatory in Berlin three years later. Messier first saw the cluster, on May 23, 1764, and added it as the fifth item on his new list. M5 shines with an integrated magnitude of 5.75.

The cluster is estimated to be 40 pc in diameter, making it the largest globular on the Messier list. The estimated age of M5 ranges

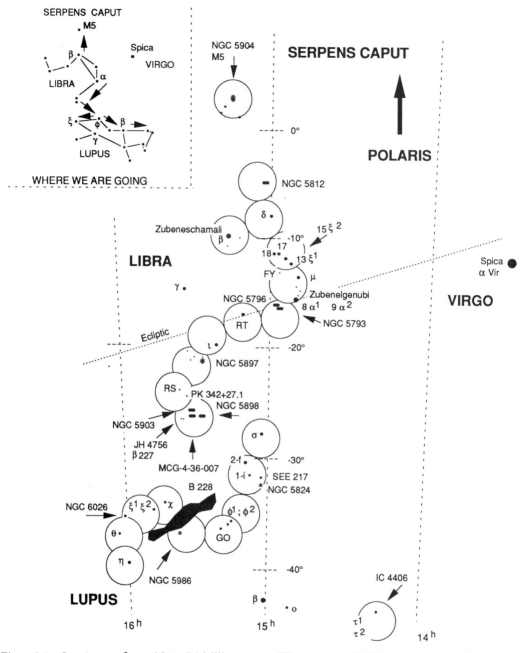

Figure 8.1 Star-hop in Libra and Lupus.

from 13 to 24 billion years. The center of M5 is so compact that you cannot separate any of its individual stars. The outer members of the cluster can be separated in 4-inch or larger scopes. The half-million members of M5 are moving away from us with a radial velocity of 54 km/s.

This slightly elongated class 5 cluster contains about 100 variable stars. In 1917, Harlow Shapley noticed the rapid rise in brightness for several of the variables took about 30 minutes and could be accurately predicted. Using the variables, he tried to see if the speed of light was different at different wavelengths. He took photographs of the cluster with yellow-sensitive and blue-sensitive plates to see if there was a delay in the arrival of light in one wavelength versus the other. The time was so small, considering the distance of 7.6 kpc that the light had traveled, that he concluded the speed of light in space was constant at all wavelengths.

Back in Libra, we hop about 2 degrees north-northwest from Beta Librae to the Algol-type (EA) eclipsing spectroscopic binary variable 19 Delta Librae [15:00;–8d31']. Delta goes from its maximum brightness of 4.92 to minimum of 5.90 in a period of 2.3273 days. This spectrum B9.5 V star has a spectrum G companion located about 4.5 million miles from it. The companion partly eclipses the primary with each revolution. These stars are located 73 pc from us and have a radial velocity of 39 km/s in approach.

One degree north of Delta is the E1-type galaxy NGC 5812 [15:01;–7d27']. This magnitude 11.2 galaxy appears as a fuzzy ball with no visual hint of arms.

About $3\frac{1}{2}$ degrees south-southwest of NGC 5812 are the optical pair consisting of the magnitude 5.84, wide multiple star 18 Lib (ADS 9456; Σ 1894) [14:58;–11d08'] and magnitude 6.4, spectrum A0 V, 17 Lib [14:58;–11d09']. A magnitude 10.0 comes is located 19 arcsec from 18 Lib at position angle 39°. The magnitude 11.3 "C" star for this system is located 169 arcsec from the primary at position angle 41°.

South-southwest of 17 Lib is the 5.46-magnitude, spectrum gK4 III, red giant 15 Xi^2 Lib [14:56;–11d24']. This star and the spectrum K0 III red giant 13 Xi^1 Lib [14:54;–11d53'] are both located 130 pc from us. These stars are going in opposite directions as 13 Xi^1 is coming toward us at a rate of 24 km/s; 15 Xi^2 Lib is racing away with a radial velocity of 15 km/s.

Forming an elongated triangle with Xi^1 and Xi^2 Lib is the red semi-regular giant (SRB) pulsating variable FY Librae [14:57;–12d26']. This variable is located about 1 degree south-southeast of Xi^1 Lib. The magnitude for FY Librae fluctuates between 7.06 and 7.46 during a period of about 45 days.

Two degrees south-southwest of FY Librae is the very close multiple system of 7 Mu Librae (ADS 9396; β 106) [14:49;–14d08']. The spectrum A0, magnitude 5.4 primary and the "B" comes are 1.8 arcsec apart. Burnham first measured this system in 1875. In 1889, the magnitude 14.5 "C" star and the magnitude 13.9 "D" star were discovered. These stars are located 15 and 25 arcsec, respectively, from the primary. The magnitude 12.5 "E" star was first observed in 1875. The system is located 22 pc from us and getting closer at a rate of 4 km/s.

The only other star in Libra to have a common name is actually a pair: 8 Alpha^1 [14:50;–15d49'] and 9 Alpha^2 Librae [14:50;–16d02']. These stars were known to the Arabs as Zubenelgenubi, meaning "The Southern Claw of the Scorpion." Zubenelgenubi (pronounced zoo-BEN-el-je-NEW-be) usually refers only to Alpha^2 and comes from *Al Zubān al Janūbiyyah*. These stars, along with Beta Librae, were considered by Hindu astronomers to be the celestial gateway, because the Sun, Moon and planets passed between them as if they were portals in the sky. The Sun is at this point on the ecliptic (225°) on November 8. Alpha[2] is the brighter of these stars. It shines as a magnitude 2.57 blue-white subgiant of spectral-type A3 IV. The fainter yellow, spectrum F4 IV, subgiant Alpha[1] is located 231 arcsec from Alpha[2]. Alpha[1] has a radial velocity of 23 km/s in approach and Alpha[2] is also coming toward us at 10 km/s. These stars are 22 pc from us.

About $1\frac{1}{2}$ degrees southeast of Alpha Lib is a pair of galaxies NGC 5796 [14:59;–16d37'] and NGC 5793 [14:59;–16d42']. The E0-type NGC 5796 glows softly at an apparent magnitude 13.2 whereas its absolute magnitude is estimated to be -20.91. This galaxy appears as an almost perfectly round fuzzy star located 41.9 Mpc from us. Due south of NGC 5796 is the magnitude 14.3 Sb-type galaxy NGC 5793. This object is small with a narrow core which makes it hard to see, except in a large aperture scope.

Hop $1\frac{3}{4}$ degrees southeast of NGC 5793 to the Mira-type long-period variable RT Librae [15:06;–18d44']. During its period of 251.74 days, this star's magnitude fluctuates between 8.2 and 14.6.

Continue about one more degree in the same direction to reach the multiple binary system of 24 Iota Librae (ADS 9532) [15:12;–19d47']. Sir William Herschel first measured the "B" star in 1782. At position angle 111°, the 9.4-magnitude, spectrum A0 comes is located 57.8 arcsec from the spectrum B9 IV blue-white subgiant, magnitude 4.54 primary star. The magnitude 5.6 "a" star was first measured in 1940 and found to be in retrograde motion when compared to the other members of the system. Due to the great disparity in brightness between the primary and the "a" star, this very close pair (less than 1 arcsec) can only be separated in very large scopes. Burnham first measured the "C" star (β 618) in 1878. The Iota Lib system is 93 pc from us and has an average radial velocity of 12 km/s in approach.

Heading in the same direction southeast from Iota Lib is the very open globular cluster NGC 5897 [15:17;–21d01']. Look for an arc of three eighth-magnitude stars northeast of the cluster. The magnitude 8.55, type 11 cluster seems to be in a stage between being an open cluster and a compact globular. The core is not very well compressed, and in a large scope you can detect many individual stars on a night of good seeing. The cluster is estimated to be as bright as 39,000 Suns, is located 12.1 kpc from us, and has a radial velocity of 10 km/s in recession.

Still heading southeast, we hop to the Mira-type variable RS Librae [15:24;–22d55']. This star changes from spectral-type M7e to M8e during its 217.65-day period. The star's magnitude ranges from 7.0 to 13.0.

About 1 degree south-southwest of RS Librae is the magnitude 11.6 planetary nebula PK 342+27.1 [15:22;–23d38']. This type II planetary is located 4.3 kpc from us and has a radial velocity of 46 km/s in recession. Spectral study indicates the gas shell is expanding from the magnitude 16.3 central star at a rate of 13 km/s. This plane-

Photo 8.1 Globular cluster NGC 5897 in Libra.

tary is very small, and a prism will help you to pick it out from the field of stars surrounding it.

Hopping 1 degree southwest brings us to a double-double consisting of magnitude 7.5 Burnham 227 (ADS 9579; β 227) [15:19;–24d16'] and to the east, magnitude 7.9, JH 4756 (ADS

145

9586) [15:19;–24d16']. The very close double β 227 was first observed in 1876 by Burnham and had a separation of 1.9 arcsec. The separation between the spectrum A2 V primary and its magnitude 8.8 comes has remained the same, but the secondary has moved from position angle 182° then to be near 168° now. When JH 4756 was measured in 1876, the magnitude 8.1 secondary was at position angle 330° with a separation of 1 arcsec. By 1959, the comes had moved to position angle 290° and the separation had closed to 0.7 arcsec. The contrast between the white and yellow main-sequence primaries makes this an interesting pair to observe even if you are unable to separate their close companions.

A triangle of three faint elliptical galaxies are to the northwest of JH 4756 and will be in the same finder view. The northern object is the E1-type galaxy NGC 5903 [15:18;–24d04']. This almost round galaxy glows at magnitude 11.48. Its absolute magnitude is -20.88. This island universe is 35.9 Mpc from us. Due south of NGC 5903 is the E-type galaxy MCG-4-36-007 [15:18;–24d07']. This galaxy is hard to see being of magnitude 13.5, less than $\frac{1}{2}$ arcsec in diameter, and has very low surface brightness. It looks just like an out of focus star. Its intrinsic brightness of -18.67 is also fainter than the other two galaxies in the area. West of MCG-4-36-007 is the easier to locate E1-type galaxy NGC 5898 [15:18;–24d06']. This galaxy glows at magnitude 11.52. Of these three galaxies, NGC 5898 is closest to us at 32.6 Mpc.

The last stop in Libra is at the spectrum M2.5-M3 III, red giant and SRB-type pulsating variable 20 Sigma Librae [15:04; -25d16']. The magnitude of Sigma fluctuates between 3.20 and 3.36 during a period of 20 days. Sigma is about 15 times brighter than the Sun. The emission spectrum for this star shows strong lines of ionized iron (Fe-II) at ultraviolet wavelengths. This star is located 51 pc from us and has a radial velocity of 4 km/s in approach. Sigma has also been known as Brachium, Cornu, Zuben el Genubi, Zubān al Akrab, and Zuben Hakrabi. All of these names are out of use today.

From Sigma Librae, we hop south into the constellation of Lupus the Wolf. On the fringe of the Milky Way, Lupus is dominated by young hot blue giants embedded in rich star fields. Most of these bright stars belong to a 100-plus member cluster called the Scorpius–Centaurus Stellar Association. These early B spectral-type stars all have about the same proper motion, which is taking them at a rate of 25 km/s in the direction of Beta Columbae.

To the Arabs, this constellation was known as *Al Fahd* "The Leopard," *Al Asadah* "The Lioness," and *Al Sabu`* "The Wild Beast." Although most modern scholars believe the name of Lupus as a wolf came about by a mistranslation of the Arabic name, the idea of the stars representing either a wolf or a fox was known to the Greeks in the first century A.D. The Romans also considered this asterism to represent a wild animal, but not necessarily a wolf. The name of Fera Lupus appeared in the Latin version of the *Almagest* and as Lupus in the Alfonsine Tables. None of the stars in Lupus has a common name that has survived the ages. Lupus covers 333.683 square degrees of the sky and ranks as the 51st constellation in size.

A very weak link between Lupus and the cannibalistic King Lycaon of Arcadia appears in Greek mythology. According to the legends about Lycaon, the son of the pious King Pelasgus, Lycaon held a sacrificial banquet in honor of the gods. The name Lycaon comes from the Greek "Lycaeus" meaning "of the she-wolf" or "of the light." Wearing a disguise, Zeus Lycaeus came to the banquet. As a test to determine which of his guests were gods, Lycaon mixed the flesh of a child in with the food. Zeus became enraged at the horrible act and caused thunderbolts to strike the king and members of his large family, which included 50 sons. Lycaon fled into the forest and tried to hide. Zeus found him and transformed Lycaon into a wolf.

Starting at Sigma Librae we hop south to the magnitude 4.34, spectrum K0 IIIa: Fe-1, red giant 2-f Lupi [15:17;−30d08'] at the border between Lupus and Libra. This variable star is located 55 pc from us and has a radial velocity of 4 km/s in approach.

Hop south-southwest from 2-f Lupi, past yellow giant, magnitude 4.91, spectrum F1 II, 1-i Lupi [15:14;−31d31'] to the open double star SEE 217 [15:03;−32d39']. This double was first measured by American astronomer Thomas Jefferson Jackson See (1866–1962), in 1897. T. J. J. See joined the staff of the Lowell Observatory in 1895, and traveled to Mexico to observe the 1896/7 opposition of Mars. While in Mexico, he observed many double and multiple systems. In 1898, he published his error-filled catalogue *Discoveries and Measures of Double and Multiple Stars in the Southern Hemisphere Made with the 61-cm Refractor of the Lowell Observatory*. SEE 217 consists of a magnitude 5.44, spectrum B4 IV, blue subgiant and a 13.7-magnitude secondary star located 36.1 arcsec away in position angle 119°.

About $\frac{1}{2}$ degree southeast of SEE 217 is the class 1 globular cluster NGC 5824 [15:04;−33d04']. This tightly compacted cluster shines at magnitude 9 and is located 23.7 kpc from us. A few scattered stars can be seen on the fringe of the cluster, but the core is too tightly compressed to view any individual members.

As we cross declination -35°, we enter that portion of the sky in which very few of the stars received a Flamsteed number in addition to its Bayer Greek letter designation. Most of these stars are below the horizon as seen from the UK and were therefore unknown to Flamsteed. If you are observing from north of latitude +35° these stars may also be very low in your horizon and can be affected by atmospheric extinction. Try to observe the southern portion of Lupus when it is on your southern meridian.

About 4 degrees south-southeast are the stars designated Phi[1] and Phi[2] Lupi. The northern of these stars, Phi[1] (SEE 229) [15:21;−36d16'], is actually a wide multiple system consisting of a magnitude 3.56, spectrum K4-5 III, yellow-orange giant and its magnitude 15 "B", and 14.5-magnitude "C" stars. In 1897, See measured both the "B" and "C" stars. The "B" star is located 16.7 arcsec from the primary in position angle 240°. The "C" star is in position angle 119° and is separated 17 arcsec from the primary. The primary star is a strong emitter of ionized iron in the ultraviolet range. The system has a radial velocity of 29 km/s in approach.

Phi² [15:23;–36d51'] is a spectrum B3 IV, magnitude 4.54, blue-white subgiant. Like the rest of the stellar association, Phi² is moving toward us with a radial velocity of 1 km/s. This star is also believed to be a distant member of the Pleiades.

One degree southeast of Phi² is the red giant SRB-type semiregular variable GO Lupi [15:28;–37d22']. This spectrum M4 III star's magnitude fluctuates between 6.98 and 7.21.

About 3 degrees east of GO Lupi is the class 7 globular cluster NGC 5986 [15:46;–37d47']. This nice binocular-observable cluster is the largest cluster in Lupus and is situated among a triangle of sixth- and seventh-magnitude stars. The cluster shines at an integrated magnitude of 7.12. Its absolute magnitude is calculated at -8.78. The core is loosely compacted and some individual stars are seen in large scopes.

About 2 degrees north of NGC 5986, you will notice a band of darkness running northwest to southeast. The long (240 arcmin) and narrow (20 arcmin) dark nebula B 228 obscures most of the stars behind it. The few reddish stars you can see are either foreground stars or background stars that have been reddened by the absorption of their light by the dark nebula and the dense interstellar dust between us and the stars.

To the north of the dark nebula are the magnitude 3.95, spectrum B9 IV, blue giant and spectroscopic binary 5 Chi Lupi [15:50;–33d37'], and the white binary doubles Xi¹ and Xi² Lupi. The spectrum A3 V northeastern star, Xi¹ [15:56.53;–33d57'], shines at magnitude 5.15 and has a radial velocity of 10 km/s in approach. The spectrum B9 V southern star, Xi² [15:56.54;–33d57'], is slightly fainter at magnitude 5.57. It has a radial velocity of 12 km/s in approach. These make a fine double to observe. The companion of Chi has an orbital period of 15.2565 years.

Line up Chi and Xi to point you to the planetary nebula NGC 6026 (PK 341+13.1) [16:01;–34d32'], located about 1 degree southeast of Xi. This type IV planetary is located 2.3 kpc from us and is racing our way at 112 km/s. The gas shell for this object glows at magnitude 12.9. In a large scope, the magnitude 13.29 central star is visible slightly off center.

One of the best color-contrast double systems in the area is Eta Lupi [16:00;–38d23']. This magnitude 3.41, spectrum B2 V, main-sequence star is a member of the stellar association with its radial velocity of 7 km/s in recession. The magnitude 7.5 comes is located 15.2 arcsec from the primary in position angle 20°. John Herschel first measured this pair in 1834. No change in position has been detected in over 100 years. A magnitude 9.3, spectrum F5, whitish-yellow "C" star is located 115 arcsec from the primary at position angle 248°. Look for Eta Lupi 4½ degrees south-southwest of NGC 6026.

In the southern portion of Lupus is a fine example of a Wolf–Rayet (W–R) ring-type planetary nebula: IC 4406 (PK 319+15.1) [14:22;–44d09']. This planetary is located about 4½ degrees north-northwest of magnitude 2.30, spectrum B1 III, BCEP-type variable,

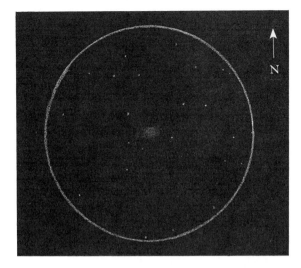

Eyepiece Impression 8.1 Planetary nebula IC 4406 in Lupus.

Alpha Lupi [14:41;–47d23']. The filamentary shell for this type IV+III planetary glows at magnitude 10.6. The W–RC central star has a visual magnitude of 14.7. W–RC are exceptionally hot, massive, and luminous stars with dominant spectral lines of carbon ions. These stars are a subclass of the regular Wolf–Rayet stars. The spectra for these stars have broad emission lines of carbon, nitrogen, helium, and oxygen. To locate IC 4406, line up Alpha Lupi with fourth-magnitude stars Tau[1] Lup (bluish) [14:26;–45d13'] and Tau[2] Lup (yellowish) [14:26;–45d22']. IC 4406 is about 1 degree north-northwest of Tau[1] Lup.

Wolf–Rayet stars are thought to be very massive, luminous objects, with intense spectra lines that indicate they are probably the hottest individual stars in the universe. The theory about them is that they are so unstable they have a tendency to expel massive amounts of glowing gas, but they do not flare-up like a nova or explode like a supernova.

If you are observing from a position south of latitude +35°, you may be able to explore the doubles, clusters, and star fields in the southern portion of Lupus. These deep-sky objects are generally too low (less than 5 degrees) above the horizon for mid-northern observers to get good distortion-free views of.

While hopping around Libra and Lupus, stop and admire the patterns created by the stars and dark nebulae in these very rich Milky Way star fields. Use binoculars or a low-power or wide-field eyepiece to get the best overall views of the star fields.

9 *July*

Scorpius, Sagittarius, and Scutum:
The Scorpion, the Archer, and the Shield of John Sobieski

A woman in the shape of a monster
a monster in the shape of a woman
The skies are full of them

a woman   'in the snow
among the Clocks and instruments
or measuring the ground with poles'

> Adrienne Rich (1929–),
> "Planetarium: Thinking of Caroline Herschel
> (1750–1848) astronomer, sister of William; and
> others."
> (lines 1–6),
> 1968

Without a doubt, Scorpius (Sco), Sagittarius (Sgr), and Scutum (Sct), are very popular to observe. They are laden with easy to locate objects emersed in the hazy stellar glow from the nuclear bulge of our Galaxy.

From -45° to -6° declination, Scorpius, Sagittarius, and Scutum radiate along the Milky Way. These constellations dominate an area of the sky honoring men, monsters, and objects. Surrounding these constellations are Ophiuchus the Serpent Bearer, Libra the Balance Scales, and Lupus the Wolf to the north and east of Scorpius. South of the Scorpion are Norma the Square, and Ara the Altar. The Southern Crown of Corona Australis sparkles on the southern border of Sagittarius and southeastern boundary of Scorpius. Telescopium, Microscopium, and Capricornus the Goat are south and east of Sagittarius the Archer. Aquila the Eagle and Serpens Cauda round out the neighborhood of our main objects this month.

Star-hop in Scorpius

We will start the three star-hops for this month in Scorpius, the 33rd largest constellation, as shown in Figure 9.1. Scorpius is tied to the legend of Orion the Hunter. The armored scorpion, Scorpio, was sent by Mother Earth to destroy Orion after he boasted that he would kill all of the wild beasts on Earth. Orion failed to kill Scorpio and both

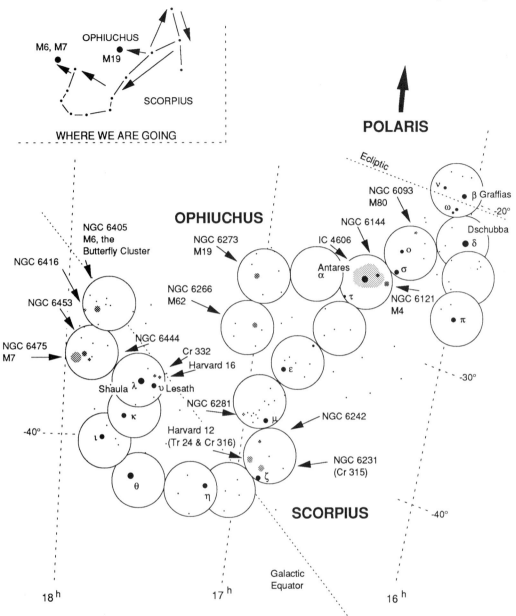

POLARIS

Figure 9.1 Star-hop in Scorpius.

wound up being placed in the sky by the gods. Though about 180 sky degrees behind the great hunter, the Scorpion stalks him through infinity.

Face south and look for the curving fishhook asterism of Scorpio. We will hop first around the western end of the constellation, then follow the spine of the Scorpion to observe the Messier objects near his stinger.

The red supergiant "Heart of the Scorpion" known as Antares is the predominant star in the area so it is the logical starting point on this star-hop. Magnitude 0.96, spectrum M1 Ib, 21 Alpha Sco (ADS 10074) [16:29;–26d25'] is a visual and spectroscopic binary, and a variable. The Greeks considered this star to be the equal of Ares and

151

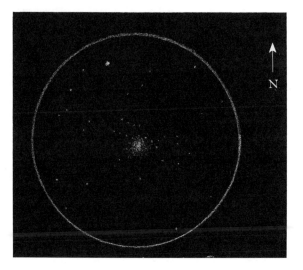

Eyepiece Impression
9.1 NGC 6093
(M80) in Scorpius.

the constellation to be the house of their god of war. Ptolemy listed it under the name we know it by, which means the "Rival of Ares." Antares (pronounced an-TIAR-ease) is the 15th brightest star in our nighttime skies. The exact distance to Antares is in doubt, with estimates ranging from 70 to 159 pc (230 to 520 light-years). If the greater distance is correct, then its diameter may be as large as 600 million miles, which is about two-thirds the size of the orbit of Jupiter around the Sun. Though 700 times larger than the Sun, Antares contains only about 15 times the mass. A magnitude 5.4, spectrum B3, green companion is 2.9 arcsec to the west of Antares in position angle 275°. This comes is difficult to separate due to the brightness of the primary. The green secondary has an orbital period of 878 years.

The semiregular variability of Antares places this pulsating star in the variable type of SRC. This type of variable consists of supergiants whose magnitude fluctuates about 1 magnitude during periods that can be as short 30 days or as long as several thousand days. These stars are of spectral classes M, C, S, Me, Ce, or Se. Antares fluctuates from magnitude 0.88 to 1.80 during a period of 1,733 days. This star is the brightest member of the Scorpius–Centaurus Stellar Association. The foggy mist-like reflection nebula IC 4606 encircles Antares.

About $1\frac{1}{2}$ degrees west of Antares is the class 9 globular cluster NGC 6121 (M4) [16:23;–26d32']. This magnitude 5.93 cluster was discovered by Swiss astronomer and mathematician Philippe Loys de Chéseaux (1718–51) sometime prior to August 6, 1746. On that day, a letter from him to his grandfather in which he listed 20 nebulous objects, was read into the records of the French Academy of Sciences. The list was not published elsewhere. Messier observed M4 on May 8, 1764. It contains about 10,300 stars located 2.4 kpc from us. This cluster has a radial velocity of 64 km/s in recession. You can almost detect individual stars near the core due to the loose concentration of stars.

About $\frac{1}{2}$ degree northwest of Antares is the magnitude 9.07, class 11 globular cluster NGC 6144 [16:27;–26d02']. This tiny cluster, located 7.6 kpc from the Sun, has a radial velocity of 130 km/s. NGC 6144 can be hard to locate due to the glare of Antares.

Protecting the heart of the Scorpion is *Al Niyāṭ*, the wide double and variable 20 Sigma Scorpii (ADS 10009) [16:21;–25d35']. To the Arabs, this star represented the blood vessels coming out of the heart. The primary is a yellowish, spectrum B1 III, giant. The magnitude 8.9 comes is spectral-type B9 and is located 20 arcsec from the primary in position angle 273°. The primary is a Beta Cephei (BC) (also Beta Canis Majoris) type of pulsating variable. These variables are spectral classes O8 to B6 giants and subgiants. They have slight variations of magnitude of 0.03 to 0.3 during a period of 0.6 to 1 day. Sigma fluctuates from magnitude 2.94 to 3.06 in about 6 hours. Sigma is surrounded by the emission/reflection nebula Sh2-9. A spectroscopic companion orbits Sigma in a period of 34.23 days.

About 1 degree north of Sigma is magnitude 4.55, spectrum A5 II, 19 Omicron Sco [16:20;–24d10']. This star leads us to the class 2 globular cluster NGC 6093 (M80) [16:17;–22d59']. Messier discovered this magnitude 7.31 knot of stars, on January 4, 1781. This cluster is very compact and appears almost like a bright star with ragged edges in moderate scopes. This cluster has an absolute magnitude of -8.8. M80 is located 8.3 kpc distant and has a radial velocity of 13 km/s in recession. In 1860, Dr. Auwers discovered a nova near the center of this cluster. The nova is designated T Scorpii (N 1860). For a few days, the star was as bright as magnitude 7.

M80 is on the western edge of a dark nebula that obscures a vast area of the Milky Way. Sir William Herschel thought this blackness was a "Hole in the Heavens" that allowed him to see deep into space.

The middle of the Scorpion's head is marked by the spectroscopic multiple system of magnitude 2.32, spectrum B0.5 V, Dschubba (7 Delta Scorpii) [16:00;–22d37']. The separation between blue-white Dschubba (pronounced JUBB-ah) and its three faint companions is less than 0.2 arcsec. This main-sequence star is 3,500 times brighter than the Sun and about seven times more massive. Dschubba is 170 pc from us and has a radial velocity of 14 km/s in approach. Its name comes from the Arabic *Al Jeb'hah* meaning "The Front" or "The Forehead."

The northern end of the Scorpion's head is marked by the triple system of Graffias (8 Beta Scorpii; ADS 9913) [16:05;–19d48']. The name seems to come from the Greek word for crab. This is a nice pair and is easy to separate. Graffias (pronounced GRAF-ee-as) consists of a magnitude 2.5, spectrum B0.5 V, spectroscopic binary primary (Beta[1]) and the MK standard spectrum B2 V, magnitude 4.9 secondary "C" star (Beta[2]). The purplish "C" star is located in position angle 21°. The separation is 13.6 arcsec. The bluish primary has a radial velocity of 7 km/s in approach whereas the "C" star is approaching at 5 km/s.

To the southeast of Graffias is the beautiful color-contrast visual pair of 9 Omega[1] [16:06;–20d40'] and 10 Omega[2] Scorpii [16:07;–20d52']. Omega[1], the MK standard for spectrum B1 V stars, is to the northwest and shines blue-white at apparent magnitude 3.96.

Its absolute magnitude is -3.56. Omega[2] is a spectrum gG2, yellow-white variable star that shines at magnitude 4.32. Its absolute magnitude is 0.4. Omega[1] is located 250 pc from us and has an approach radial velocity of 4 km/s. Omega[2] is close at 53 pc. Its radial velocity is 5 km/s in approach.

Northeast of Graffias is the double-double of Scorpius, 14 Nu Scorpii (ADS 9951) [16:11;-19d26']. The two primaries ("A" and "C") in this four-star system are easy to separate from each other being 41.4 arcsec apart. The two secondaries ("B" and "D") are very close to their primaries. The magnitude 6.8 "C" star is about 1 arcsec due north of, and is very difficult to separate from, its magnitude 4.00, spectrum B2 IV-V, primary. The "A" star is also a spectroscopic binary system with an orbital period of 5.9222 days. The Nu system is surrounded by the extremely faint reflection nebula IC 4592. Dark skies and a high-contrast filter are needed to even glimpse this object.

The southern star marking the scorpion's head is the double-lined spectroscopic binary 6 Pi Scorpii (ADS 9862; β 662) [15:58;-26d06']. This magnitude 2.89, spectrum B1 V, star is on the edge of the irregular-shaped nebula Sh2-1.

Eight degrees due east of Antares is the class 8 globular cluster NGC 6273 (M19) [17:02;-26d16'] in Ophiuchus. Messier discovered this "Nébuleuse sans étoiles" on June 5, 1764. This oval-shaped object shines at magnitude 7.15. The core is rather open and many stars can be resolved. M19 is located 10.5 kpc from us and has a system radial velocity 121 km/s in approach.

About 4 degrees south of M19 is the class 4 globular cluster NGC 6266 (M62) [17:01;-30d07']. This magnitude 6.60 cluster is located on the boundary between Ophiuchus and Scorpius. Messier found this asymmetrical cluster, on June 7, 1771. He rechecked its position on June 4, 1779, and entered it as the 62nd object on his list. M62 is 5.9 kpc from the Sun. The cluster contains a high number of RR Lyrae-type pulsating variables. These stars are radially pulsating giants in spectral classes A through F. Their magnitude ranges from 0.2 to 2 magnitudes during periods from 0.2 to 1.2 days. RR Lyrae stars were formerly known as short-period Cepheids or cluster-type variables. Most RR Lyrae stars are found in globular clusters.

Heading southeast down the body of Scorpius from Antares we first come to magnitude 2.82, MK standard spectrum B0 V, 23 Tau Scorpii [16:35;-28d12']. This star is 240 pc from the Sun and has a radial velocity of 1 km/s in approach.

The next bright star along Scorpio's spine is 26 Epsilon Scorpii [16:50;-34d17']. This yellow subgiant K2 IIIb spectral star shines at magnitude 2.29 and marks the beginning of Scorpio's tail. Epsilon is one of the few bright stars in the area that is not a member of the Scorpius–Centaurus Association. Its radial velocity is 3 km/s in approach.

Three degrees south of Epsilon is the naked-eye multiple system of Mu[1] and Mu[2] Scorpii. To the west is Mu[1] at coordinates [16:51;-38d02']. This spectrum B1.5 IV star is a double-line spectroscopic, Beta Lyrae-type (EB) eclipsing binary variable. Its magnitude fluctuates between 2.80 and 3.08 during a 1.440-day period. Between eclipses in this type of variable the apparent brightness continually changes so the actual timings of eclipses cannot be accurately made. These stars are spectral types B and A. The components of Mu[1] are believed to be only six million miles apart. The "B" star moves in a near circular orbit. Mu[2] [16:52;–38d01'] is 346 arcsec to the east of Mu[1]. Some sources claim that Mu[1] and Mu[2] are a physical pair only 55,000 AU (0.88 light-years) apart, whereas other sources have them separated by as much as 50 pc (163 light-years). Mu[1] has a radial velocity of 25 km/s in approach and Mu[2]'s velocity is 2 km/s in recession.

Hop $2\frac{1}{3}$ degrees east to the open cluster NGC 6281 (II2pn) [17:04;–37d54']. This 220 million-year old cluster shines at magnitude 5.4. The cluster contains about 30 stars located 600 pc distant.

From Mu, hop $1\frac{1}{2}$ degrees south-southeast to the open cluster NGC 6242 (I3m) [16:55;–39d30']. This cluster is slightly larger than NGC 6281, but is fainter at magnitude 6.4. Several of its brightest members appear yellowish. NGC 6242 is about 51 million years old and is 1,200 pc from us.

One degree south brings us to the visually overlapping clusters Tr 24 (IV2pm) [16:57;–40d40'] and Collinder 316 (Cr 316) (I2m) [16:55;–40d50']. Together, these open clusters are also known as Harvard 12. Cr 316 is the brighter and more scattered of the clusters. Its integrated magnitude is 3.4. Tr 24 is harder to pick out from the surrounding star field and its magnitude is much fainter at 8.6. The two clusters contain about 170 stars. IC 4628, a seventh-magnitude emission nebula spreads across the northern stars of Tr 24.

Near the center of Harvard 12 is the EB-type eclipsing binary variable V861. This blue-white giant star fluctuates between magnitude 6.07 and 6.69 during a period of 7.848 days.

The bright knot of stars south of Harvard 12 is the magnitude 2.6 open cluster NGC 6231 (Cr 315) (I3pn) [16:54;–41d48']. This beautiful young (3.2 million years) cluster is 1,900 pc from us and has a radial velocity of 22 km/s in approach.

Embedded in the heart of NGC 6231 is the multiple system van den Bos 1833 (vdB 1833). This eight-member system consists of a spectrum B0, magnitude 5.6 primary and a 7.1-magnitude "B" star only 0.4 arcsec apart. Included in the system are a magnitude 13.8 "C" star, and a magnitude 13.7 "D" star (SEE 293) in position angle 128°. The magnitude 13 "E" star is 20.9 arcsec from the primary in position angle 282°. The spectrum B0 "F" star shines at magnitude 7.3, 56.6 arcsec from the "A" star in position angle 21°. The "G" star (SEE 294) is another 13th-magnitude star. The "H" star (h 4892) shines at magnitude 10.3.

The area around Harvard 12 and NGC 6231 is encased in an extensive and faint nebula. Use your nebula filter to see its wispy filaments. Be sure to scan the area in binoculars to enjoy the rich Milky Way star fields as the galactic equator cuts through this area of the sky.

From NGC 6231, you can either follow the loop of stars marking Scorpio's tail or just hop east to our next target, the visual pair of 35 Lambda and 34 Upsilon at the stinger. The area enclosed by the loop of Scorpio's tail contains several open and globular clusters and planetary nebulae. This U-shaped ring of stars was known to the Pacific islanders as the Fishhook of Maui.

With a variable magnitude of 1.59 to 1.65, Lambda Scorpii [17:33;–37d06'] is the 24th brightest star and is known as Shaula (pronounced SHAW-la). The name for this blue-white, spectrum B2 IV, subgiant comes from the Arabic word *Al Shaulah* which means "The Sting." Shaula is suspected of being a spectroscopic binary and a short-period pulsating variable of type Beta Cephei (also called Beta Canis Majoris). The variables of this type have visual magnitude amplitudes of 0.03 to 0.3 and are spectral types O8 through B6. Their periods of maximum to minimum brightness range from 14 to 24 hours. They can be giants or main-sequence stars. Shaula's period is 0.214 day. This star appears to be at the beginning of its evolutionary move off of the main-sequence. A distance of 84 pc separates us from Shaula, which dims its absolute magnitude of -3.3.

West of Lambda is the magnitude 2.69 stinger Lesath (34 Upsilon) [17:30-37d17']. The origins of the name Lesath (pronounced LESS-ath) are in doubt. Some sources claim it comes from the Arabic word *Al Las`ah* meaning "The Sting" whereas other sources credit French astronomical writer Joseph Justus Scaliger (1540–1609) with applying the latinization of the same Arabic *Al Shaulah* used for Shaula. Lesath is a blue-white, spectrum B3 Ib, bright supergiant located 480 pc from the solar system. This spectroscopic binary system is receding at a rate of 18 km/s.

Just to the north of Lesath is the tight open cluster Cr 332 (IV1p) [17:30;–37d05'] and the larger and brighter open cluster Harvard 16 (IV2p) [17:31;–36d51']. Cr 332 consists of about 12 stars with an integrated magnitude of 8.9. About 70 stars make up Harvard 16. Due to the rich Milky Way star field around them, these clusters are hard to distinguish from their neighbors.

Line up Upsilon and Lambda to point you in the direction of the large and bright open cluster NGC 6475 (M7) [17:53;–34d49'] about 4 degrees northeast of them. This Trumpler class II2r cluster has been known since ancient times as Ptolemy included it in the *Almagest* as a nebula among the unfigured stars in the vicinity of Scorpius. On May 23, 1764, Messier came across this southernmost of his objects.

M7 consists of about 80 stars brighter than tenth magnitude, the brightest of which are third magnitude. The majority of the cluster's stars are spectral types B and A. The cluster contains at least six spectroscopic binary systems and several difficult to resolve visual pairs. In the southwestern portion near the center of the cluster is the very close double star SEE 342. The components are both mid sixth-magnitude, spectrum K0, stars separated by 0.4 arcsec. In the heart of the cluster is the very close double β 1871. The primary shines at magnitude 6.5 and the comes at 7.3 in position angle 111°. Another very close double is β 1123 in the center of the cluster. The magnitude 6.9,

Photo 9.1 Open clusters NGC 6475 (M7) and NGC 6405 (M6).

spectrum B9, components are separated by 0.1 arcsec. You will need a large scope to separate any of these doubles. M7 is located 240 pc from us and has a system radial velocity of 14 km/s in approach.

To the west of M7 are the tiny class 4 globular cluster NGC 6453 [17:50;–34d36'] and the open cluster NGC 6444 (III2m) [17:49;–34d49']. NGC 6453 shines at magnitude 9.77, and being small looks very similar to a planetary nebula, except in a large scope or under high power. This object is 10.7 pc from us. The eight members of open cluster NGC 6444 are hard to pick out from the surrounding star cloud. The brightest members are 11th magnitude.

Three and a half degrees northwest of M7 is the open cluster NGC 6405 (M6) [17:40;–32d13']. Some sources state that M6, along with

157

M7, were listed as one object in the 1551 edition of the *Almagest*. Hodierna included this object on the list he created prior to 1654. Messier first spotted it on May 23, 1764. The spread of the brighter stars gives the cluster the appearance of a butterfly's wings; hence its name, the Butterfly Cluster. This 100 million-year old cluster has an integrated magnitude of 4.2, and its brighter members are sixth-magnitude blue main-sequence stars.

On the northeast side of M6 is the one of its brightest stars, which is an exception being a yellow, spectrum K0-K3, giant. This star is also known as the SRD-type pulsating variable BM Scorpii. SRD-type variables are semiregular supergiants and giants in spectral classes F, G, and K. Their periods range from 30 to 1,100 days. These stars show a wide magnitude amplitude of from 0.1 to 4. BM Sco fluctuates between magnitude 6.8 and 8.7 during a period of 815 days (?). On the western fringe of M6 is the eruptive variable V862 Sco. This star is generally in the magnitude range of 6.6 to 6.8, but has been detected as bright as magnitude 2. M6 is 600 pc from us and stands out as an easy to locate target since few stars surround it.

About $\frac{1}{2}$ degree east of M6 is the bright open cluster NGC 6416 (IV1p) [17:44;–32d21']. This magnitude 5.7 cluster contains about 50 stars located 800 pc from us.

The area of the sky bounded by M7, M6, 19 Delta Sgr [18:20; -29d49'] and 3-x Sgr [17:47;–27d49'] is the realm of the planetary nebula. You may be able to locate several dozen of these fuzzy star-like objects within this box of sky. Many of these are very faint and you will need to employ a prism to separate them from the non-planetary nebulae in the area.

Star-hop in Sagittarius

The second leg for the star-hops this month, as shown in Figure 9.2, takes us into the 15th largest constellation, Sagittarius. This constellation covers 867.432 square degrees of the sky. The brighter eight stars for this constellation form the famous teapot with steam (the Milky Way) coming out of its spout. Scattered throughout the Sagittarius Star Cloud are numerous bright and dark nebulae, open and globular clusters, and planetary nebulae. Since we are looking in the general direction of the hub of our own Galaxy, we are unable to view the universe beyond the 100 billion stars that lie at the center of our Galaxy.

The mythology for this constellation has been in dispute since ancient times. Some ancient writers considered the teapot to be the human torso of Chiron, the wisest of the centaurs, holding his bow. Chiron is more commonly referred to as being the constellation Centaurus. Many pre-19th-century sky maps show a half-man half-horse creature as Sagittarius. Eratosthenes wrote that Sagittarius is the satyr Crotus, the son of Pan and Eupheme, the Nurse of the Muses. Crotus invented archery and is shown with drawn bow and an arrow aimed at Scorpio's heart. He lived with the Muses on Mount

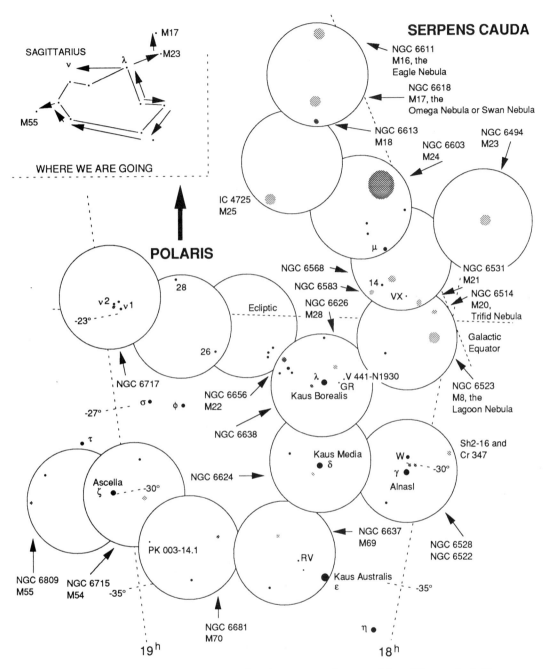

SERPENS CAUDA

NGC 6611
M16, the
Eagle Nebula

NGC 6618
M17, the
Omega Nebula or Swan Nebula

NGC 6613
M18

NGC 6603
M24

NGC 6494
M23

IC 4725
M25

NGC 6568 →

NGC 6583 →

NGC 6626
M28

NGC 6531
M21

NGC 6514
M20,
Trifid Nebula

Galactic
Equator

λ
.V 441-N1930
'GR
Kaus Borealis

NGC 6656
M22

NGC 6523
M8, the
Lagoon Nebula

NGC 6638

Kaus Media
δ

Sh2-16 and
Cr 347

W
γ
Alnasl

NGC 6624 →

NGC 6717

NGC 6637
M69

NGC 6528
NGC 6522

PK 003-14.1

.RV

Kaus Australis
ε

NGC 6809
M55

NGC 6715
M54

Ascella
ζ

NGC 6681
M70

η

POLARIS

ν
λ
M17
M23

M55

WHERE WE ARE GOING

Figure 9.2 Star-hop in Sagittarius.

Helicon and brought them many hours of joy. They asked Zeus to place their friend in the sky. The stars of Sagittarius guided Jason across the seas to the location of the Golden Fleece.

We will start this star-hop at the tip of the teapot spout 10 Gamma Sgr, hop around the teapot, and follow its steamy flow along the Milky Way toward Scutum. The yellowish magnitude 2.99, spectrum K0 III, and spectroscopic binary star 10 Gamma Sgr [18:05;–30d25'] also marks the head of the archer's arrow and is known as Alnasl

159

"The Point." Alnasl (pronounced AL-na-zel) is about 36 pc from us and has a radial velocity of 22 km/s in recession. The thick mass of stars north of Gamma is the "Large Sagittarius Star Cloud." You can see many foreground dust lanes dividing this cloud up into dozens of what appear to be separate clouds. Explore this area with binoculars for the best view.

About $\frac{1}{4}$ degree north-northwest of Gamma are two globular clusters: NGC 6528 and NGC 6522. These clusters can be seen together in the same low-power eyepiece field. To the west in your eyepiece is NGC 6522 [18:03;–30d02']. This class 6 cluster is larger than NGC 6528 and is slightly brighter at magnitude 8.75. NGC 6522 is located 6.3 pc from us and has a radial velocity of 8 km/s in recession. The class 5 globular NGC 6528 [18:04;–30d03'] is located 7.3 kpc away and is getting farther from us at a rate of 160 km/s. This object shines at magnitude 9.67. Stars deep into the core of NGC 6522 are easy to resolve, but are more difficult to see as individuals in NGC 6528.

This area of the Milky Way is laden with thin obscuring dust clouds. Estimates are that visual magnitudes for objects farther away than 1 kpc, like the globular clusters, are reduced by about 1.5 to 2.5 magnitudes. The more distant an object is from us, the greater the drop in visual magnitude.

About $\frac{1}{4}$ degree north of NGC 6528 is the multiple and Delta Cephei (DCEPS) type variable W Sgr (Gamma[1] Sgr; ADS 11029; SEE 346) [18:05;–29d35']. This open system consists of a spectrum F8-G1 Ib, fluctuating magnitude 4.30 to 5.08, primary and two 13.5-magnitude secondaries. The primary star's magnitude changes in a cycle of 7.5947 days. The two companions are an occultation double located 33 and 48 arcsec from the primary star. T. J. J. See first measured these stars, in 1897. DCEPS-type variables are thought to be stars in the beginning stages of being a pulsating star as it moves off the H–R diagram main-sequence. Their magnitude changes are less than 0.5 during a period of 7 days or less. The radial velocity for W Sgr is 29 km/s in approach.

Hop about $4\frac{1}{4}$ degrees to the west of W Sgr to find the emission nebula Sh2-16 [17:46;–29d18']. This gaseous cloud is energized by the open cluster Cr 347 (III2pn) embedded in it. This double object is located less than $\frac{1}{4}$ degree south-southeast of the Galactic Center. Cr 347 contains about 40 stars 1,500 pc from us.

About $3\frac{1}{2}$ degrees to the east of W Sgr is the middle of the Archer's bow marked by the orange-yellowish magnitude 2.70, spectrum K2.5 IIIa: Cn 0.5, Kaus Media (19 Delta Sgr; ADS 11264; SEE 350) [18:20;–29d49']. The name Kaus Media (pronounced koss ME-dee-ah) comes from the Arabic *kaus* meaning bow and Latin *media* or *meridionalis* meaning middle. Sicilian astronomer Giuseppe Piazzi (1746–1826) named the three bow stars early in the 19th century. See first measured this open system in 1896 and found the bluish magnitude 14.4 "B" star 25.8 arcsec from the primary in position angle 276°. The magnitude 14.9 "C" star is 40.1 arcsec away in position angle 165°. The "D" star shines at magnitude 12.9 and is located in position angle 221° and is separated from the "A" star by

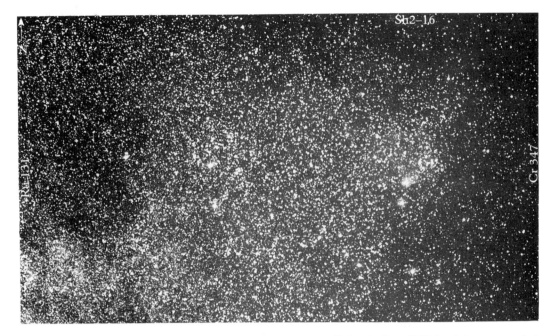

Photo 9.2 Sh2-16, Ru 131, and Cr 347 in Sagittarius.

58.1 arcsec. The system has a radial velocity of 20 km/s in approach. Delta is about 60 times brighter than the Sun.

About $\frac{1}{2}$ degree southeast of Delta Sgr is the class 6 globular cluster NGC 6624 [18:23;–30d22']. This nice magnitude 8.31 cluster is a strong X-ray source located 7.5 pc from us. This cluster is also getting farther away at a rate of 69 km/s. At the southern end of the bow is the blue-white spectrum B9 IV subgiant open double star Kaus Australis (20 Epsilon Sgr; SEE 351) [18:24;–34d23']. At magnitude 1.85, Kaus Australis (pronounced koss oss-TRAY-lis) is the brightest star in Sagittarius and 35th brightest in the sky. In 1896, See found the magnitude 14.2 secondary star 32.5 arcsec from the primary in position angle 295°.

About $\frac{3}{4}$ degree northeast of Epsilon is the Mira-type variable RV Sgr [18:27;–33d19']. This star goes in a cycle from magnitude 7.2 to 14.8 during a period of 317.51 days.

Hop another degree in the same direction from Epsilon Sgr to locate the class 5 globular cluster NGC 6637 (M69) [18:31;–32d21']. While at the Cape of Good Hope in 1751–2, De Lacaille first noted this apparent magnitude 7.79 (-7.90 absolute) cluster. Messier included it in his catalogue on August 31, 1780, as a "Nébuleuse sans étoile, dans le Sagittaire." M69 appears nearly round with a few streamers of stars along its ragged edges. The cluster is estimated to be 21 pc in diameter and 10.4 kpc from us. Every second, M69 moves 50 km farther away.

After observing M69, the easiest way to find our next Messier object is to let the sky drift through your eyepiece for 12 minutes until the class 5 globular NGC 6681 (M70) [18:43;–32d18'] comes into view. Messier discovered this magnitude 8.18 object on the same evening as he first located M69. There appears to be a swarm of stars surrounding the main clump that makes up this cluster. The distance to M70 is uncertain with values given from 10.8 to 20 kpc.

Let your scope drift for 12 more minutes to allow the planetary nebula PK 003-14.1 [18:55;–32d16'] to enter your eyepiece. This type II planetary shines at magnitude 10.9. The central star is a magnitude 13.9, white, spectrum O-type object, with a radial velocity of 65 km/s in approach.

Two degrees north of PK 003-14.1 is the class 3 globular cluster NGC 6715 (M54) [18:55;–30d29']. Messier discovered this magnitude 7.61 cluster on July 24, 1778. It appears almost round with a few faint scattered stars near it. This cluster is larger in both apparent and actual size than M69 and M70 even though it is more than twice as distant at 21.5 kpc. Its absolute magnitude is calculated at -9.41 and its radial velocity is 131 km/s in recession. Over 80 variables have been detected in the cluster. Most of these are RR Lyrae types.

The bright star $1\frac{1}{2}$ degrees north-northeast of M54 is the magnitude 2.5, visual and spectroscopic binary Ascella (38 Zeta Sgr; ADS 11950) [19:02;–29d53']. Ascella (pronounced as-SELL-ah) marks the southern tip of the "Milk Dipper," which consists of the teapot handle stars Zeta, 40 Tau, 34 Sigma, and 27 Phi Sgr. The handle for the Milk Dipper is Lambda. Zeta's primary is a magnitude 3.2 blue-white, spectrum A2 III, giant. Circling the "A" star in a period of 21.138 years is a blue-white, spectrum A2 V, subgiant of magnitude 3.4. A magnitude 9.9 "C" star located 75.5 arcsec from the "AB" stars is also part of this system. These stars are 24 pc from us and have a radial velocity of 22 km/s in recession.

To observe the class 11 open cluster NGC 6809 (M55) [19:40;–30d58'], we have to follow a trail of stars about 7 degrees to the east of Zeta Sgr. The cluster is bright at magnitude 6.33 and was first seen by De Lacaille, in 1751–2, from the Cape of Good Hope. Messier was unable to resolve this very loose globular when he verified De Lacaille's catalogue, on July 24, 1778. The brighter members are of 13th and 14th magnitude. The cluster has a radial velocity of 167 km/s in recession and is 5.7 pc from the Sun. Its estimated age is 14 billion years.

Sweep back to Zeta Sgr then northwest to magnitude 2.81, spectrum K1 IIIb, Kaus Borealis (22 Lambda Sgr) [18:27;–25d25'] at the top of the teapot. Kaus Borealis (pronounced koss BORE-ee-ALICE) also marks the northern tip of the Archer's bow. This yellow giant is coming toward us at a rate of 21 km/s. Lambda is located 30 pc from us.

In the same wide-field, low-power eyepiece as Lambda is the 13.3 kpc distant, class 6, globular cluster NGC 6638 [18:30;–25d30']. This small cluster is located about $\frac{1}{2}$ degree southwest of Lambda and shines at magnitude 9.03.

About $1\frac{1}{4}$ degrees west of Lambda are two novae-type cataclysmic variables; GR Sgr (Nova 1924) [18:23;–25d35'] and V441 (Nova 1930) [18;22;–25d29']. Novae are close binary stars that experience a rapid increase in brightness followed by a long (days, years, or decades) gradual decline in magnitude. The component of the binary system that undergoes this magnitude change is a hot dwarf. These stars can have a brightness change of as much as 7 to 19 magnitudes. The orbital periods for novae range from 0.05 to 230 days. GR Sgr became as bright as magnitude 7.6 in 1924 and has faded to 16th magnitude. V441 Sgr is an Na-type nova. This type of nova has both a rapid increase and decrease in luminosity. The nova outburst period lasts 100 days or less. V441 Sgr was at magnitude 8.2 in 1930 and is now 16th magnitude.

About $\frac{3}{4}$ degree northeast of V441 Sgr is the bright globular cluster NGC 6626 (M28) [18:24;−24d52']. On the evening of July 27, 1764, Messier saw this object and was unable to resolve its mass of stars. This class 4 globular cluster is 5.8 kpc from us and shines at apparent magnitude 6.99, but its absolute magnitude is -8.1. Many stars can be resolved around its dense bright core.

Three degrees to the northeast of M28 is the spectacular class 7 globular cluster NGC 6656 (M22) [18:36;−23d54']. The actual discoverer of this magnitude 5.07 cluster has been in doubt, but the credit is now given to an obscure German amateur astronomer named M. J. Abraham Ihle, who noted the cluster while he was observing Saturn, in 1665. Messier verified its existence on June 5, 1764. At 3.0 kpc, less than a handful of globular clusters are closer to us than M22. This accounts for its brightness and large apparent size. The central mass of stars is estimated to have a diameter of about 50 light-years. The cluster appears as if someone has splattered white paint on a black wall. Thousands of faint stars can be seen all around the bright core.

In the *Almagest*, Ptolemy listed the class 8 globular cluster NGC 6717 (Pal 9) [18:55;−22d24'] along with the double stars 32 Nu[1] and 35 Nu[2] as "The Nebular and double star in the eye" of the Archer. This appears to be the earliest listing of a double star. NGC 6717 is located about 4 degrees northwest of M22. Nu[1] [18:54;−22d44'] is a spectrum K2 I bright red giant and double system the Arabs named *Ain al Rāmī* meaning "The Archer's Eye." Nu[2] [18:55;−22d40'], like Nu[1], is also a red giant double system. This star's spectrum is K3 III: Ba0.8; CN2.

A very popular Messier object is the Lagoon Nebula (NGC 6523; M8) [18:03;−24d23']. This object is made up of 9 Sgr, an open cluster (NGC 6530), a large emission nebula, and three dark nebulae (B 88, B 89, and B 296). Flamsteed recorded the existence of this object around 1680. De Chéseaux, Le Gentil, and De Lacaille all noted it in their separate listings around 1750. Messier took his first look at this magnitude 5.8 object, on May 23, 1764. The spectrum O8 If, magnitude 5.97, star 9 Sagittarii is considered to be the source of the radiation that causes the gas cloud to emit its soft glow. The H–R diagram of the stars in the open cluster NGC 6530 (II2m) [18:04;−24d20'] indicates that this is a young cluster (two million years old). Many of the stars in this cluster are erratic variables, which are not yet stable members of the main-sequence. Use a wide-field eyepiece to view all of this great object. A nebula filter helps to see the wispy nature of the cloud. The magnitude 5.35, spectrum F3 III, giant 7 Sgr is a foreground star.

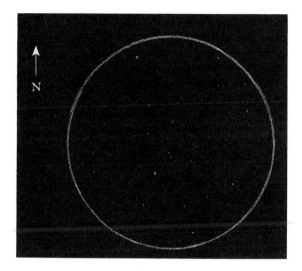

*Eyepiece Impression
9.2   NGC 6514
(M20); the Trifid
Nebula.*

The next few objects are very close to the galactic equator. A favorite at star parties is the magnitude 7.63, quad-segmented, Trifid Nebula (NGC 6514; M20) [18:02;–23d02']. M20 is both an emission and reflection nebulae. The dark region that criss- crosses the nebula is the dark nebula B 85. La Gentil first charted this object sometime prior to 1750. Messier added it to his growing list on June 5, 1764. John Herschel gave it the name Trifid. Near the center of the nebula is the multiple star HN 40 (ADS 10991). This system consists of at least seven stars. The brightest is a magnitude 7.6, spectrum O7, giant. The magnitude 10.7 "B" star is 6 arcsec from the primary in position angle 22°. The brighter "C" star (HN 6) shines at magnitude 8.7 only 10.8 arcsec from the primary in position angle 212°. The other components of the system are 10th- to 14th-magnitude stars. The magnitude 6.3 open cluster NGC 6514 is a part of the nebula.

In the same low-power field as, and $\frac{3}{4}$ degree northeast of, M20 is the open cluster NGC 6531 (M21) [18:04;–22d30']. This Trumpler (I3m) class cluster consists of about 70 members that have an integrated magnitude of 5.9. Messier discovered this cluster, on June 5, 1764. Distance estimates range from 680 to 1,300 pc. The brightest members are eighth- and ninth-magnitude B0-type giants.

Midway between M21 and reddish magnitude 5.44, spectrum gK0, 14 Sagittarii [18:14;–21d42'] is the SRC-type variable VX Sgr [18:08;–22d13']. This star cycles from magnitude 6.5 to 12.5 in a period of 732 days.

About $\frac{1}{2}$ degree northeast of 14 Sgr is the magnitude 8.6 open cluster NGC 6568 (III1m) [18:12;–21d36']. About 30 members of this cluster are eighth-magnitude and another 20 members are fainter than ninth-magnitude.

In the same low-power field as NGC 6568 is the naked-eye multiple system of 13 Mu[1] (ADS 11169) and Mu[2] [18:13;–21d03']. Mu[1] is a spectrum B9 Ia bright giant spectroscopic binary and eclipsing EA-type variable. Its magnitude fluctuates from 3.79 to 3.92 during a period of 180.45 days. Four other faint companions can be seen near Mu[1].

About 3 degrees northeast of Mu^1 Sgr you will see a large and dense congregation of stars. This is the "Small Sagittarius Star Cloud" more commonly known as M24 [18:16;–18d29']. This object, first seen by Messier on June 20, 1764, has no NGC number, but some of the editors of Messier's catalogue identified M24 as being the open cluster NGC 6603 (I1r)[18:18;–18d25'] embedded in the star cloud. The 100-plus members of this rich cluster are difficult to differentiate from the star cloud, since the cluster's brightest members are 14th magnitude. The cluster is 2,880 pc from us. To see the cluster requires a scope of at least eight inches and medium power. The two dark holes in the northwestern edge of the cloud are the nebulae B 93 and B 94.

Sweep 4 degrees east from the southern end of M24 to observe the large 220 million-year old open cluster NGC 6494 (M23) (III1m) [17:56;–19d01']. About 150 stars, the brightest of which are ninth-magnitude make up this cluster Messier discovered, on June 20, 1764. The cluster is 660 pc from us and is easy to pick out from the surrounding star fields.

Head east and sweep past M24 at the same declination as M23 to locate the open cluster IC 4725 (M25) [18:31;–19d15']. M25 is 35 minutes (about 9 degrees) east of M23, so you can also let the sky drift through your eyepiece for that length of time until M24 appears. De Chéseaux first saw this loose Trumpler (I2p) class cluster, in 1746. Messier entered this magnitude 4.6 cluster in his logbook, on June 20, 1764. The cluster is estimated to be 89 million years old and is 580 pc from us. Near the center of the cluster is a box of four bright stars. The cepheid variable and multiple star U Sagittarii (ADS 11433; β 966) is the northeastern of these four stars. This spectrum F5 Ib star's magnitude range is from 6.34 to 7.08 during a period of 6.744925 days. U Sgr is accompanied by 13 faint comes and was first measured by Burnham, in 1879. These stars range from eighth to tenth magnitude. As members of the M25 cluster, these stars all seem to be simply physical companions rather than binary systems.

Return to M24. A pair of sixth- and seventh-magnitude stars at the northeastern tip of M24 point you in the direction in which to hop 1 degree north to the small open cluster NGC 6613 (M18) (II3p) [18:19;–17d08']. The 20 members of M18 shine with an integrated magnitude of 6.9 from a distance of 1,200 pc. Look for two triangles of stars, one much brighter than the other.

One degree north brings us to the magnificent emission nebula NGC 6618 (M17) (III2p) [18:20;–16d11']. This glowing cloud of gas is also known as the "Omega Nebula," the "Horseshoe Nebula," and the "Swan Nebula." De Chéseaux noted the existence of this object, in 1746. Messier thought it resembled a spindle when he first looked at it, on June 3, 1764. British astronomer Sir William Huggins (1824–1910) announced in 1866 his discovery that M17 is a cloud of glowing gas instead of a mass of unresolved stars. The cloud contains enough material to make 800 Sun-size stars. About 35 stars of open cluster NGC 6618 are embedded in the cloud and are the radioactive sources for its glow.

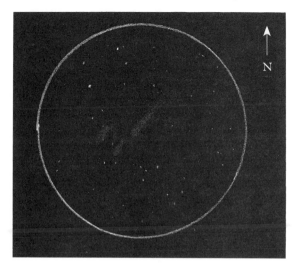

Eyepiece Impression
9.3 NGC 6618
(M17); the Swan
Nebula.

Before going on to hop around Scutum we will stop to observe the Eagle Nebula (NGC 6611; M16; IC 4703) (III2p) [18:18;–13d47'] in Serpens Cauda, the Serpent's Tail. Visually, M16 appears as an open cluster in small scopes. Larger scopes used at a dark site reveal the wondrous nebulosity enveloping this gathering of spectrum O and B hot blue baby stars. The age of the cluster is estimated at only three to four million years. Use a nebula filter to see the wispy gas cloud and the black nebulae cutting into it. The gas cloud is estimated to be about 21.4 pc (70 light-years) in diameter. No accurate distance to M16 has been established due mainly to the unknown amount of interstellar dust between M16 and us, which makes it very hard to get a reliable measurement.

Also, take some time to look at the class 1 globular cluster NGC 6864 (M75) [20:06;–21d55'] on the eastern fringe of Sagittarius, as shown in Figure 9.3. This 8.55-magnitude ball of stars is difficult to locate, since it is in a rather barren area of the sky with few guide stars near it. You will find M75 about 8 degrees southwest of magnitude 3.08, spectrum F8 V, Dabih (9 Beta Capricorni) [20:21;–14d46'] and about 14 degrees north-northeast of magnitude 3.77, spectrum gG8 IV, 39 Omicron Sgr [19:04;–21d44']. Méchain found M75 on the night of August 27, 1780. After a couple of attempts to locate it, Messier found it on October 18, 1780, and added it to his list. About 11 cluster variables are known to be members of M75. From the tremendous distance of 18 kpc away, M75 is greatly dimmed from its absolute magnitude of -8.30. The radial velocity for M75 is 195 km/s in approach.

Star-hop in Scutum

Our third star-hop this month takes us to the small yet celestially rich constellation originally known as Scutum Sobiescianum (Sobieski's Shield). Today, the 84th constellation in size is known simply as

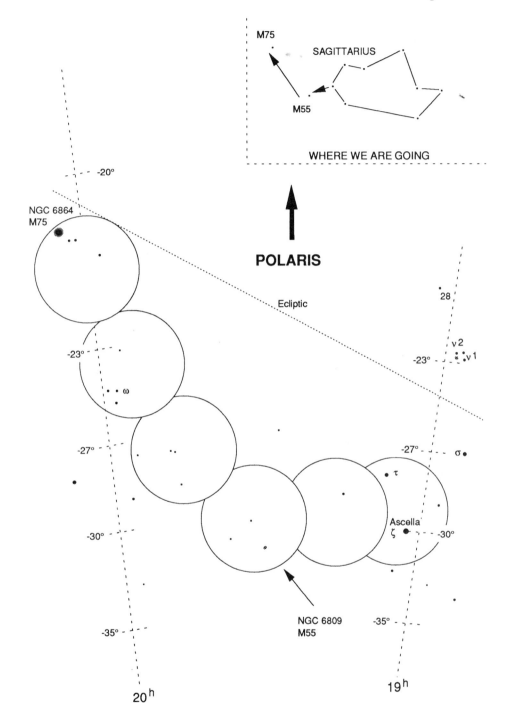

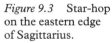

Figure 9.3 Star-hop on the eastern edge of Sagittarius.

Scutum. Elected Polish King, John III Sobieski (1629–96) defeated an advancing Ottoman army at the gates of Vienna, on September 12, 1683, in one of the last great battles between Christian Europe and the followers of Mohammad. His fame as the "Vanquisher of the Turks" in prior wars had led to his May 1674 election as King of Poland.

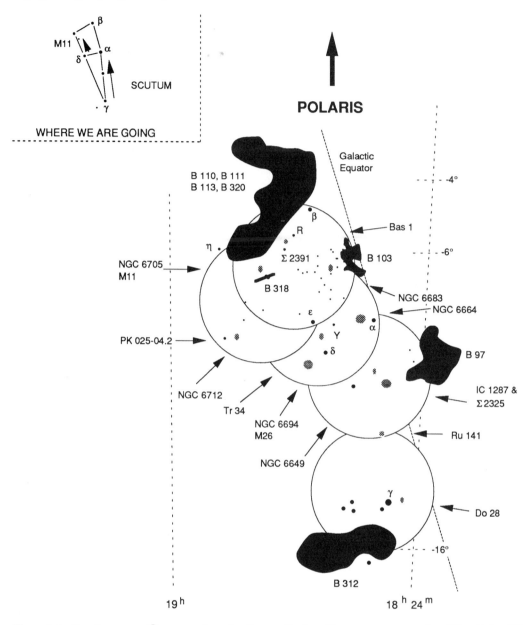

Figure 9.4 Star-hop in Scutum.

Compared to the fame of other European monarchs, King John III Sobieski is a rather obscure figure to most people today. Why then a constellation in his honor and not some other famous European rulers? At the time of the 1683 battle, Johannes Hevelius lived in Danzig. In addition to being an astronomer, Hevelius was a very loyal subject of his king. In his 1690 catalogue Hevelius outlined the constellation and considered its fourth-magnitude stars as a representation of the cross on Sobieski's shield.

Scutum is one of those constellations in which stars for two constellations share designations. The brightest stars in the new constellation were labeled using the method of identifying stars with a Greek letter. Though Hevelius's atlas was probably known by John

Photo 9.3 Scutum region. The bright triangle of three stars in the center of the photo are α Scuti, ∂ Scuti, and Σ2325.

Flamsteed, the early non-acceptance of the constellation left some of Scutum's stars to share their designations with the Flamsteed numbers assigned to a few of the stars in Aquila the Eagle.

We will start our tour of Scutum, as shown in Figure 9.4, at the magnitude 4.70, spectrum A2 III, giant star, Gamma Scuti [18:29;–14d33'], which is about $2\frac{1}{2}$ degrees (about 19 minutes of

R.A.) east of M16. About $\frac{1}{2}$ degree southeast is the black patch of the dark nebula B 312 [18:30;–15d08']. This nebula straddles the boundary between Scutum and Sagittarius. B 312 is elliptical in shape being almost 100 by 30 arcmin. This black patch covers an area of about 1 square degree of sky. A few foreground stars shine brightly between us and the cloud.

Star-hop about 1 degree west of Gamma Sgr to the open cluster Dolidze 28 (Do 28) (IV2p) [18:25;–14d39']. The 20 faint stars of this small cluster are hard to pick out from the surrounding star field.

Look 2 degrees north-northeast for the open cluster Ruprecht 141 (Ru 141) (III2m) [18:31;–12d19']. The 20 stars of this cluster appear scattered with no central concentration. The cluster appears about 11th magnitude with the brightest members being 12th-magnitude stars.

About $1\frac{1}{4}$ degrees north of Ru 141 is the reflection nebula IC 1287 [18:31;–10d50']. Almost smack in the middle of the nebula is the wide double star Σ 2325. A dust lane bisects the nebula southeast of the double star. Σ 2325 consists of a 5.8-magnitude, spectrum B3, white primary and a 9.1-magnitude secondary located 12.3 arcsec from the primary. Look for the secondary star in position angle 257°. F. G. W. von Struve first measured this pair, in 1829.

About $\frac{1}{4}$ degree northeast of IC 1287 is the open cluster NGC 6649 (II2m) [18:33;–10d24'] contains about 50 members in a tight grouping located 1,300 pc from us. This 8.9-magnitude cluster appears to be the hub for several spokes of dark dust lanes radiating through the star field.

About 1 degree northwest of NGC 6649 is the irregular-shaped dark nebula B 97 [18:29;–9d56']. This dark patch of gas and dust covers an area of about 1 square degree. The galactic equator cuts through the southeastern edge of the nebula.

The vast concentration of stars to the east of B 97 is known as the Scutum Star Cloud. This area is visible to the naked eye, but shows best with binoculars. In some parts of the cloud, the stars are so densely packed you can hardly see black space between them. This is one area of the sky you should take your time to look over and savor the view. Compare the over-abundance of stars in the Scutum Cloud with the star-less areas north of it. Near the center of the constellation is the magnitude 3.85, spectrum K3 III, red giant Alpha Scuti (1 Aquilae) [18:35;–8d14']. About $\frac{1}{4}$ degree east of Alpha is NGC 6664 (III2M) [18:36;–8d13'], the largest open cluster in Scutum. The 50 members of NGC 6664 are spread out and shine with an integrated magnitude of 7.8. The brightest members are about tenth magnitude. The cluster is 1,639 pc from the solar system.

On the eastern fringe of NGC 6664 is the pulsating Delta Cephei-type variable, Y Scuti. This star's magnitude fluctuates from 9.2 to 10.0 during a period of ten days. Another variable member of NGC 6664 is the short-period Delta Cephei-type (DCEPS) star; EV Scuti.

About $\frac{1}{2}$ degree southeast of NGC 6664 is the small open cluster Trumpler 34 (Tr 34) [18:39;–8d29']. This Trumpler class II2m cluster contains just eight members and shines with an integrated magnitude of 9.4.

About $\frac{1}{2}$ degree south-southeast of Tr 34 is the short-period pulsating variable Delta Scuti (2 Aquilae) [18:42;–9d03']. Delta Scuti is the prototype for short-period variables that show very slight fluctuations in their visual brightness during a period lasting less than five hours. In the Delta Scuti class, this range is about 0.20 or less of a magnitude. Delta's magnitude fluctuates from 4.98 to 5.16. The changes in your own local "seeing" during the night can cause you to see this much of a difference in the star's apparent magnitude, so you will have to compare Delta to the stars around it to see if they are also changing in brightness. If these other stars do not appear to dim or brighten while Delta is changing, then you are observing Delta during one of its pulsations.

Delta Scuti variables are all young stars and fit within spectral classes A5 to F5. Unlike many of the other types and classes of variable stars, the spectral type for a Delta Scuti star remains constant during the star's pulsations.

About 100 Delta Scuti-type stars have been detected. Since these stars are Population I giant stars, they are all located in or near to the Milky Way and generally are members of open clusters. Many of these stars also appear to be short-period spectroscopic binary stars. This leads astronomers to think the rapid motion and gravitational effects of the nearby companion star may be a cause of the pulsations in the primary star.

Delta Scuti has two visual companions. The magnitude 9.2 "C" star was first measured in 1879. This star is located 52 arcsec from the primary ("A") star in position angle 130°. The fainter (magnitude 12.2) "B" star was first measured in 1938. This star is at position angle 46° and has a separation from the primary of 15 arcsec. The Delta Scuti system is about 49 pc from us and has a radial velocity of 45 km/s in approach.

On the evening of June 20, 1764, Messier noticed a cluster of stars near Delta Scuti and added it as the 26th object on his list. NGC 6694 (I1m) (M26) [18:45;–9d24'] looks like a wishbone or tuning fork with four bright stars near the center. Two curving arms of fainter stars can be detected to the north and south of the cluster's center. M26 has an integrated magnitude of 8.0, and its brightest member is a magnitude 11.9 blue giant. The 100 stars that make up M26 are located 1,550 pc from the Sun. This cluster is receding from us at a rate of 4 km/s.

Star-hop about 2 degrees north-northeast from M26 to the class 9 globular cluster NGC 6712 [18:53;–8d42']. In small scopes, this magnitude 8.13 cluster appears somewhat nebulous with no individual stars visible. In larger instruments (13 inches or greater), individual stars can be resolved around a very dense core. The cluster is located 7.4 kpc from us and is coming closer at a rate of 124 km/s.

To the southeast in the same field of view as NGC 6712 is the magnitude 15.0 planetary nebula IC 1295 (PK 025-04.2) [18:54;–8d50']. Two 13.5-magnitude stars appear to be surrounded by a thin gas shell. Compared to the majority of planetary nebulae, this planetary is large at about 86 arcsec in apparent diameter. The gas shell has low surface brightness, so use a high-contrast-type filter. Once you have IC 1295 centered in a low-power eyepiece, switch to high power to get the best view of it. This object is 1.3 kpc from the solar system.

In the northern area of Scutum is the largest of its dark nebulae, consisting of B 110, B 111, B 113, and B 320 [18:50;–5d00']. This crescent-shaped nebula covers about 3 square degrees of sky as it curves its way through a rich star field.

On the southern end of B 111 is the bright open cluster NGC 6705 (M11) (II2r) [18:51;–6d16']. While verifying the discoveries of his predecessors, Messier located this dense cluster on the night of May 30, 1764. He added it as the 11th object on his growing list. Gottfried Kirch discovered this object, in 1681. About 600 stars brighter than magnitude 14.8 can easily be seen in M11. The cluster contains hundreds of fainter stars and is rich in yellow and red giants. The stars are so tightly packed at the cluster's center they may be less than 1 light-year from each other. The cluster is estimated to be 500 million years old and located 1,720 pc from us. M11 is not part of the Scutum Cloud, but is a foreground object. Admiral William Henry Smyth, RN (1788–1865) wrote that this cluster resembled a flight of wild ducks. Since then, M11 has been known as the Wild Duck Cluster. The thin black streak south of M11 is the dark nebula B 318.

The nice small open cluster Basel 1 (Bas 1) (I2m) [18:48;–5d51'] is located about $\frac{3}{4}$ degree north-northwest from M11 along the edge of B 111. This cluster consists of 15 stars and shines with an integrated magnitude of 8.9. The cluster is 58 million years old and is 1,460 pc from us.

About 1 degree northwest of M11 is the radially pulsating supergiant R Scuti [18:47;–5d42']. This RVA-subtype of RV Tauri-type pulsating variable stars fluctuates from magnitude 4.45 to 8.20 during a period of 140.05 days. The variability of this star was first noticed by the British astronomer Edward Piggot (1753–1825), in 1775. R Scuti is in the class of variables whose prototype is RV Tauri. These variables appear to have double waves of alternating minima. This causes the star to have various magnitudes each time it reaches its extreme faintness phases. Most of the time, R Scuti is a fifth- or sixth-magnitude star. About every 146 days however, the star fades to eighth magnitude and slowly rises back in brightness. R Scuti is a large star and may have a diameter as large as 100 times that of the Sun.

Halfway between M11 and R Scuti is the open multiple star Σ 2391 (ADS 11696) [18:48;–6d00']. The primary is a magnitude 6.5, spectrum A2 star. The 9.8-magnitude comes ("B") is located 37.9 arcsec from the primary in position angle 332°. This is a nice double and is easy to separate even in small scopes. A third member of the system is a magnitude 14.3 star located 12 arcsec from the "B" star.

Sweep $1\frac{1}{2}$ degrees west from M11 to the U-shaped open cluster NGC 6683 (I2p) [18:42;–6d17']. This cluster is hard to pick out from the edge of the Scutum Star Cloud. The 20 stars in this cluster

Photo 9.4 Dark
Nebula Bernard 103
in Scutum.

have an integrated magnitude of 9.5 with the brightest members
being 11th magnitude.

The expansive dark cloud west of NGC 6683 is the dark nebula
Bernard 103 (B 103). This cloud covers about $\frac{1}{2}$ square degree of sky.
The eastern portions seem darker than the western side. In binoculars
you can see twisting branches of dark matter.

In this chapter we were able to observe a tiny selection of targets in
Scorpius, Sagittarius, and Scutum. Look over your star charts and
make our own additional star-hops of this fascinating area of the summer sky. The whopping array of celestial objects we did not go to on
these star-hops will keep your attention high during many long warm
summer nights.

10 *August*

Draco:
Follow the trail of the Dragon

> Press close bare-bosom'd night — press close magnetic nourishing
> night!
> Night of south winds — night of the large few stars!
> Still nodding night — mad naked summer night.
>
>> Walt Whitman (1819–92),
>> "Song of Myself,"
>> (lines 435–7),
>> 1855, 1881

Star-hop in Draco

For this month's star-hop, we will follow the winding "Trail of the Dragon" as shown in Figure 10.1, take side trips to other objects within Draco's celestial boundary, and swing by a couple of objects in Camelopardalis the Giraffe and Ursa Major. Circumpolar Draco (Dra) covers portions of 11 right ascension hours and 30° of declination. Encompassing 1,082.952 square degrees of the sky, this fire-spewing reptile ranks eighth in size. Draco looks something like an upside-down question mark winding through the sky between the two celestial bears.

I have read books in which the authors discourage amateurs from looking over Draco by claiming it to be a telescopic desert. They may have written that because many of the constellation's stars are faint, and except for Thuban there are no famous stars or *real* Messier objects there (M102 is thought to be one of Messier's missing objects). As we star-hop from the tail of the dragon to his head, you will find Draco to be a starry treasure trove of galaxies, double and multiple star systems, and nebulae that will entice you to look them over again and again. Being faint, Draco is best viewed in dark skies that allow you to see a rainbow of color, since the Dragon's stars glow like iridescent scales that cover the entire spectra of stellar colors.

A fire-breathing, evil dragon in the night sky is an ancient concept, but not the only way in which this constellation was viewed. The Egyptians pictured the constellation as either a hippopotamus or a crocodile. In their writings, they may also have depicted the stars as their version of a goddess of love. In ancient Persia, Draco was a man-eating serpent. The Chinese thought of this group of stars as *Tsi Kung*, "The Palace for the Heavenly Emperor." The Arabs saw a herd of camels wandering around the north pole and predators waiting to

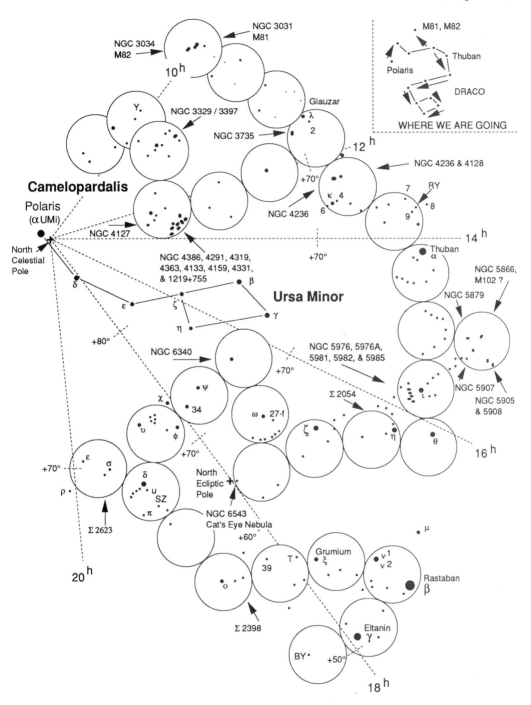

Figure 10.1
Star-hop in Draco.

kill the weaker members of the herd. Roman tales indicate that Draco was a mighty snake taken by Minerva (their equivalent to Athena) and thrown into the sky.

The name of Draco comes to us from the Greeks. They seem, however, to have had a tough time figuring out just how Draco came to be. They considered the constellation as a dragon, but sometimes it was also referred to as a snake. Draco was also thought to have been

175

the serpent-like feet for the Titan giants. After the battle between the Titans and the Olympian gods, Athena grabbed one of the Titans and hurled him into the sky. Other stories tell of Draco being the dragon or the sea monster that was slain by Perseus when he rescued Andromeda from her terrible fate. Cetus the Sea Monster is also supposed to also be this monster.

I have no idea how many times one dragon can be slain, but Draco may also have been the forked-tongued serpent slain by Cadmus the Tyrian at the Spring of Ares on the plain of Panope. After slaying the dragon, Cadmus offered a sacrifice to Athene. She appeared and told Cadmus to sow the dragon's teeth in the ground near the battle site. Immediately, the teeth sprouted and grew to become the Sown Men. After a battle among these mighty Sparti, only five remained alive. These five warriors helped Cadmus to found his city-state at Thebes (present day Thívai, Greece).

For this month, we will take one long hop along the back of Draco. The easiest way to start this month's star-hop is to locate Polaris then hop across a section of Camelopardalis to reach the tip of Draco's tail. The tip of Draco's tail is located about 8 degrees southeast of Polaris. You can find this undesignated magnitude 4.29, spectrum K3 IIIa, star at coordinates [9:39;+81d22'] by making about three hops from Polaris heading slightly west of the line between the Big Dipper pointer stars (Dubhe and Merak) and Polaris.

In the same field as our first star is the M-type variable Y Draconis [9:42;+77d51']. The magnitude for this spectrum M5e star ranges from 7.8 to 14.5 during a period of 326 days. If you are using a scope larger than eight inches, look carefully around this area for it is populated by several faint galaxies.

After you have studied the vicinity around the last star in Draco's tail, hop southeast to NGC 3329 (also listed as NGC 3397) [10:44;+76d49']. This Sa-type galaxy is 30.3 Mpc from us and appears as a faint blob with a bright core in a field of stars.

From NGC 3329, sweep about 4 degrees east into Camelopardalis (Cam) to a star grouping and Sbc-type galaxy NGC 4127 [12:08;+76d48'], then $\frac{1}{2}$ degree south to the galaxies NGC 4159 and NGC 4331 in Draco. From 29.8 Mpc away, NGC 4127 glows at an apparent magnitude of 12.5. Its absolute magnitude is -19.87. NGC 4331 is a magnitude 13.8 Imp-type irregular galaxy 26.6 Mpc from us.

Locate the white, magnitude 5.38, spectrum A2, star [12:18;+75d09'] about one degree south of NGC 4331. To the east of this star is a cluster of five galaxies. The northernmost cluster member is the small 12.8-magnitude, E6/SO-type galaxy, NGC 4386 [12:24;+75d32']. About 9 arcmin southwest of NGC 4386 is NGC 4291 [12:20;+75d22'], an E1-type galaxy 29.4 Mpc from us. This 12.4-magnitude galaxy looks like a faint globe of whiteness. Six arcmin southeast of NGC 4291 are two very close galaxies: NGC 4319 and 1219+755. NGC 4319 is another very small Sab-barred spiral with a bright core. At 13th magnitude, this galaxy can be difficult to locate.

Quasi-stellar object (QSO or quasar) 1219+755 [12:21;+75d18'] is typical of these extremely distant, high radio-emitting objects. Quasars were detected during radio surveys of the sky conducted in the 1950's. The emission lines of atomic hydrogen in the spectrum of these objects are highly shifted to the red indicating the source of the spectrum is at a great distance and receding at tremendous speed. Only about 10 percent of the known quasars are high radio emitters. The rest have extremely high luminosities. On the average, quasars have absolute luminosities 100 times brighter than our nearby extragalactic neighbors. QSOs are very difficult to see visually due to their small angular and actual size. These objects are very compact galaxies with active nuclei. Many are within the grasp of large amateur scopes. Quasar 1219+775 has a visual magnitude of 14.5 and appears as a tiny star.

Magnitude 3.84 Giauzar (1 Lambda Dra) [11:31;+69d19'] is our next stop. Giauzar (pronounced GUY-u-zar) usually marked the tail of the Dragon on old star charts instead of the star we began this hop at. This spectrum M0 III, orange star was known by several names from *Al Juzā'* meaning "The Twins" to *Al Jauzah*, "The Central One" indicating its place midway between Polaris and the pointers in Ursa Major. Giauzar means "the Poison Place" in Persian and refers to the mistaken belief that where the Moon crossed the ecliptic at the head and the tail of the Dragon were poisonous places in the sky. About $\frac{1}{2}$ degree east of Lambda is 2 Dra [11:36;+69d19']. The name *Al Jauzah* refers to this, spectrum K0 III, star as Lambda's twin. In the same field as Lambda is the edge-on Sb spiral galaxy NGC 3735 [11:36;+70d32']. This faint streak of 12th-magnitude light is located north of Lambda.

From Lambda, sweep about three hops west into Ursa Major to view the Sb-type galaxy NGC 3031 (M81) [9:55;+69d04']. Johann Bode discovered this faint patch of light on December 31, 1774. Méchain saw it in August 1779. Messier added it to his list on February 9, 1781. M81 was the first spiral in which rotation of its arms was detected with spectroscopic analysis. In 1914, Maximillian Wolf (1863–1932) discovered the linear rotational velocity at the edge of the galaxy was about 300 km/s. This magnitude 6.8 oval appearing galaxy is inclined 35 degrees to our line of sight, and in a large scope the hint of spiral structure is visible extending from its bright core. The absolute magnitude for this galaxy is -18.29. With no bright stars in the area of M81, you will first find a 2 degree long by 1 degree wide box of fourth- and fifth-magnitude spectrum F and G stars. At the northwestern corner of this box is the magnitude 4.56, spectrum G2 IV, yellow subgiant 24 DK Uma. M81 is about 2 degrees south-southeast from 24 DK Uma.

To the north in the same low-power view of M81, you will also have the irregular galaxy NGC 3034 (M82) [9:55;+69d41'] in view. This magnitude 8.41 streak of light is about 11 arcsec long running west to northeast. Bode discovered this galaxy on the same evening he found M81. Messier reported that Méchain had seen it sometime in August 1779, and he first saw it on February 9, 1781. M82 was one of the first Seyfert-type galaxies found. In 1962, studies showed this

Photo 10.1 NGC 3031 (M81) and NGC 3034 (M82).

strong radio-emitting galaxy had experienced a massive explosion 1 to $1\frac{1}{2}$ million years ago. This explosion expelled polarized gaseous material away from its nucleus at speeds of 160 km/s. Synchrotron radiation causes this material to shine. The absolute magnitude for M82 is -19.42. The light from it has traveled 5.2 Mpc to reach us.

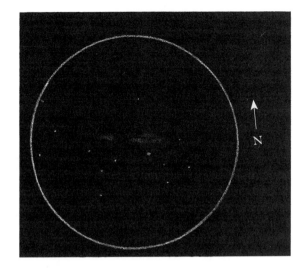

*Eyepiece Impression
10.1* NGC 3034
(M82) in Ursa
Major.

Hopping due east of Lambda Dra, we come to NGC 4236 [12:16;+69d28'], a large bright Sdm spiral with very loose structure. This galaxy glows at magnitude 9.47, but being elongated it appears much fainter with only a hint of brightness at its core. Use averted vision and a low-power eyepiece for the best views of this object.

The next star along Draco's tail is the variable and spectroscopic binary 5 Kappa Draconis [12:33;+69d47']. Kappa is a spectrum B6 IIIpe star shining at magnitude 3.87.

South of Kappa is a triangle formed by fifth-magnitude stars 7, 8, and 9 Draconis. Located about midway between 8 and 9 is the deep red, spectrum G3, SRB-type variable, RY Draconis [12:56;+66d00']. This star fluctuates between magnitudes 9.4 and 11.4 during a period of 172.5 days.

Two hops east from RY Draconis brings us to Thuban (11 Alpha Dra) [14:04;+64d22']. Thuban (pronounced thoo-BAN) was the pole star about 5,000 years ago, being closest to the pole in 2,830 B.C. The Arabs used the name Thuban for the entire constellation. Thuban is 71 pc away, but coming toward us with a radial velocity of 16 km/s. This golden yellow giant is the MK luminosity standard for spectrum A0 III stars. It shines at magnitude 3.65 and is actually about 135 times brighter than the Sun. Alpha is a spectroscopic binary with the comes only about 20 million miles from the primary. Much speculative writing has been published over the years linking the descending passageway into Pharaoh Khufu's (also known as Cheops; reigned 26th century B.C.) Great Pyramid in Egypt to a visual alignment with Thuban's ancient pole star position. Through modern calculations of Thuban's precession, and analysis of Egyptian archeological records dating the period of the pyramid's construction, this alignment between the star and passageway is now believed to be unrelated. The passageway never pointed directly at Thuban during its time as the pole star.

The area between Thuban and orange, magnitude 3.29, spectrum K2 III, 12 Iota Dra [15:24;+58d57'] contains few guide stars. Iota is also known as Ed Asich (pronounced ed-ahs-ITCH). This name is derived from the Arabic *Al Dhīh*, "The Male Hyena." Iota is 48 pc from us and is coming toward the Sun at a rate of 11 km/s. In early January each year, the area near Iota is the radiant point for the Quadrantid meteor shower.

A side trip about 3 degrees southwest of Iota will take us to one of Messier's missing objects: M102. In 1781, Méchain discovered a nebula in the area and added it to Messier's list. Two years later, Méchain admitted this sighting was probably a mistake. Since then, astronomers have speculated that what Méchain saw was one of several galaxies: NGC 5866, NGC 5879, NGC 5907, or NGC 5908.

Most Messier lists show NGC 5866 [15:06;+55d46'] as the prime candidate for being M102. At magnitude 10.0, this E6p-type galaxy is the brightest of the four mentioned galaxies. NGC 5866 is easy to locate, even in small scopes. This object is 15.3 Mpc from us and has an absolute magnitude of -20.8. NGC 5866 appears as an elongated oval with a dark streak running from end to end. Second on the probability list is NGC 5879 [15:09;+57d00']. This Sbc-type galaxy is also easy to see in small scopes, but is much fainter than NGC 5866, since its magnitude is listed at 11.5. NGC 5907 [15:15;+56d19'] is an edge-on viewed Sb galaxy. It appears long and thin and fairly bright at magnitude 10.38. Last on the list is NGC 5908 [15:16;+55d25']. This Sb galaxy is smaller than the other M102 candidates and is harder to locate since its magnitude is 11.9. In the same field as NGC 5908 is NGC 5905 [15:15;+55d31']. This tiny smudge of light is actually larger than NGC 5908, but is slightly dimmer at magnitude 13.1. It has a bright core, but its arms are very thin and extremely hard to see. Its thinness makes it appear smaller than NGC 5908, yet its apparent diameter on photographs is larger.

Northeast of Iota Dra are the galaxies NGC 5976, NGC 5976A, NGC 5981, NGC 5982 and NGC 5985. NGC 5976 and NGC 5976A are very small and faint so you probably will not be able to see them as galaxies, unless you are using a 20-inch or larger scope. Otherwise they appear as very faint fuzzy stars.

About $2\frac{1}{4}$ degrees southeast of Iota is 13 Theta Draconis [16:01;+58d33']. This spectrum F8 IV-V spectroscopic binary shines at magnitude 4.01. It never received an Arabic name, but was known to the Chinese as *Hea Tsae*, "the Lowest Steward" in the Palace of the Heavenly Emperor. Theta generally marks the location where Draco's body and tail meet. From here we will turn and head $2\frac{1}{2}$ degrees northeast to 14 Eta Dra.

14 Eta Dra (ADS 100058) [16:23;+61d30'] is a standard double, but, since the comes is a much fainter dwarf companion, these stars will appear as a single yellow-white star. The comes is 5.2 arcsec southeast of Eta. Being 26 pc from us, Eta's 2.74 apparent magnitude and G8 III spectral/luminosity classification means the primary is actually about 40 times brighter than the Sun. Eta is coming toward us with a radial velocity of 14 km/s. Together with 22 Zeta Dra [17:08;+65d42'], these two stars were known to the Arab world as *Al Dhī'bain*, "The Two Hyenas" or "Wolves." These two stars were

Photo 10.2
Planetary nebula
NGC 6543 (PK
096+29.1) in Draco.

lying in wait to kill the baby camel, *Al Ruba`*, 21 Mu Dra [17:05;+54d28'].

North of, and in the same field as, Eta is the very close double Σ 2054 (ADS 10052) [16:23;+61d42']. The primary is spectrum G5 and shines at magnitude 6.0. The comes is a 7.2-magnitude star about 1 arcsec from the primary in position angle 355°.

22 Zeta Dra was also known as *Al Dhī'bah* along with its other Arabic name tying it to Eta. Zeta shines at magnitude 3.17 and is a spectrum B6 IIIp star. At a distance of 97 pc from us, Zeta has an absolute magnitude of -3.2 and is about 1,500 times brighter than the Sun. Its approach radial velocity is 14 km/s.

About $1\frac{1}{2}$ hops northeast of Zeta Dra is the small, but bright, blue-greenish planetary nebula NGC 6543 (PK 096+29.1) [17:58;+66d38']. Though the "Cat's Eye Nebula" is listed as being magnitude 8.1 it is best viewed in a telescope larger than eight inches and under high power. The ellipsoidal gas shell around the super-hot, Wolf–Rayet O7 spectral-type white dwarf central star appears to be two separate layers of doubly ionized oxygen gas, which indicate the star experienced two explosive events, many years apart. Its magnitude 10.9 yellowish central star is partly obscured, but is bright and easy to see. On August 29, 1864, Sir William Huggins, using NGC 6543, made the first spectroscopic analysis of a planetary nebula. Until he conducted his study of NGC 6543, planetaries were thought to be tiny clusters of stars. His work indicated this type of nebula was a cloud of very low pressure gas instead of unresolved stars.

About $1\frac{1}{2}$ arcmin east of NGC 6543 is the point in the sky marking the North Ecliptic Pole. This spot is 90 degrees north of the ecliptic.

We get back on the main trail at 27-f Dra [17:31;+68d08'], a spectrum K0, double star system. In the same field is 28 Omega Dra [17:31;+68d45'], which is a 4.80-magnitude, spectrum F4 V, spectroscopic binary star. Together, these stars were called *Al Athfār al Dhi'b*, "The Hyena's Claws." They were also called *Al Dhīh*, "The Wolf."

One hop northwest of Omega Dra and in line with 21 Eta Ursae Minoris [16:17;+75d45'], is the magnitude 11.03 Scp-type galaxy NGC 6340 [17:10;+72d18']. This object is small, but has a bright core. NGC 6340 is 22 Mpc from us.

About $3\frac{1}{4}$ degrees north of Omega, 31 Psi[1] and Psi[2] Dra (ADS 10759; Σ 2241) [17:41;+72d09'] form an open double star system known to the Arabs as *Dsiban*. The Chinese called this pair *Niu She*, the "Palace Governess." The white F5 IV-V primary (Psi[1]) shines at magnitude 4.58. The yellow 6.1-magnitude Psi[2] is also an F5 star, located at position angle 15°.

To the Chinese, the very close multiple and spectroscopic binary 43 Phi Dra (ADS 11311; OΣ 353) [18:20;+71d20'] was known as *Shaou Pih*, the "Minor Minister." Phi Dra shines at a combined magnitude of 4.22 with a spectra that classifies it as an A0 p: Si star. The magnitude 6.1 "B" star is 0.2 arcsec from the primary in position angle 304°. The magnitude 12.7 "C" star is easier to separate being 70.8 arcsec from the "A" in position angle 115°. The binary star's orbital period is 26.768 days.

Al Tais, "The Goat," is the Arabic name for the open double 57 Delta Dra [19:12;+67d39']. This magnitude 3.07 star also marks the second convolution in the Dragon's body and was known by the Latin name *Nodus Secundus*, the "Second Knot." Delta also marks the northwest corner of the Chinese "Heaven's Kitchen," *Tien Choo*. Delta is 36 pc from us and is a yellow, spectrum G9 III, star. Its absolute luminosity is about 75 times brighter than the Sun. The magnitude 12.3 "B" star is 88.1 arcsec from the primary in position angle 352°. This system has a radial velocity of 25 km/s in recession.

Southwest of Delta Dra are two variables: U Dra and SZ Dra. U Dra [19:10;+67d17'] is an M-type variable with a magnitude range of 9.1 to 14.5 during its 316.42-day period. SZ Dra [19:10;+66d20'] is an irregular M5 spectral class with a magnitude range of 8.5 to 9.5. No set period has been established for this variable. SZ Dra is about $\frac{1}{8}$ degree north of white, magnitude 6.2, spectrum A2V, 55 Dra [19:09;+65d58'].

About $6\frac{1}{2}$ degrees south-southwest of SZ Dra is 47 Omicron Dra (ADS 11779; Σ 2420) [18:51;+59d23']. This star is a G9 III: Fe-0.5 spectral-type open double and spectroscopic binary system. The primary shines at magnitude 4.8 and its magnitude 7.8 comes is 34 arcsec away at position angle 326°. These stars appear orange and blue, respectively.

In the same field as Omicron Dra and about 1 degree west of it is the wide double Σ 2398 (ADS 11632) [18:43;+59d33']. This system consists of two DM-type spectral class cool red dwarfs located 3.4 pc from us. This distance makes these 8.9 and 9.7 apparent magnitude stars about the closest double system to us. Both stars are only about one-quarter the mass of the Sun. Their absolute magnitudes are +11.2 for the primary and +12.0 for its companion. The comes is 15.3 arcsec from the primary at the 160.8° position angle.

Another interesting object in this area is the close multiple 39-b Draconis (ADS 11336; Σ 2323) [18:23;+58d48']. The primary for this system is a spectrum A1 white giant shining at magnitude 5. The first comes is an F5 yellow, eighth-magnitude star located 3.7 arcsec away at position angle 351°. The third star in the system is a yellow, 7.5-magnitude, F8 star 89 arcsec from the primary at position angle 21°.

The base of Draco's head is marked by the open double Grumium (32 Xi Dra) [17:53;+56d52']. Grumium (pronounced GRUM-e-yum) is a spectrum K2 III orange giant that shines at magnitude 3.75. Located 316 arcsec from the primary in position angle 290° is a magnitude 14.9 "B" star.

In the same field as, and about one degree north of, Xi Dra, is the long-period variable T Draconis [17:56;+58d13']. T Dra was classed as an NOe spectral type, but is now classed as a M7e red carbon star. Its magnitude ranges from 7.2 to 13.5 during a period of 421.22 days. T Dra has a companion that was thought to be a variable and was given the designation UY Dra, but its variability is now in doubt.

About $2\frac{1}{4}$ degrees southwest of Xi Dra are blue-white 24 Nu[1] and 25 Nu[2] Draconis (ADS 10628) [17:32;+55d10']. These two stars form a wide double system in which both of its visible components are white A5 spectral types nearly equal in brightness at 4.87 and 4.88 magnitudes. These stars are the faintest of the four stars that make up the quadrangle outlining the Dragon's head. Being 61.9 arcsec apart, these stars are easy to separate in binoculars. Both stars are also spectroscopic binaries. Depending on the dragon's orientation on old mythological maps of the heavens, Nu Dra represents either the Dragon's fiery tongue or one of his eyes. To the Arabs, this star was one of the mother camels protecting their faint foals inside the quadrangle.

Close double yellow giant Rastaban (23 Beta Dra; ADS 10611; β 1090) [17:30;+52d18'] is the southwestern point forming the Dragon's head. Rastaban (pronounced RAS-tah-ban) comes from the Arabic *Al Rās al Thu'bān* meaning "The Dragon's Head." The Arabs also saw this star as one of the mother camels. This MK standard for spectrum G2 Ib-IIa stars is coming toward us with a radial velocity of 200 km/s. Though shining at magnitude 2.79, Rastaban is intrinsically about 600 times brighter than the Sun.

The last star on this star-hop along the trail of the main stars of the Dragon is the multiple system of Eltanin (33 Gamma Dra; ADS 10923; β 633) [17:56;+51d29']. Eltanin (pronounced el-TAY-in) is a magnitude 2.99, MK spectrum standard K5 III, yellow giant and usually marks the Dragon's forehead. Gamma Dra is on the zenith circle for Greenwich and used to be known as the Zenith star. All six comes in this system are fainter than magnitude 11.

Gamma Draconis was used by James Bradley to calculate the aberration of starlight. Over the period of a year (December 1725 to

December 1726), Bradley observed the changing positions of Eltanin. He recorded an astonishing 1 arcsec shift in the star's position during a three-day period. He knew that this was too great a change to be the result of parallax. Confounded by what he had observed, he studied about 200 more stars during the following few years. By comparing his figures to the speed of light, he discovered the stellar positional changes were caused by the speed of the Earth's motion. Bradley published his findings in 1729.

While trying to figure out the cause of the shifts observed in Gamma Draconis, Bradley also discovered an annual change in declination in many other stars and concluded that these shifts were due to the Earth's slight uneven movement, which he proved was caused by the gravitational pull of the Moon. This Earth motion is called nutation. In 1732, he published these further findings. Bradley's work became the basis for one of the proofs of the Copernican theory that the Earth rotates and moves through space, and he also confirmed the speed of light.

The last star on this hop is also one of the more interesting stars in Draco: the rotating class variable BY Draconis [18:33;+51d44']. This star is the prototype star for the type of variable which consists of dwarf stars with non-uniform brightness due probably to the axial rotation of the star combined with spots or other changes in its atmosphere. These stars fit within spectral classes DG to DM, or DGe to DMe. Stars of this type show very slight changes in magnitude, with periods of less than a day to about 120 days. BY Draconis has a period of 3.81 days, but its brightness changes less than half a magnitude, from 8.07 to 8.48, so you probably will not notice much of a change. When you look at yellow, K7 Ve spectral type, BY Dra you are seeing what the Sun would look like from deep space, for the Sun is a BY Draconis-type variable star. To locate BY Draconis, either hop two to three fields of view due east of Gamma Dra, or let the sky drift in your scope for 37 minutes. BY Dra is also a spectroscopic binary with an orbital period of 5.9760 days.

You should feel a great sense of accomplishment after traveling from the tip of Draco's tail to his head. Now see if you can go back the way you came and end up at Polaris. I think you will now also agree that Draco is anything but a telescopic desert.

11 *September*

Cygnus, Lyra, Vulpecula, and Sagitta: The Swan, the Lyre, the Fox, and the Arrow

> For every one, as I think, must see that astronomy compels
> the soul to look upwards and leads us from this world to
> another.
>
> Plato (*c.* 428 – *c.* 348 B.C.)
> *The Republic*, Book VII

With the warm midsummer nights giving way to the cooler autumn evenings of September, we still have many great opportunities to continue to stay out into the early morning hours to observe the vast array of objects embedded in the Milky Way. This month we will first star-hop around the tail and wings of Cygnus the Swan. Cygnus is flying south along the Milky Way heading toward Scorpius the Scorpion. Our second hop will take us from a former pole star, Vega (3 Alpha Lyrae), in Lyra the musical Lyre, through Vulpecula the Fox to Sagitta the Arrow. These four constellations are located along the summer Milky Way, midway between Sagittarius the Archer on the southern horizon and Perseus to the north. During the evening, they slowly sweep westward passing directly overhead at mid-northern latitudes.

Star-hop in Cygnus

Seen as a bird since ancient times, Cygnus (Cyg) was known by various names in different lands. The Arabs considered it to be "The Flying Eagle" *Al Tā'ir al Arduf* or "The Hen" *Al Dajājah*. Until the 1700's the constellation was widely known as Gallina, a forlorn hen.

The Romans gave the asterism the name we know it by today, after Cycnus a mythological offspring of Mars (Ares). The constellation's lineage is somewhat in dispute as no less than five mythological characters were also named Cycnus. Most of the mythological sources consider Cycnus to be the son of Mars as the most likely source of the constellation's name. But the son of Mars has two mothers, and which lady is the correct mother depends on which myth is used. Cycnus's mother was either Pelopia or Pyrene. The legends surrounding Cycnus the son of Mars and Pyrene and Cycnus the son of Mars and Pelopia are very similar for, in each case, Cycnus becomes involved in a battle with the hero Heracles.

The son of Pelopia was known as a brigand and used to kill travelers. He used the money he took from his victims and offered sacrifices to his father. This activity enraged Apollo, who in turn sent Heracles to slay Cycnus. Ares came to avenge his son's death, but Athena intervened and the battle ended in a draw.

The son of Pyrene also ran into Heracles on the hero's way to find the golden apple tree in the Garden of Hesperides; his 11th Labor for Eurystheus. When Heracles crossed the Echedorus stream in Macedonia, this Cycnus challenged the hero to a fight. This tale ends when Zeus stops the fighting with a thunderbolt. In neither of these tales is Cycnus, the son of Mars, a swan, nor does he become one.

In the mythology of a Cycnus who was the son of Apollo and Thyria (Hyria) there *is* a connection between legend and a swan. This Cycnus was a handsome man, but he was capricious and lost all of his friends, except Phylius. Cycnus sent Phylius on several errands, which resulted in the displeasure of Zeus when he led a wild bull into his temple. Dishonored, Cycnus threw himself into a lake. His mother joined him in death. Apollo transformed them into swans, so placing his son among the stars.

One more Cycnus emerges from the depths of Greek mythology. This Cycnus was the son of Poseidon and was raised by a swan. Given the protection of invulnerability to swords and spears, this Cycnus became a mighty warrior for the Trojans. He came into combat with Achilles. Achilles, unable to kill his foe with one of his weapons, grabbed Cycnus and threw him to the ground. Achilles piled rocks on Cycnus in an attempt to smother him. Poseidon rescued his son and turned him into a swan.

Pick the Cycnus you like best and fly with it.

The brightest stars in the 16th largest constellation form an easy to locate cross of stars, as shown in Figure 11.1, with the base of the cross pointing toward Sagittarius. Cygnus is commonly referred to as the Northern Cross. I see the group of stars (18 Delta Cygni, 64 Zeta Cygni, 37 Gamma Cygni, and 50 Alpha Cygni) that form his wings and tail creating a numeral 4. Cygnus covers 803.983 square degrees of sky, most of which is bathed in the soft glow of the Cygnus Arm of the Milky Way. This can make finding some objects difficult, so be sure you are looking at your intended target. Almost everywhere you look in Cygnus you will see clusters, nebulae, and rich star fields. The greater portion of these objects lay along the line from Deneb (50 Alpha Cygni; ADS 14172) located at coordinates [20:41;+45d16'] to Albireo (6 Beta Cygni) [19:30;+27d57']. Binoculars will give you the best views of these fantastic star fields.

We will start this month's first star-hop off at Deneb. The 19th brightest star is the northeastern angle of the Summer Triangle. The other members of the triangle are Vega [18:36;+38d47'] and Altair (53 Alpha Aquilae) [19:50;+8d52']. Blue- white, MK standard spectrum A2 Ia, supergiant Deneb (pronounced DEN-ebb) marks the tail of the Swan. The name is derived from the Arabic *Al Dhanab al Dajājah* meaning "The Hen's Tail." Deneb shines at magnitude 1.25 and has an absolute magnitude of -7.5. If this star was at a distance of 10 pc

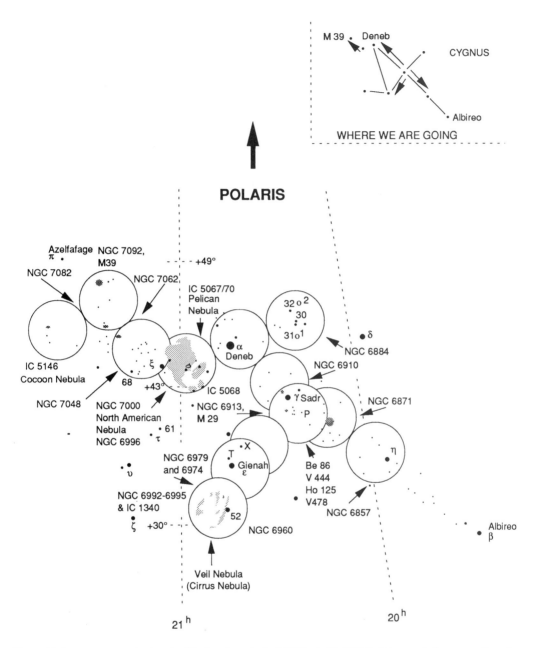

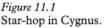

Figure 11.1
Star-hop in Cygnus.

instead of being located at 560 pc (1,600 light-years) from us, the star would appear brighter in our skies than does Venus. Deneb is about 70,000 times brighter than the Sun and about 25 times more massive.

Deneb is the primary for a wide double star. Sir William Herschel listed it as the 73rd object in his 1821 catalogue of double stars. The 11.7-magnitude comes is located 75.4 arcsec from the primary at position angle 106°. The glare of Deneb makes detection of the comes difficult in small telescopes.

Deneb is also a pulsating-class variable and is the prototype for the ACYG (Alpha Cygni) type of pulsating variables. ACYG variables

187

Photo 11.1 Central Cygnus region.

display non-radial pulsations as their atmospheres expand and contract on an irregular basis with cycles of a few days to several weeks.
As the pulsations occur, their spectrums shift back and forth and so
does their radial velocity. They are supergiants in spectral classes Beq
to Aeq Ia. Deneb's pulsations result in a small irregular radial velocity

Photo 11.2 NGC 7000; the North American Nebula, IC 5068, Deneb, and IC 5067/70; the Pelican Nebula.

overlaying its larger radial velocity of 5 km/s in recession. Its magnitude fluctuates by about 0.05 during its 11-day cycle.

From Deneb, we will first head east to view some objects then return to the vicinity of Deneb. About 3 degrees east of Deneb is the 3.72-magnitude, MK standard spectrum K4.5 Ib-II, supergiant variable and spectroscopic binary 62 Xi Cygni [21:04;+43d55']. This star has an absolute magnitude of -4.4 and is located 290 pc distant, but is coming closer at a rate of 20 km/s.

Use Xi to locate the emission nebulae just west of it. The closest to Xi is the faint emission nebula NGC 7000, commonly known as the "North American Nebula" [20:58;+44d20']. Due to its large size (three times the Moon's apparent diameter), you will need a low-power or wide-field eyepiece for the best view of the nebula. Use a high-contrast or O-III filter to enhance your view. You can see this object in binoculars and naked-eyed on a dark night from a dark site. The nebula is laced with wispy patches of glowing gas and thousands of stars, including the scattered open cluster NGC 6996 [20:56;+44d38']. The overall magnitude for the nebula is 5.96. Most references give Deneb as the source of the energy that lights the nebula. Deneb and the nebula are relatively close, being only approximately 70 light-years apart, and Deneb is one of the brightest stars in the region. Take your time to study this object from coast to coast and from Canada to Central America.

To the southwest of NGC 7000 are the faint emission nebulae IC 5070 [20:50;+44d21'] and IC 5067, commonly called the "Pelican Nebula." This nebula is fainter than its larger neighbor. Look for it off the Atlantic Coast of the North American Nebula.

South of the Pelican are the three patches of the emission nebula IC 5068 [20:50;+42d31']. This nebula is very faint and difficult to separate from the surrounding glow of the Milky Way. Use a high-contrast or O-III filter to enhance your view.

Follow the trail of fifth- and sixth-magnitude stars northeast to the planetary nebula NGC 7048 (PK 088-01.1) [21:14;+46d16']. This 3b-type nebula is elongated and glows at visual magnitude 12.1. Unless you are using a large scope, you probably will not see its 18.3-magnitude central star. The gas shell is expanding at a rate of 11 km/s.

Due east of NGC 7048 is the open cluster NGC 7062 (III1p) [21:23;+46d23']. The compact cluster contains about 30 stars, most of which are fainter than 11th magnitude. The apparent visual magnitude for the 100 million-year old cluster is 8.3. NGC 7062 is located 1,900 pc from the Sun.

Head about $\frac{1}{2}$ degree east to magnitude 5.24, spectrum K0 III, 71-g Cygni [21:29;+46d32'] then north to the open cluster NGC 7082 (IV2p). This cluster is brighter than NGC 7062 with its apparent magnitude of 7.2, but with its scattered 25 stars it blends into the surrounding Milky Way star field, which makes it harder to pick out as a cluster. Use a low-power or wide-field eyepiece to view this cluster.

With your finder centered on NGC 7082, you will have the open cluster NGC 7092 (M39) (III2p) [21:31;+48d26'] on the northeastern edge of the field. The brighter dozen or so stars of this naked-eye object's 30 members stand out in the star field around it. M39 may have been discovered by Le Gentil, in 1750, but, since he did not give its coordinates, credit for its discovery is generally given to Messier. He claimed to have first seen it on October 24, 1764. Most of the stars appear white with one yellow and a few blues mixed into the cluster.

Back at Deneb, we will hop northwest toward 18 Delta Cygni [19:44;+45d07']. Located about midway between Deneb and Delta are the neighbors 30 Cygni [20:13;+46d48'], 31 Omicron[1] Cygni (ADS 13554; V695) [20:13;+46d44'], and yellow giant 32 Omicron[2] Cygni (V1488) [20:15;+47d42']. Both of the omicron stars are EA-type eclipsing variables and spectroscopic binary systems. In EA-type eclipsing variables, the light from the stars remains constant until one member of the pair is eclipsed by the other. The total light from the combined stars dims as the orbiting star passes behind the extended atmosphere of the primary star.

An excellent color-contrast visual triple consists of yellow-orange, spectrum K2 II, magnitude 3.79, giant 31 Omicron[1]; its 6.7-magnitude bluish, spectrum B4 V, "C" star, and 30 Cygni. Greenish giant 30 Cygni is a 4.83-magnitude, spectrum A3 III, star located 338 arcsec northwest of Omicron[1].

The "B" star of the Omicron[1] multiple system is a faint 13.1-magnitude blue star located 36.6 arcsec from the primary. We see this pair almost edge on, so every 10.42 years the blue star passes behind the primary and is eclipsed. The "a" star is less than 0.5 arcsec from the primary and is visible only on photographs. The "D" star is a white A2 star located 337 arcsec from the primary. The 13th-magnitude "E" star is close to the "D" star. The whole Omicron[1] system is 160 pc from us and has a radial velocity of 7 km/s in recession. The "A" and "B" star pair have an absolute magnitude of -2.2, which means together they put out 500 times more light than the Sun.

About $\frac{1}{3}$ degree west of Omicron[1] is the planetary nebula NGC 6884 (PK 088+07.1) [20:10;+46d28']. This type IIb planetary appears bluish and is difficult to separate from the rich star field it is embedded in. Use a high-contrast filter and a high-power eyepiece to pull it out from the surrounding stars. The magnitude 15.2 Wolf–Rayet central star is lost in the glow of the gas shell. The gas shell is expanding at a rate of 23 km/s. The nebula is 5.1 pc from us.

From NGC 6884, hop to the 2.20-magnitude white giant Sadr (37 Gamma Cygni; ADS 13765) [20:22;+40d15'] at the central point of the cross. Sadr (pronounced sad-der) is the MK standard for spectrum F8 Ib stars and is located 230 pc from us. The star's name comes from the Arabic *Al Sadr al Dajājah*, "The Hen's Breast." The area surrounding this star is rich in bright and dark gaseous nebulosity, star fields and open clusters. Use a high-contrast or O-III filter to bring the faint emission nebulae out from the star fields. Their wispy nature is spectacular to observe. Sadr is also a spectroscopic binary and variable star.

Northeast of Gamma Cygni is the open cluster NGC 6910 (I2p) [20:23;+40d47'] embedded near the edge of a portion of the faint emission nebula, Gamma Cyg OB1. The cluster consists of about 50 stars with magnitudes ranging from 9th to 11th magnitude. The apparent magnitude of the cluster is 7.4. This cluster is estimated to be about ten million years old and is 1,650 pc from us. The cluster is spread out with strings of stars making it look like the letter Y.

About $1\frac{1}{2}$ degrees south-southeast of Sadr and $\frac{1}{2}$ degree west of 40 Cygni [20:27;+38d26'] is the compact open cluster NGC 6913 (M29) (III3p) [20:23;+38d32']. Messier discovered this ten million-year old cluster on July 29, 1764. Though in the Cygnus star field, M29 is easy to find, even in binoculars. M29 consists of about 50 stars and has an apparent magnitude of 6.6. The core of the cluster is made up of four eighth- and ninth-magnitude stars with twin patches of stars branching off toward the north. The cluster is 1,250 pc from us and has a radial velocity of 28 km/s in approach.

We will branch off once again from Cygnus's body and head east along his wing. Our first stop is at the multiple star 54 Lambda Cygni (ADS 14296; OΣ 413) [20:47;+36d29']. In 1842, this 4.8-magnitude, spectrum B5 V, star was first noted as a double by Otto Struve (1819–1905), the son of F. G. W. von Struve. Objects listed in Otto Struve's 1843 catalogue are designated as OΣ or STT. The visual and spectroscopic binary system OΣ 413 is located 130 pc from us and has a radial velocity of 23 km/s in approach. The 6.07-magnitude comes takes 391.3 years to orbit the primary star. In 1863, James South recorded the position of a 9.9-magnitude "C" star (S 765) located 85.7 arcsec from the binary pair at position angle 105°.

About $\frac{1}{2}$ degree south-southwest from Lambda is the Delta Cepheid-type variable X Cygni [20:43;+35d35']. This bright supergiant fluctuates from spectral class F7 Ib to G8 Ibv as its magnitude changes from 5.87 to 6.86 during a 16.36-day period.

Look about 1 degree south-southeast of X Cygni for the visual multiple and spectroscopic binary system of Gienah (53 Epsilon Cygni; ADS 14274; β 676) [20:46;+33d58']. This Gienah (pronounced GEE-nah) is the second star we have come to with this Arabic name. The other Gienah is in Corvus. Epsilon Cygni is a spectrum K0 III yellow giant located 25 pc from us with a radial velocity of 26 km/s in approach. The components of this system were first measured by S. W. Burnham, in 1878. The 11.5-magnitude "B" star is 55 arcsec from the "A" star.

Less than $\frac{1}{4}$ degree north-northeast of Epsilon Cygni is the pulsating class, Lb-type variable and multiple star T Cygni (ADS 14290; β 677) [20:47;+34d22']. T Cygni fluctuates 0.5-magnitude. The variability of this star is uncertain as not enough study has been done on it. This spectrum K3 III star has a 9.9-magnitude companion gravitationally bound to it 10 arcsec away in position angle 121°. An 11.2-magnitude optical companion is also part of this system, which is located 100 pc from us. The system has a radial velocity of 23 km/s in approach.

The next bright star south of Epsilon Cygni is 4.22-magnitude 52 Cygni [20:45;+30d43']. This spectrum K0 IIIa yellow giant is visually embedded in the western streamer (NGC 6960) of the "Bridal Veil Nebula" (also known as the Veil Nebula, and the Cirrus Nebula). The entire Veil Nebula consists of NGC 6960, NGC 6974, NGC 6979, NGC 6992, NGC 6995, IC 1340 [20:56;+30d43'], and several undesignated segments. The western branch of the nebulosity extends north and south through 52 Cygni which is actually in the foreground and not a physical part of the nebula.

The Veil Nebula was discovered by Sir William Herschel, in 1784. Use an O-III or high-contrast filter with a wide-field, low-power eyepiece to get the best view. On a night of good seeing from a dark site, you may be able to view it through binoculars.

The Veil is thought to be the remnant of a supernova explosion that occurred 30 to 40 thousand years ago. No central star has been

Photo 11.3 52
Cygni and NGC
6960; the Veil
Nebula.

detected, though several candidates in the vicinity of the center of the bubble have been studied. A top candidate is the 6.2-magnitude, spectrum K2, double star β 67 (ADS 14355) [20:50;+30d55']. Burnham first classified this star as a double, in 1875.

The distance to this expanding bubble of glowing gas is uncertain, but estimates put it at about 460 pc (1,500 light-years). It glows softly at an integrated magnitude of 5.69. The nebula is about four-and-a-half times the apparent size of the full Moon, but only the outer western and eastern limbs are easily visible, visually. The eastern portion of the bubble is thicker and brighter than the western side.

Back at Sadr, we hop along the back of Cygnus to explore some clusters and a star considered to be the most luminous single object in the sky: P Cygni (34 Cygni) [27:17;+38d01']. Begin this part of the tour by heading southwest, following the trail of stars to the variable V444 [20:19;+38d44']. This EA-type eclipsing variable consists of a spectrum O6 primary star and an extremely hot Wolf–Rayet-type, spectrum WN5.5, star. The combined magnitude of these stars fluctuates from 8.3 to 8.6 in a period of 4.2 days. The total mass of this two-star system is estimated to be 53 times that of the Sun. These stars are located 1,500 pc from us and have a radial velocity of 3 km/s in recession.

In the same eyepiece field are the open cluster Berkeley 86 (Be86) [20:20;+38d42'], the close double star Ho 125 [20:19;+39d00'], and the EA-type eclipsing variable V478 [20:19;+38d20']. Berkeley 86 (I3p) is a compact cluster of about 30 stars. Though small, it is bright at magnitude 7.9, which makes it stand out in the rich field around it. Ho 125 consists of a 6.9-magnitude, spectrum K2, primary and a 10.6-magnitude comes located 3.1 arcsec away at position angle 194°. The four-magnitude difference between the components makes this a difficult double to separate. V478 consists of a pair of spectrum B0 V, white main-sequence stars. Their magnitude fluctuates from 8.9 to 9.3 in a period of 2.8808 days.

To the west of V478 is a small triangle of stars. The brightest one, at magnitude 4.81, is the intense-blue hypergiant P Cygni (also known as Nova Cygni 1 or Nova 1600) [20:17;+38d01']. This MK standard B1.pe-type star is located 44 pc from us and has a radial velocity of 9 km/s in recession. This star has retained its now obsolete Roman-letter designation. P Cygni was the fourth new star discovered in historic times, having been first seen as a third-magnitude star on August 8, 1600, by Dutch astronomer Willem Janszoom Blaeu (1571–1638). Within a few years, the star had faded to beyond naked-eye visibility, then returned to third magnitude, in 1655. Again it faded and has oscillated around fifth magnitude for the last 200 years.

P Cygni fits into the SDOR type of eruptive variable. This type of variable contains highly-luminous stars which fall into the spectral types of Bpeq to Fpeq. Their magnitude fluctuations are irregular and can vary as much seven magnitudes. They are usually found in association with diffuse nebulae and their spectra indicate they are surrounded by expanding shells of gas. Their absolute magnitudes are as bright as -10. P Cygni is the prototype for stars with broad emission lines in the violet range of their spectra with sharp absorption lines on the border of the violet range.

Before you leave Cygnus, take a look at the double star 61 Cygni A and B (ADS 14636) [21:06;+38d44']. These yellow stars are the fourth closest ones to us at a calculated distance of 3.4 pc. Magnitude 5.23, 61 Cygni A is the MK spectrum standard for K5 V stars. The B component is the MK standard for K7 V spectrum stars. These stars were the first whose parallax was determined. This was done by W. F. Bessel, in 1838. The orbital period for the 6.00-magnitude B component is 653.3 years. You will find this system about 2 degrees north-northwest of yellow, magnitude 3.72, spectrum F2 IV, Delta Scuti-type variable and seven-star system of 65 Tau Cygni (ADS 14787) [21:14;+3802'].

Star-hop in Lyra, Vulpecula, and Sagitta

For our second star-hop this month, we will travel from the fifth brightest star in the sky, Vega in Lyra, south to the constellations of Vulpecula (Vul), and Sagitta (Sge), as shown in Figure 11.2. These three constellations are high overhead for mid-northern observers during the summer months.

Lyra, the Lyre or Harp, is an ancient constellation due partly to the fact Vega was the pole star during the dawn of civilization in the Middle East. The name we know it by comes to us from the Greeks. They named it after the musical instrument invented by Hermes, the son of Zeus and Maia. As a day-old infant, Hermes escaped from his bandages and stole a herd of cattle from his half-brother Apollo. While living in a cave on the island of Cyllene, Hermes found a tortoise at the entrance to his cave. He cleaned the shell and stretched strips of cattle intestines across it. Apollo came to the cave looking for his herd and saw Hermes's lyre. Apollo liked the instrument and traded the stolen herd for it. Apollo in turn gave the instrument to his son-in-law Orpheus. As a crewmember on Jason's *Argo*, Orpheus sang and used the music from the lyre to calm the crew during a storm and to overcome the temptations of the Sirens. Orpheus was a lover of young men. This led to his being killed and dismembered by the women of his homeland of Thrace. Upon his death, his lyre was placed among the constellations.

Vulpecula the Fox is a young constellation, having been named by Johannes Hevelius in his 1690 star catalogue. Covering 268.165 square degrees of the night sky, Vulpecula ranks as the 55th constellation in size.

Sagitta was well known as an arrow to several ancient civilizations long before Ptolemy gave it the name we know it by in his *Almagest*. The Greeks called it *Herculea* as they thought it was the arrow shot eastward by Hercules. It has also been associated with being one of Cupid's arrows. None of Sagitta's stars were named by the Arabs, even though they also called the four-star asterism *Al Sahm* meaning "The Arrow." Sagitta ranks 86th in size and covers 79.923 square degrees of sky. Only Equuleus the Foal (87th) and Crux the Southern Cross (88th) are smaller constellations.

Our jumping off point on this star-hop is magnitude 0.03, blue-white Vega (3 Alpha Lyrae; ADS 11510; H V-39) [18:36;+38d47']. Vega (pronounced VEE-ga) was first used in the Alfonsine Tables as a derivative of the Arabic *Al Naṣr al Wāki'* meaning "The Swooping Eagle."

Vega is the MK spectrum standard for A0 V main-sequence stars and serves as the standard for calibrating magnitude measuring equipment. This bright star is located 8.1 pc from us. The Sun and solar system are heading in the direction of Vega at a rate of 19 km/s. This point in the sky is called the "Apex of the Sun's Way." Vega has a radial velocity of 14 km/s in approach. On the evening of July 16/17, 1850, William Cranch Bond (1789–1859), then director of the Harvard College Observatory, aimed a camera-equipped telescope at Vega and captured for the first time the light of a star on a photographic plate.

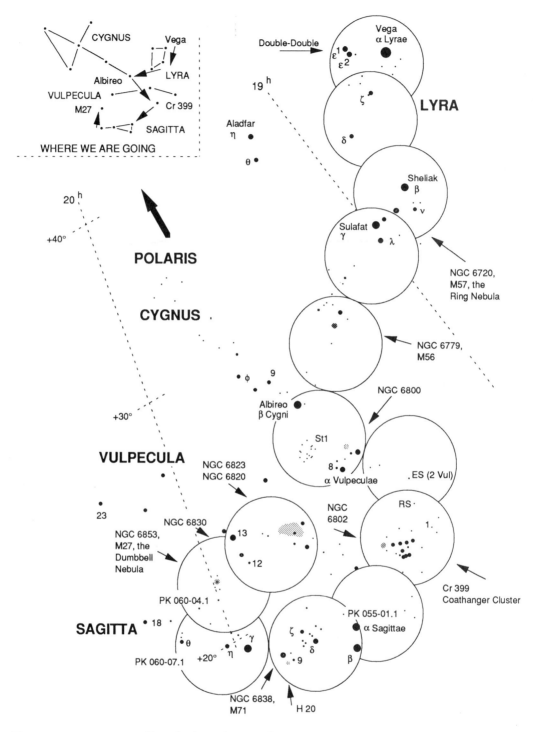

Figure 11.2
Star-hop in Lyra,
Vulpecula, and
Sagitta.

Vega is the primary of a wide multiple-star system that includes at least three secondary stars. The secondary stars appear to be only visual companions as the proper motion of Vega is causing the distances between it and these companions to gradually increase.

About $1\frac{1}{2}$ degrees northeast of Vega are two stars designated as 4

Epsilon[1] (ADS 11635; Σ 2382) [18:44;+39d40'] and 5 Epsilon[2] Lyrae (ADS 11635; Σ 2383) [18:44;+39d36']. This pair of stars is known as the Double-Double as each star has its own very easy to separate companion. This system was first observed by Sir William Herschel, in 1779. The northern pair consists of whitish, 4.76-magnitude, spectrum A3 V, Epsilon[1] (the "A" star in the system) and its 6.02-magnitude, spectrum A1 V, "B" comes 2.3 arcsec to the north. Due to its elliptical orbit, as we view it, the comes is slowly getting closer to the primary. The orbital period for the comes is 1,165 years. The "A" star is also known to be a spectroscopic binary system. Two hundred and eight arcsec to the south is magnitude 5.1, spectrum A5 V, Epsilon[2] (the "C" star) and its 5.47-magnitude, spectrum F0 V, companion ("D"). The "D" star is 2.2 arcsec from the "B" star in position angle 82°. The two primary stars do not seem to be an actual pair as no orbital motion has been detected. Both systems are 38 pc from us, but Epsilon[1] has a radial velocity of 31 km/s in approach whereas Epsilon[2] is a little slower at 24 km/s in approach.

Two degrees to the south is another open multiple system, 6 Zeta[1] Lyrae (ADS 67321) [18:44;+37d36'] and 7 Zeta[2] Lyrae (ADS 11639) [18:44;+37d35']. Several members of this system are fainter than magnitude 11 and are lost in the glare of 4.36-magnitude Zeta[1] and 5.73-magnitude Zeta[2]. You should be able to separate the brighter stars.

At the southwestern corner of the Lyre is the bright eclipsing binary variable Sheliak (10 Beta Lyrae; ADS 11745) [18:50;+33d21']. The name Sheliak (pronounced SHELL-e-ack) comes from the Arabic *Al Shilyāk* meaning "The Tortoise." The variability of the star was first noticed by British amateur astronomer John Goodricke (1764–86), in 1784. During a 12.9-day period, Beta's magnitude fluctuates between 3.34 and 4.34. You can tell when Beta is at or near its maximum by comparing its brightness to that of its 3.24-magnitude neighbor, Gamma Lyrae. Beta is located 92 pc from us. Astrophysicists are not sure exactly what is going on with this star, but the theory is that matter is flowing from the larger star to the smaller one at a rate of 300 km/s. The rapid rotation of the comes around the primary is also distorting both stars. Some scientists believe the secondary star is a black hole, sucking in any matter that comes near it.

Early in January 1779, French astronomer Antoine Darquier de Pellepoix (1718–1802) discovered an oval nebulous object appearing to be about the size of Jupiter midway between Beta and Gamma Lyrae. On January 31, 1779, Messier independently found the same object and entered it as the 57th object on his list. He thought the gaseous shell of the planetary nebula was tiny stars which he could not resolve. The Ring Nebula (NGC 6720; M57; PK 063+13.1) [18:53;+33d02'] is probably the easiest planetary nebula to locate. The type IV+III bluish-green shell of gas around its 14.8-magnitude central star is very bright at magnitude 9.7. The shell glows from the strong emissions of doubly ionized oxygen. The central star is a very dense and extremely hot (100,000 K) sub-dwarf. Spectroscopic study shows that the gas is expanding at the rate of 30 km/s. The gas shell has been described as being like a smoke ring or a doughnut about 30,000 AU in diameter. The distance to the Ring Nebula is uncertain, but is believed to be about 0.6 kpc.

To the east of the Ring Nebula is the 3.24-magnitude, spectrum B9 II, star Sulafat (14 Gamma Lyrae; ADS 11908; AGC 9) [18:58;+32d41']. Sulafat (pronounced SU-la-faat) was first thought to be a double when Alvin G. Clark entered it as the ninth double in his catalogue, in 1868. The spectrum A0, magnitude 12, companion is now known to be only an optical companion.

About $4\frac{1}{2}$ degrees southeast of Gamma Lyrae and about midway between Gamma and 6 Beta Cygni is the class 10 globular cluster NGC 6779 (M56) [19:16;+30d11']. Messier discovered this cluster on January 23, 1779. At magnitude 8.25, M56 looks like a tight knot of light planted in a rich field of faint stars. M56 is about as bright as 90,000 Suns. The cluster is 9.5 kpc from us and is racing toward us at a radial velocity of 138 km/s.

Still heading southeast, we cross into Cygnus and stop at probably the most famous color-contrast double star, Albireo (6 Beta Cygni; ADS 12540) [19:30;+27d57']. These stars are an optical pair with the spectrum K3 II, golden-yellow giant primary star 120 pc from us. The deep blue "B" spectrum B9.5 V, main- sequence star is 300 pc from us. The B" star is 34 arcsec from the "A" star in position angle 54°. The name Albireo (pronounced al-BIR-ee-oe) seems to have come from a typographical error in the 1515 edition of the *Almagest*. The Arabs knew this star as *Al Minḥar al Dajājah* meaning "The Hen's Beak."

Heading south from Albireo we cross into the constellation of Vulpecula the Fox. About $3\frac{1}{2}$ degrees from Albireo is the very loose open cluster Stock 1 (IV2p) [19:35;+25d13']. The 40 members of this 5.3-magnitude cluster appear as a slight condensation within the rich star field they inhabit. The cluster has an apparent diameter of about the same size as the Moon, so your best views of the entire cluster will be in your finder or binoculars.

South of Stock 1 is the optical double consisting of magnitude 4.44, M0 IIIa spectral-type, 6 Alpha Vulpeculae [19:28;+24d39'] and magnitude 5.81, gK0 spectral-type, 8 Vul [19:28;+24d46'].

Use the bright red giant Alpha Vul as your guide to the loose open cluster NGC 6800 (III2p) [19:27;+25d08'] about $\frac{1}{3}$ degree to the west. NGC 6800 consists of about 20 to 25 members, most of which are fainter than tenth magnitude. The cluster stands out from the rich field of faint stars around it.

About a finder field distance to the south-southwest is the BCEP-type pulsating variable ES Vul (2 Vul) [19:17;+23d02']. ES Vul goes from magnitude 5.4 to 5.46 in a period of 14 hours 30 minutes and changes from being an O8 IV to a B0.5 IV spectral type.

Half a degree south of ES Vul is the EA-type eclipsing binary variable RS Vul [19:17;+23d06']. This variable, spectrum B5 V, star fluctuates from magnitude 6.9 to 7.6 during a period of 4.47766 days.

The spectrum B3 IV, magnitude 4.77, double 1 Vul (ADS 12243) [19:16;+21d23'] is about $\frac{1}{2}$ degree south of RS Vul. The 11.8-magni-

Photo 11.4 Cr 399; the Coathanger Cluster in Vulpecula.

tude B5 spectral-type secondary is located 39 arcsec from the primary in position angle 13°. 1 Vul is located 210 pc from us and has a radial velocity of 17 km/s in recession. An optical companion of 1 Vul is magnitude 5.46, spectrum A3, 1 Sagittae [19:15;+21d13']. The distance to 1 Sge is uncertain, but it has a radial velocity of 23 km/s in recession.

South-southeast of 1 Vul is the interesting open cluster Collinder 399 (III2p) [19:25;+20d11']. This cluster is also known as Brocchi's Cluster or the Coathanger Cluster. The latter name is a good description, for six of the brighter stars in the cluster form a slightly curving line similar to the arm of a coathanger with a hook-like string of four more stars south of the main line. The bottom of the hook is marked with the multiple star 4 Vulpeculae, a magnitude 5.16, spectrum K0 III, red giant (ADS 12425) [19:25;+19d47']. The cluster is estimated to contain about 40 stars. The best view of this easy to locate magnitude 3.6 cluster is in your finder or binoculars, because it is too large to fit into an eyepiece field. The 200 million-year old cluster is embedded in a rich field of faint stars. The cluster is calculated to be is 130 pc from us.

At the north eastern edge of the Coathanger Cluster is the faint open cluster NGC 6802 (III1m) [19:30;+20d16']. This tiny cluster (3.2 arcsec in diameter compared to 60 arcsec for the Coathanger) is hard to pick out as a cluster from the stars around it. NGC 6802 consists of about 50 members, most of which are fainter than 12th magnitude, but the cluster has a combined magnitude of 8.8. This cluster is about 1.7 billion years old and is 990 pc from us.

Our next stop is in the rich star field of Sagitta the Arrow. Everywhere you look in this small constellation you will find interesting Milky Way stars and asterisms to observe. Use binoculars to explore this area of the summer sky. Our first stop in Sagitta is at its lucida, the double star 5 Alpha Sagittae (ADS 12766) [19:40;+18d00']. Alpha and 6 Beta Sge [19:41;+17d28'] form the base of the arrowhead. Alpha is a yellow-orange giant shining at magnitude 4.37, with a G1 II spectrum, and is the primary in a wide multiple system. The magnitude 13.2 "B" star is 31 arcsec from the primary at position angle 179° (in 1960). The "C" star is much harder to separate being dim at magnitude 14.9. Alpha is 190 pc from us and coming closer with a radial velocity of 2 km/s in approach.

About $\frac{3}{4}$ degree north of Alpha is the tiny type II planetary nebula PK 055-01.1 [19:40;+18d49']. This is one of those planetary nebula you will need a diffraction grating or prism to separate it from the background stars. It has a bright core and appears as a star that you cannot focus into a sharp spot.

Beta Sge is a spectrum G8 IIIa: Cn0.5, yellow-orange giant 200 pc from us, but it is racing away from us at a rate of 22 km/s. Its magnitude is equal to Alpha's at 4.37.

In the middle of the arrow is the magnitude 3.82, spectral-type M2 II, red giant 7 Delta Sge [19:47;+18d32']. This star is 170 pc from us and coming closer at a rate of 3 km/s.

Use the alignment of Alpha Sge, Delta Sge, and magnitude 6.3, spectrum O8.f, bluish 9 Sge [19:51;+18d40'] to point you to the globular cluster NGC 6838 (M71) [19:53;+18d47']. This object has also been considered to be a dense open cluster. Studies show the cluster lacks many main-sequence stars typical of an open cluster and it has many red giants that would make it a globular, but it also has numerous metal-rich stars, which are generally not found in a globular cluster. These contradictory findings make this one deep-sky object astronomers are still uncertain about: is it a globular or an open cluster? There is also some uncertainty about who first recorded an observation of this object. It appears as probably the 13th object on Philippe de Chéseaux's 1746 list. The uncertainty arises, because De Chéseaux did not include the coordinates for this object, which he listed as being "very close to Sagitta." Sometime between 1772 and the summer of 1779, Johann Koehler independently located a cluster and recorded its coordinates as the seventh object on his list of 20 nebulous objects. Pierre Méchain came across the same cluster on June 28, 1780, and reported the find to Messier. Glowing at magnitude 8.3, M71 appears as a knot of stars with no tightly packed central core typical of globulars. You can resolve many of its central stars. The cluster resembles an arrowhead pointing toward 9 Sge. The cluster is located about 4 kpc from us and has a radial velocity of 19 km/s in recession.

About $\frac{1}{2}$ degree south-southwest of M71 is the poor open cluster Harvard 20 (H 20) (III2p) [19:53;+18d20']. The 15 members of this cluster are mostly 11th magnitude and fainter, except for two ninth-magnitude stars near the center of the cluster. The stars of the cluster appear only slightly brighter than those of the Milky Way star field.

The red giant 12 Gamma Sge [19:58;+19d29'] marks the tip of the arrow. This MK standard spectrum M0 III star is located 51 pc from us and shines at magnitude 3.47.

Photo 11.5 NGC 6853 (M27); the Dumbbell Nebula.

Forming a right triangle with 16 Eta Sge [20:05;+19d59'] and the multiple system 17 Theta Sge [20:09;+20d54'] is the type III+II planetary nebula PK 060-07.1 [20:11;+20d20']. The central star for this magnitude 16 planetary is an irregular spectrum variable known as FG Sge. The spectral type for this object has changed from being a white B4 star to a yellow G8. For a time, the star reddened to be a K spectral type. In 1890, the star was a photographic magnitude 13.7 and had brightened to 9.5 by 1967. The gas shell for this 6,000-year old irregular-shaped nebula is expanding at a rate of 34 km/s, which is

Photo 11.6 Open cluster NGC 6823 and emission nebula NGC 6820 in Vulpecula.

slightly slower than the planetary's radial velocity of 39 km/s in approach. Spectral analysis of the shell indicates the star recently expelled another layer of gas, which is undergoing its own nuclear reactions. Use a diffraction grating to pinpoint this planetary as it is situated south-southeast of two 12th-magnitude stars, and several other very faint stars are in the same field of view.

The easiest way to find the type III+II planetary nebula PK 060-03.1, which is also known as the Dumbbell Nebula (NGC 6853; M27) [19:59;+22d43'] is to hop north 3 degrees from Gamma Sge. M27 was first observed by Messier on the evening of July 12, 1764. This 7.3-magnitude planetary nebula appears as an oval smudge of light in small scopes and binoculars. In large scopes, the famous wide edges and narrow middle band of greenish gas are easy to see. The bluish dwarf, spectrum O7, 13.9-magnitude, central star is actually located near the western edge of the gas shell. At 85,000 K, this star is one of the hottest objects detected and is a strong emitter of ultraviolet radiation. Estimates for its location range from 150 to 300 pc. The gas shell is expanding at a rate of 30 km/s. The shell is estimated to be 0.76 pc ($2\frac{1}{2}$ light-years) in diameter.

If you have access to a large-aperture scope, the very faint planetary nebula PK 060-04.1 (Abell 68) [20:00;+21d43'] is located 1 degree south of M27. This type III planetary has a visual magnitude of 15.2 for the gas shell and the central star is 18th magnitude. As a comparison of size, M27 is greater than 348 arcsec in diameter whereas Abell 68 is about 39 arcsec. The use of a diffraction grating will help you to pinpoint this planetary.

About $\frac{1}{2}$ degree north of spectrum B3 V, 4.95-magnitude, 12 Vulpeculae [19:51;+22d36'] is the open cluster NGC 6830 (II2p) [19:51;+23d04']. The 20 members of this loose cluster glow at magnitude 7.9. The 100 million-year old cluster is slightly brighter than the star field in which it is embedded.

Look about 1 degree west of NGC 6830 for the open cluster NGC 6823 (I3p) [19:43;+23d18'] and the emission nebula NGC 6820 surrounding the cluster. Most of the 30 members of this cluster shine at magnitude 9 or fainter. The cluster has a magnitude of 7.1. NGC 6823 is visible in binoculars, but the irregular-shaped nebula requires a large scope or high-contrast filter to detect its soft background glow.

I enjoyed creating this star-hop and hope you enjoyed taking it. Though small, these last three constellations are packed with many more deep-sky objects for you to search for during warm summer nights under dark skies.

12 *October*

Andromeda and Perseus:
The Chained Maiden and her rescuer

Now Time's Andromeda on this rock rude,
With not her either beauty's equal or
Her injury's, looks off by both horns of shore
Her flower, her piece of being, doomed dragon's food.

Gerard Manley Hopkins (1844–89),
"Andromeda" (lines 1–4),
(1876–89)

Fall is upon us now, and with its chill we get crisp evenings with steady air and usually good seeing, but the nights are not too cold to force us in early. Andromeda (And) and Perseus (Per) share a common border in the sky and a place in Greek mythology. These two mythological characters are tied to the constellations depicting Andromeda's parents, King Cepheus of Joppa and his Ethiopian queen Cassiopeia. Cetus the Sea Monster and Pegasus the Winged Horse are also components of their tale.

An oracle told Perseus's grandfather, King Acrisius of Argos, that he would be killed by his grandson. The king had Perseus and his mother, Danae, sealed in a box then cast out to sea. Being the offspring of Zeus, Perseus along with his mother were saved by the fisherman Dictys. King Polydectes was the brother of Dictys and he wanted Danae for his wife. Perseus did not want his mother to marry the king. Thinking he could get rid of Perseus by tricking him into believing he was going to marry Hippodameia instead, Polydectes sent Perseus on a mission to get him the head of the Gorgon Medusa for a wedding present.

Using a shield of polished bronze given to him by Athene and Hephaestus' diamond sword, Perseus was able to slay the feared demon without looking directly at her. When he sliced off her snake-covered head, the winged horse Pegasus and the warrior Chrysaor emerged from her body. Perseus began to fly back to the island of Seriphos with the Gorgon's head in a sack.

While Perseus was slaying the Gorgon, Cassiopeia boasted that she and her daughter, Andromeda, were more beautiful than the sea nymphs. This boast infuriated Poseidon. He cast a raging flood and a sea monster to destroy the queen, Andromeda, and Joppa (present day Tel Aviv, Israel). Some mythological sources consider this monster to be Draco the Dragon while others claim is was Cetus the Sea Monster, but scholars are not sure which is the correct tormentor.

Witnessing the destruction to their homes, the people of Joppa pleaded with Cepheus to find a way to put an end to it. He contacted the Oracle of Ammon and was told he would have to sacrifice his virgin daughter. Following the Oracle's orders, Andromeda was chained to the cliffs overlooking the sea.

Perseus came upon the maiden chained to boulders at the sea shore. The monster was closing in on her. Seeing Andromeda's plight and falling in love with her, Perseus offered to rescue her, but only if her father would give her hand in marriage. Cepheus quickly agreed. In a ferocious battle, the gallant Perseus beheaded the sea monster, then unchained Andromeda.

Cepheus's brother Phineus had previously been offered Andromeda's hand in marriage. Intent on fulfilling his prior claim to her, Phineus led his army into the wedding ceremony in order to kill Perseus. Outnumbered, Perseus uncovered the ugly head of the Gorgon, thereby turning Phineus and his army to pillars of stone.

Perseus and his bride returned to his island. There he found his mother and Dictys hiding from King Polydectes. Enraged, Perseus stormed into a party being given by the king. He pulled the head from the sack and turned the king and his party into stone. Athene received the head of Medusa from Perseus and placed it in her shield.

Star-hop in Andromeda

We will start this month's star-hop tour in Andromeda, as shown in Figure 12.1, then star-hop around her husband's constellation. Consisting of 722.278 square degrees of sky, Andromeda ranks as the 19th constellation in size.

Tying Andromeda and Pegasus together is their common use of the blue-white, magnitude 2.06, spectrum B8 IVp: Mn; Hg, wide double and spectroscopic binary star Alpheratz (21 Alpha Andromedae; ADS 94; [formerly known as Delta Pegasi]) located a coordinates [0:08;+29d05]. Alpheratz (pronounced al-FEE-rats) marks the northeastern point of the Great Square of Pegasus and Andromedae's head. The magnitude 11.3 secondary is located 82 arcsec from the primary in position angle 280°. The spectroscopic secondary star orbits Alpheratz in a period of 96.6960 days. This system is about 22 pc from us and has a radial velocity of 12 km/s in approach.

About 2 degrees northeast of Alpheratz is a knot of nine faint galaxies: NGC 67, NGC 68, NGC 69, NGC 70 (IC 1539), NGC 71, NGC 72/72A, NGC 74, and NGC 76. These galaxies make up the group of interacting galaxies known as VV 166. The brightest, at magnitude 12.96, is the SO-type galaxy NGC 68 [0:18;+30d04'].

The brightest stars in Andromeda also form the hind legs for Pegasus the Winged Horse. We will first follow the eastern (left) leg from Alpheratz to Almach (57 Gamma And) and observe NGC 1039 (M34) in Perseus. We will then view objects near the western (right) leg from 29 Pi And to 51 Andromedae.

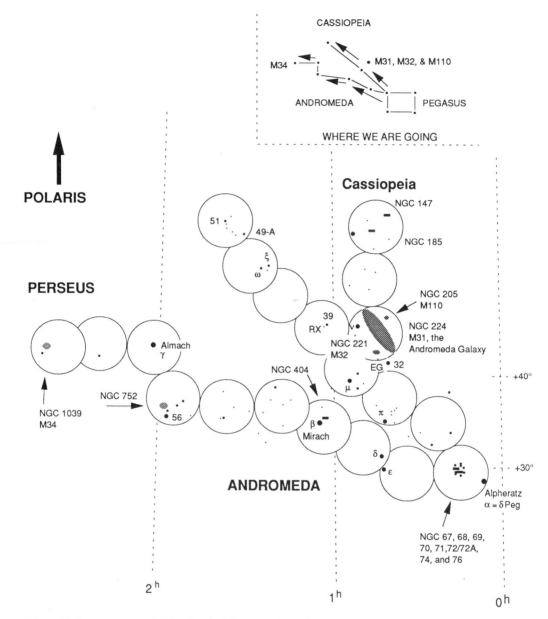

CASSIOPEIA

M34

M31, M32, & M110

ANDROMEDA | PEGASUS

WHERE WE ARE GOING

Cassiopeia

NGC 147

51 · 49-A

NGC 185

ξ
ω

NGC 205
M110

39
RX · ν

NGC 224
M31, the
Andromeda Galaxy

NGC 221
M32

POLARIS

PERSEUS

Almach
γ

NGC 404

EG · 32

μ

- · · +40°

NGC 752

π

NGC 1039
M34

56

β
Mirach

δ

ε

- · · +30°

ANDROMEDA

Alpheratz
α = δ Peg

NGC 67, 68, 69,
70, 71,72/72A,
74, and 76

2ʰ

1ʰ

0ʰ

Figure 12.1
Star-hop in
Andromeda.

The first bright star along the eastern leg is the reddish, MK standard spectrum K3 III, wide multiple system 31 Delta And (ADS 548; β 491) [0:39;+30d51']. This star marks Andromeda's left breast. The deep red dwarf, magnitude 12.4, spectrum M2 "B" star was first measured by S. W. Burnham, in 1878, and found to be 27.9 arcsec from the "A" star in position angle 299°. By 1934, the "B" star had moved 1 degree to position angle 298° and a separation of 28.7 arcsec. A 15th-magnitude star is also part of the system, which is located 49 pc from us.

The next bright star in the leg is the yellow-orange wide multiple Mirach (43 Beta Andromedae; ADS 949) [1:09;+35d37'], which consists of five stars. Mirach (pronounced MY-rack), the magnitude 2.06, spectrum M0+ IIIa primary and its 14th-magnitude "B" dwarf companion are an easy pair to separate. The "B" is 27 arcsec from the primary in position angle 202° and was discovered by E. E. Barnard, in 1898. The spectra for Mirach shows an abundance of absorption lines for neutral metals. A 13.4-magnitude "b" star is located 24 arcsec from the "B" star. Located about 90 arcsec from the primary are two optical companions. The "C" and "D" stars are easy to separate being so far from the glare of the bright primary.

In the same low-power field with Mirach is the elliptical galaxy NGC 404 [1:09;35d43']. This E0-type galaxy is 6.5 arcmin north-west of Mirach and can be lost in the star's glow. The magnitude 10.11 galaxy appears almost star like. This galaxy has a radial velocity of 178 km/s in recession. Use moderate power to get Mirach and its glare out of your eyepiece in order to have a better view of this galaxy.

About 14 degrees east-northeast from Mirach is the open cluster NGC 752 (III1m) [1:57;+37d41']. The 60 members of this spread-out magnitude 5.7 cluster form a twisted "X" that is visible in the finder. The cluster is located 400 pc from us and has a radial velocity of 4 km/s in approach.

To the southwest of NGC 752 is the magnitude 5.67, spectrum K0 III, open multiple system 56 Andromedae [1:56;+37d15']. The primary and secondary stars are separated by 190 arcsec. Look for the reddish magnitude 6 secondary in position angle 300°.

About 4 degrees north of NGC 752 is the interesting multiple system of the golden-yellow giant Almach (57 Gamma[1] And; ADS 1630; Σ 205; and Gamma[2] And; OΣ 38) [2:03;+42d19']. Magnitude 2.18, spectrum K3 Iib, Almach (pronounced al-MAK) was discovered to be a double star by Johann Tobias Mayer in either 1778 or 1788. The magnitude 5.03, spectrum A0p, greenish-blue "B" star (Gamma[2]) is located 9.8 arcsec from the primary in position angle 63°. The "B" is a visual binary with a blue-white, magnitude 6.3, spectrum A0 V, "C" that orbits in a period of 61.1 years. The "A" star is also a spectroscopic binary with a period of 2.67 days. Gamma And is 37 pc from us. Gamma[1] has a radial velocity of 12 km/s in approach and Gamma[2] is a little faster at 14 km/s. This is an excellent system to view, even in binoculars.

About 9 degrees (38 minutes of R.A.) to the east is the open cluster NGC 1039 (M34) (II3m) [2:42;+42d47'] in Perseus. Messier discovered this magnitude 5.2 cluster, on August 25, 1764. About 60 stars are associated with this 190 million-year old cluster. Estimates place M34 at 440 to 468 pc from us with a system radial velocity of 10 km/s in approach. The brighter stars, most of which are white, spectrum B8, giants, are in the center of the cluster and form a box-shaped asterism. On the southeastern corner of the box you can find the wide double star h 1123, first measured by John Herschel. Both components are white, spectrum A0, magnitude 8.0 stars, which have a separation of 20 arcsec. About 6 arcmin to the southeast from h 1123 is the double star OΣ 44. Otto Struve discovered this pair and, in 1850, listed their separation as being 1.4 arcsec. The primary is a spectrum B9 star with a magnitude of 8.4. The comes shines at magnitude 9.1 in position angle 55°.

Photo 12.1 NGC 224 (M31); the Andromeda Galaxy and companion galaxies NGC 221 (M32), and NGC 205 (M110).

Return to Alpheratz then hop to the first bright star in the western leg, which is 29 Pi And (ADS 513) [0:36;+33d43']. This is a wide multiple system with a blue-white primary of spectrum B5 V and magnitude 4.4. The magnitude 8.16 "B" star is located 35.9 arcsec from the primary in position angle 173°. The 13.0 magnitude "C" star is 55.2 arcsec from the primary in position angle 357°.

The next bright star along the body of Andromeda is the multiple system of magnitude 3.8, spectrum A5 V, 37 Mu And (ADS 788) [0:56;+38d30']. The magnitude 12.9 "B" star shines 42.2 arcsec from the primary in position angle 307°. The magnitude 11.4 "C" star is in position angle 126°, 34.2 arcsec away from the "A" star. A "D" star is also part of the system, which is 25 pc from us.

Mu And is at the eastern apex of a triangle with blue 32 And [0:41;+39d27'] and yellow 35 Nu And [0:49;+41d04'] at the base angles. Along the base of the triangle is the elongated glow of the great Sb I-II-type spiral Andromeda Galaxy, NGC 224 (M31) [0:42;+41d16']. After factoring in our own galactic and solar system movements of about 200 km/s, M31's blueshifted spectrum indicates it is coming toward us at a speed of 59 km/s.

The faint glow of the Andromeda Galaxy has been known since at least the early Middle Ages as it appears on several maps. The earliest surviving record of it is on a chart contained in the A.D. 964 book *Suwar al-Kawākib al-thābitah* ("Book of the Fixed Stars") by al-Sufi. The first to view it through a telescope was the German astronomer Simon Marius (also Mayer, Mair, or Meyer) (1570–1624), on December 12, 1612. This former student of Tycho Brahe described it as "like a flame of a candle seen through horn" (referring to horn lanterns) in his booklet known by its short title *Mundus Jovialis* ("Jupiter's Universe"). Marius gave the four bright moons of Jupiter their names of Io, Europa, Ganymede, and Callisto. In the intervening years between the sighting by Marius and when Messier saw it on August 3, 1764, several prominent astronomers also recorded their observations of it in their logbooks.

A bright star was seen in the great nebula in August 1885. For a short time this object reached an estimated magnitude of 6. The estimates are that as it exploded the star reached an absolute magnitude of -18.2 and was as luminous as 1.6 billion suns. We now know the observers had witnessed a rare supernova, since then designated S Andromedae (SN-1885). This object long ago faded from view.

In December 1924, Hubble announced his discovery of pulsating cepheid variables in the nebula and his use of them to determine the distance to the nebula as being about 900,000 light-years. This was the first time the theory of the existence of galaxies besides our own Milky Way was proven to be a reality.

A 1953 study of M31 made with photographs taken by the 200-inch Hale telescope on Mount Palomar showed the nuclear bulge contains faint old red and yellow Population II stars whereas the spiral arms are loaded with dust and young hot luminous blue giant Population I stars. The study also discovered the Population I cepheids are brighter by 1.5 magnitude than the older ones in the core of the galaxy. With this discovery, the distance to the galaxy was revised to be 2.2 million light-years (675.26 kpc).

M31 can be disappointing to see through an eyepiece after admiring photographs of it. You can easily see the bright nucleus and the arms, but you will probably not be able to actually see any of the detail in the arms as shown in a photograph. A large scope and a dark sky are needed to observe the thin dust lanes in the outer reaches of the galaxy. After the Magellanic Clouds in the southern skies, this is the largest, in apparent size, extragalactic object and is generally considered to be the most distant object that can easily be seen naked-eye.

M31 is accompanied in space by four gravitationally bound companion galaxy systems: NGC 147, NGC 185, NGC 205 (M110), and NGC 221 (M32). NGC 147 [0:33;+48d30'] and NGC 185 [0:39;+48d20'] are visually located about 7 degrees north of M31. These two ninth-magnitude elliptical galaxies are about 1 degree west of magnitude 4.54, spectrum B5 IIIe, 22 Omicron Cassiopeiae [0:44;+48d17'].

To the southeast of M31 is the magnitude 8.21 E2-type galaxy NGC 221 (M32) [0:42;+40d52']. Le Gentil discovered this nebula, on October 29, 1749. Messier claimed that he saw it in 1757, but did not add it to his list until August 3, 1764. The total mass of M32 is calculated to be equivalent to three billion solar masses. Its absolute magnitude is low for a galaxy at -16.5. Owing to its motion around M31, M32 has a radial velocity of 21 km/s in recession. It appears as an egg-shaped patch of light under moderate power.

NGC 205 (M110) [0:40;+41d41'] is located about $\frac{1}{4}$ degree northwest of M32, and its inclusion on Messier's list has invoked much discussion over the years. Messier claimed to have first seen it in 1773 and he included it on his drawing of M32 published in 1807.

Caroline Herschel claimed to have discovered it on August 27, 1783. Messier did not include it on any of his original lists. Since 1966, NGC 205 has been generally accepted as the last item in the Messier catalogue. This E6-type galaxy glows as a circular hazy patch of light at magnitude 8.01. It can be seen in the same low-power field with M32 if the western side of M32 is placed on the edge of the field. Its radial velocity is 1 km/s in recession.

To the east of M32 is EG And [0:44;+40d41'], a Z Andromedae-type (ZAND) cataclysmic variable. This type of star is actually a symbiotic binary system consisting of a hot star, a cool red giant, and an envelope of glowing gas being pulled from the cool star to the other. EG And fluctuates between magnitude 7.1 and 7.8 on an irregular basis that may be as short as 40 or as long as 470 days.

From 35 Mu And, hop 3 degrees east to magnitude 5.98, spectrum A7 V, blue-white wide double, 39 And (ADS 863) [1:02;+41d20']. In the same field as 39 And is RX And [1:04;+41d18']. This faint star is a dwarf nova of the Z Camelopardalis-type (UGZ) of cataclysmic variable. UGZ stars have outbursts on a recurring basis, usually about 10 to 40 days apart. They sometimes do not return to their former magnitude for several cycles. RX And goes from magnitude 10.3 to 14.0 during a 14-day cycle. Though faint most of the time, this is the brightest star of its variable type.

Following north along the western leg we come to the long-period binary and multiple system of 48 Omega And (ADS 1152; β999 and β 82) [1:27;+45d24']. The magnitude 4.83, spectrum F5 V, primary and the magnitude 11.5 red dwarf "B" can be difficult to resolve being only 2.0 arcsec apart. The "A" and "B" stars were discovered to be a double by Burnham, in 1881. Nine years earlier, he had catalogued as β 82 the magnitude 10.2 "C" and "D" stars. The "C" star is 119 arcsec from the "A" star in position angle 111°. The "D" star is 4.9 arcsec from the "C" star. The "B" star has a luminosity of about 0.025 Suns whereas the "A" star is about 17 times brighter than the Sun. Its distance of 36 pc, system radial velocity of 11 km/s in approach, and true space motion indicate that Omega And is an outlying member of the Hyades Moving Group.

Star-hop in Perseus

The last bright star forming the western leg in Andromeda is the magnitude 3.57, spectrum K3 III, red giant 51 And [1:37;+48d37'], which is the starting point for our hop in Perseus, as shown in Figure 12.2. Less than $\frac{1}{8}$ degree west of 51 And is the difficult double A 817 (ADS 1263) [1:37;+48d43']. Robert Aitken first measured these magnitude 8.5 and 9.0, spectrum A0 stars, in 1904. Their separation is less than 1 arcsec so you will need a large scope to resolve them.

The last bright star in the western Pegasus/Andromeda leg is actually the magnitude 4.03 to 4.11, spectrum B2 Vpe, Phi Persei [1:43;+50d41']. This spectroscopic binary system is just north of the boundary between Andromeda and Perseus. This Gamma

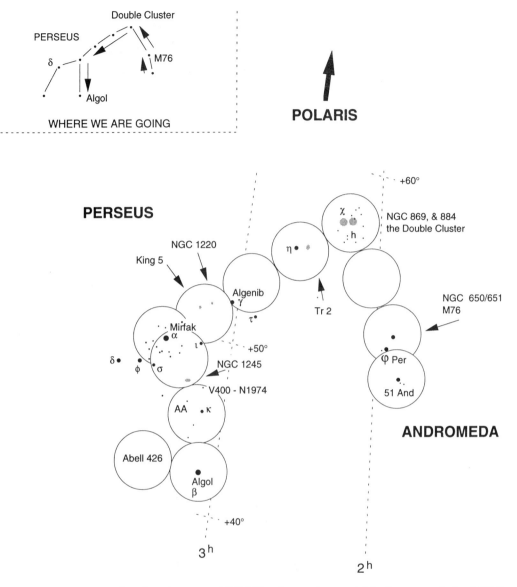

Figure 12.2
Star-hop in Perseus.

Cassiopeiae-type (GCAS) of eruptive variable has a period of 19.5 days between outbursts as the star spins on its axis. Phi Per is 350 pc from us and has a radial velocity of 19 km/s in approach.

About 1 degree north of Phi Persei is a magnitude 6.7, spectrum K5 III, red giant star. One-quarter degree west of this red giant is the type III+ VI planetary nebula NGC 650/651 (M76; PK 130-10.1) [1:42;+51d34']. Méchain found the "Little Dumbbell Nebula" on September 5, 1780, and reported his find to Messier. Messier verified the discovery on October 21, 1780, and entered this nebula on his list. In the telescopes of the 19th century this object appeared to be two separate objects so it was assigned two NGC numbers. NGC 650 is for the southern portion of the box-shaped planetary. The gas shell glows at magnitude 10.1 and the extremely hot (60,000 K) central star at magnitude 15.87. The integrated magnitude for M76 is 11.5.

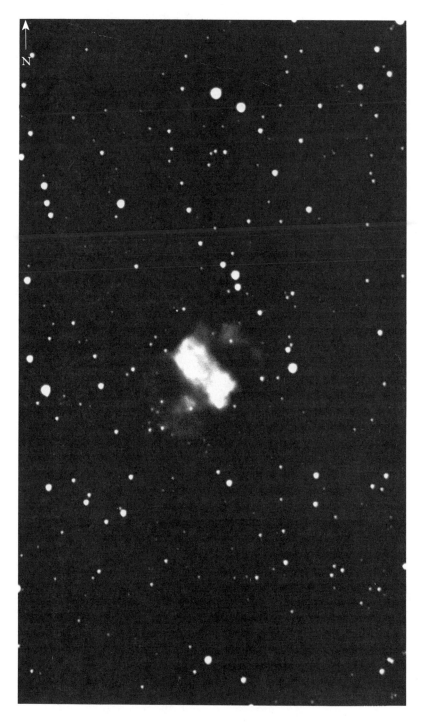

Photo 12.2 NGC 650/651 (M76); the Little Dumbbell Nebula.

The gas shell is expanding at a rate of 42 km/s. The distance to M76 is not firmly established and ranges from 537 pc to 2.51 kpc. The gas shell appears thicker and brighter at the north and south ends than in the middle. Use a high-contrast or LUMICON O-III filter for the best view of this object.

NGC 884 χ   NGC 869 χ

N

Photo 12.3 Open clusters NGC 869 and NGC 884 (χ Persei); the Double Cluster in Perseus.

To me, one of the most beautiful objects to view, especially in binoculars or a wide-field eyepiece, is the Double Star Cluster consisting of open clusters NGC 869 (h Per) [2:19;+57d09'] and NGC 884 (Chi Per) [2:22;+57d07']. Both of these Trumpler I3r class clusters are bright and very easy to locate being about 5 degrees northeast of Phi Persei. These clusters were seen as hazy patches by Hipparchus and Ptolemy, who included the clusters as cloudy spots in their ancient star lists. The Arabs named them *Mi`ṣam al Thurayya* indicating their position at the wrists of Perseus on the Pleiades side. The clusters usually mark the handle of Perseus's sword and his hand.

Both of these clusters contain hundreds of stars, many of which are double and multiple systems with practically all of the stellar colors visible. Even a few rare M-type red supergiants are cluster members. Three of these red stars are located in NGC 884. No red star has been found in NGC 869. The exact distance to the clusters has not been calculated, but knowing the brightest members of each cluster are very luminous white spectrum O, B, and A supergiants of apparent sixth- and seventh-magnitudes, a distance of about 2,200 pc has generally been accepted. With estimated absolute magnitudes in the range of -7.5 to -3.3 some of these stars may be as much as 60,000 times brighter than the Sun, but are dimmed by distance and interstellar dust. Take your time to explore these great clusters. You will come back to them many times.

Heading toward the bright string of stars that outline the great warrior's body, we first come to the small open cluster Tr 2 (III2p)

213

[2:37;+55d59'] two degrees west of the multiple, magnitude 3.6, 15 Eta Persei (ADS 2157; Σ 307) [2:50;+55d53']. This cluster is the second item in Trumpler's catalogue and contains about 20 stars with an integrated magnitude of 5.9. The cluster is 78 million years old and is 600 pc distant.

Eta Persei is a fine multiple color-contrast system to view. A yellowish or golden, spectrum K3 Ib-Iia, magnitude 3.8 primary and a bluish magnitude 8.5 "B" star dominate a field of ninth- and tenth-magnitude companions. Having a separation of 28.3 arcsec, the "A" and "B" stars are easy to resolve. The area between Eta Per and the Double Cluster is the radiant point for the Perseids meteor shower in mid-August each year.

About $2\frac{1}{2}$ degrees southeast of Eta Per is the very close binary and visual double Algenib (23 Gamma Persei; ADS 2324; h 2170) [3:04;+53d30']. Two other stars share this same name: Alpha Persei and Gamma Pegasi. The etymology for this name is hazy as some sources say it originally was used for Alpha and/or Gamma Persei and was mistakenly applied to Gamma Pegasi, yet the word is thought to derive from the Arabic *Al Janāḥ* meaning "The Wing" or *Al Janb* "The Side." These words would better fit Algenib's location at the southeast corner of the Great Square of Pegasus than in Perseus. The binary portion ("A" and "a") of Algenib (pronounced al-JEE-nib) is very difficult to separate since the two stars are less than 0.4 arcsec apart even when "a" is at its greatest elongation from the magnitude 2.9, spectrum G5 III, "A" star. We see this system as an edge-on ellipse with an orbital period of 14.647 years. A magnitude 10.6, spectrum A2 V, "B" star is located 57 arcsec from the "A" star in position angle 326°, but this star is not a true physical member of the system. This system is 34 pc from us.

One degree southeast of Algenib is the very small open cluster NGC 1220 (II2p) [3:11;+53d20']. This poor cluster contains about 15 faint stars, yet at magnitude 11.8 it can be a good test of your ability to locate difficult targets.

An easier target is the open cluster King 5 (K 5) [3:14;+52d41'], which is about $\frac{3}{4}$ degree south-southeast of Gamma Per. This cluster contains about 40 faint stars, many of which are 13th magnitude or fainter.

Heading in the same southeasterly direction, we come to the 33rd brightest star in our skies: Mirfak (33 Alpha Persei) [3:24;+49d51']. Magnitude 1.80, spectrum F5 Ib, white-yellow supergiant Mirfak (pronounced MERE-fak) is the heart of a very wide double and a moving association of about 110 stars. This star has also been known as Algenib, but Mirfak, which comes from the Arabic *Marfik al Thurayya* meaning "The Elbow," better indicates its position in the constellation than Algenib. Mirfak's 11.8-magnitude secondary is located 167 arcsec from it in position angle 196°. Young hot blue-white giants of spectral types B5 through A5 are the predominant stars in the Alpha Persei Association (Melotte 20 and Cr 39/40) [3:20;+49d], which is located mainly to the southeast of Alpha Per. You will also see a lot of yellow and orange spectral-type F0 through

G0 stars. The association is moving about at 16 km/s in the general direction of Beta Tauri. This cluster shows best in binoculars. Mirfak is located about 190 pc from us and has a radial velocity of 2 km/s in approach.

We will branch off from Perseus's body and head southwest down his left leg. About $2\frac{1}{2}$ degrees southwest of Mirfak is the magnitude 8.4 open cluster NGC 1245 (III1r) [3:14;+47d15']. The 200 members of this rich 1.1 billion-year old cluster appear to be packed close together, but not as densely as in a globular.

About $1\frac{1}{2}$ degrees west of NGC 1245 is the intermediate speed nova (NAB) V400-N1974 [3:07;+47d08']. This star flared up to eighth magnitude in 1974, then over a period of about 125 days, faded back to be about 19th magnitude. Intermediate speed novae are cataclysmic variables that decrease by at least three magnitudes in a period of 100 to 150 days after their flare-up as a nova. Nova 1974 is about $\frac{1}{4}$ degree southwest of a magnitude 6.41, spectrum A0 V, star [3:07;+47d18'].

One degree south of NGC 1245 is the SRA-type pulsating variable AA Persei [3:15;+46d35']. These SRA-types are semiregular giants of spectral types M, Me, C, Ce, S, or Se. They have small visual magnitude amplitudes (less than 2.5) and their periods can vary in a range of 35 to 1,200 days. AA Persei pulsates between magnitude 9.0 and 10.0 during a period of 9.103 days.

Marking variously the forehead or the eyes of the demon Gorgon Medusa is the bright double and spectroscopic eclipsing variable Algol (26 Beta Persei; ADS 2362; β 526) [3:08;+40d57']. Algol (pronounced AL-gall) comes from the Arabic *Rās al Ghūl*, which means the "Demon's Head." Cultures worldwide held this strange star in evil regards. The Hebrews knew it as *Rosh ha Satan* meaning "Satan's Head." The Chinese called it the "Piled-up Corpses" *Tseih She.*

Around 1669, Italian astronomer Geminiano Montanari (1632–87) was the first to study the variability of Algol. In 1782, John Goodricke noticed the changes in Algol's brightness from one minimum (magnitude 3.40) to the next occurred on a regular basis of 2 days, 20 hours, 48 minutes, and 56 seconds. He theorized this was caused by a dark companion orbiting Algol. While studying the periodic Doppler shift in spectral lines for Algol in 1889, H. C. Vogel discovered the spectral-type Am dark companion, Algol B; thus discovering the first spectroscopic binary system. This star is not really dark, it is just not visible due to the brightness of Algol, which is a white, spectrum B8 V, star with a normal magnitude of 2.12. The spectroscopic binary is believed to be a subgiant only 10.4 million kilometers from its center to the center of Algol. A third spectroscopic companion, Algol C, is also known to exist in an orbit about 80 million kilometers from the AB stars. A fourth spectroscopic star, Algol D, is believed to be part of this system.

Also part of the Algol system are two 12th-magnitude stars ("B" and "C") and a tenth-magnitude "D" star. These visual stars were first measured by Burnham in 1878–9. The Algol system is 29 pc distant and has a radial velocity of 4 km/s in recession.

About 2 degrees northeast from Algol is the center of a tight group of about 40-plus faint galaxies known as the Perseus Galaxy Cluster (Abell 426) [3:20;+41d30']. This is one of the 2,712 rich "galaxian archipelagos" American astronomer George Ogden Abell (1927–83) catalogued in a 1958 paper entitled "The Distribution of Rich Clusters of Galaxies." He gathered his data by studying the National Geographic Society Palomar Observatory Sky Survey (POSS) plates. At a distance of 72 Mpc, this is believed to be one of the closest superclusters to us after the Virgo and Centaurus clusters. Abell 426 appears as a faint chain of about 24 fuzzy star-like objects extending southwesterly from 11.59-magnitude NGC 1275 to 14.3-magnitude IC 310. Most of these galaxies are too faint and small to be seen visually in all but the largest telescopes or on photographs.

One of the unusual members of Abell 426 is the peculiar and Seyfert-type galaxy NGC 1275 (Perseus A; 3C 84) [3:19;+41d31']. Exactly what is the nature of this galaxy has intrigued astronomers for years and remains a mystery. Some think it is two galaxies colliding at great velocity (3,000 km/s), which gives off high temperatures (80 million K) and tremendous X-ray energy. Around the edges of the galaxy are numerous filaments of ejected gas. These gas jets have velocities of several thousand kilometers per second above the radial velocity of 5,361 km/s in recession for the balance of the galaxy. This galaxy has also been catalogued as a tight spiral or a large elliptical galaxy.

As in all of the star-hops in this book, more objects, such as NGC 1499, (the California Nebula), are available in Andromeda and Perseus for you to explore than I could present here.

13 *November*

Cassiopeia and Cepheus:
The Queen and her King of Joppa

Mysterious star!
Thou wert my dream
All a long summer night —

Be now my theme!
By this clear stream,
Of thee will I write;
Meantime from afar
Bathe me in light!

Edgar Allen Poe (1809–49),
"Mysterious Star!" (A new
introduction to "Al Aaraaf")
(lines 1–8),
1831

As the autumn evenings become cooler, signaling the coming of winter, we are given the best opportunity to study Cassiopeia and Cepheus. The Greek mythological Queen of Ethiopia sits on her heavenly throne, while her husband, King Cepheus, is nearby. Though Cassiopeia (also written as Cassiope) and Cepheus are visible all year for northern hemisphere stargazers, during November these constellations hover overhead at the zenith and transit the southern meridian early in the evening. In this position, we are looking through a thinner layer of the Earth's atmosphere than at other times of the year. This allows us to see faint galactic star clusters and nebulosity, and to separate close doubles. These objects can be difficult to see when you have to look through thicker layers of our atmosphere.

Due to telescope mounting interference, users of equatorial fork-mounted telescopes may have trouble trying to view these constellations at this time of year. If mounting interference is a problem for you, despair not, you can always take this month's star-hops at other times of the year.

Star-hop in Cepheus

Though Cassiopeia is the starring attraction this month, we will first take a brief tour of the constellation of Cepheus, as shown in Figure 13.1. According to the ancient tales, King Cepheus sailed with Jason

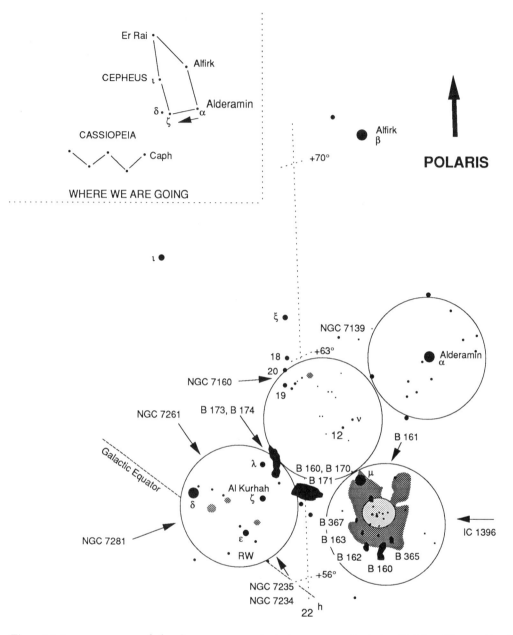

Figure 13.1
Star-hop in Cepheus.

and the Argonauts on their journey to find the Golden Fleece. In the Bible, Jonah sailed from the seaport of Joppa and wound up inside the big fish.

Being entwined with Perseus and Andromeda, I have already told you the tragic tale of Cepheus and Cassiopeia. After their deaths, Cassiopeia and Cepheus were transported to the heavens by Poseidon. Still angry with the queen because of her boast and treachery, he placed Cassiopeia in a circumpolar position. There, she would never be out of his sight, and as additional punishment, for part of the year she would hang upside down in the sky. Being weak and tall, King Cepheus was placed in an area of faint stars that stretch out over

587.787 square degrees of the sky. He ranks 27th in size among the constellations.

Cepheus (Cep) contains few bright stars, but does have several interesting and astronomically important objects to look at. We will commence our tour at the king's brightest star, Alderamin (5 Alpha Cephei; ADS 14858) at coordinates [21:18;+62d35']. To locate Alderamin, look to the north and find the prominent "W" asterism of Cassiopeia. The two bright stars forming the western leg of the asterism point you in the direction of Alderamin (pronounced al-DARE-ah-min), which is about six finder hops to the west. Alderamin is a 2.44-magnitude blue-white, spectrum A7 Van, open double star that forms the southwestern corner of the hut-shaped constellation. The other corners are defined by 21 Zeta [22:10;+58d12'], 32 Iota [22:49;+66d12'], Er Rai (35 Gamma at the peak of the roof) [23:39;+77d37'], and Alfirk (8 Beta Cep) [21:28;+70d33']. As the shape of the king is portrayed in drawings of the sky, Alderamin forms his right shoulder. His feet are near Polaris. Alderamin is derived from "Al Deraimin" in the Alfonsine Tables, which is from the Arabic *Al Dhirā al Yamīn* meaning "The Right Arm."

Alderamin's magnitude 10.2 "B" star is located 206 arcsec from the primary in position angle 22°. In 1907, Burnham discovered a close pair of 11th-magnitude ("C" and "D") stars 19.9 arcsec from the "B" star in position angle 172°. The rapid rotation of Alderamin causes its spectrum to appear hazy on spectrographs. This star is about 21 times brighter than the Sun. Alpha is located 14 pc from us and has a radial velocity of 10 km/s in approach.

Your field of view for Alderamin will include optical doubles of 10th and 11th magnitude. However, this fringe area of the Milky Way teems with actual binary systems.

About 5,000 years from now, Alderamin will be the pole star. Alderamin is presently at 26.8 degrees of declination south of Polaris. Notice how far Alderamin is from Polaris and how far on the other side of Polaris is the former pole star; Thuban (11 Alpha Draconis). This will give you some idea of how far the Earth's precession wobble causes the North Pole to shift in the sky over a period of 10,000 years.

Centering your finder scope on 10 Nu Cephei [21:45;+61d07'], about 2 degrees southeast of Alderamin, reveals a field richer in doubles than seen at the first hop stop. On the northern fringe of your view should be the small open cluster NGC 7160 (II3p) [21:53;+62d36']. NGC 7160 is about 1 degree west of magnitude 5.27, spectrum K4 III, 20 Cep [22:05;+62d47']. Binoculars or wide-field eyepieces give the best views of 12 stars that make up NGC 7160. A large field of view allows you to distinguish the cluster from the other stars near it as the cluster appears slightly brighter than the background field.

Within NGC 7160 is the EW/KE-type eclipsing variable EM Cep. W Ursae Majoris is the prototype for EW-type eclipsing variables. These variables have visual magnitude amplitudes of less than 0.8 and periods of less than a day. With multiple minima, the beginning

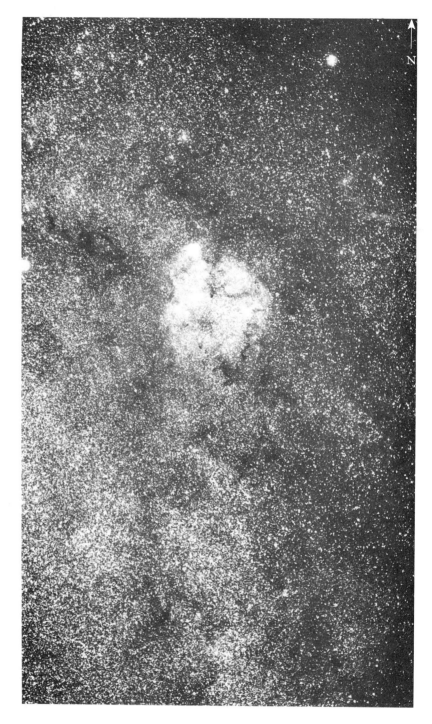

Photo 13.1 Mu Cephei, IC 1396, B 169/70, B 171, and Alderamin.

and end of each cycle cannot be determined. The KE designation means the mass of these stars extends beyond their Roche lobes and are in contact with each other across their inner Lagrangian point, the primary star is a main-sequence star generally of spectral classes F0 through K, and the secondary is a dwarf near the main-sequence. EM

Cep consists of a blue-white, spectrum B1 Ive, primary that fluctuates between magnitude 7.02 and 7.17 during a period of 0.8061 days.

Two degrees south of Nu is the bright red giant, MK standard for spectrum M2 Iae stars; Mu Cephei [21:43;+58d46']. This SRC-type pulsating star varies between magnitude 3.43 and 5.1 during a period of 730 days. Mu marks Cepheus's left elbow. This star is the brightest naked-eye red star and is estimated to be a whopping 12,000 times brighter and hundreds of times larger than the Sun. The star is commonly known as William Herschel's Garnet Star. He wrote about it, but did not discover its variability. Through a telescope, Mu seems reddish-orange to slightly yellowish-orange.

Mu Cephei is located on the northern edge of IC 1396 [21:39;+57d30']. Heart-shaped IC 1396 consists of both a faint reflection nebula and an open cluster (II3m). Dark veins are visible and appear entwined in the cluster. A prominent vein starts at the edge of the nebula, very close to Mu Cephei, and cuts an L-like path part of the way through the cloud. The very rich cluster is embedded almost dead center in our view of the nebula. Over 200 stars make up this 3.5-magnitude cluster. In the foreground are the dark nebulae B 160, B 161, B 162, B 163, B 365, and B 367. On the northeastern side of the nebula are two sixth- and seventh-magnitude double stars. IC 1396 is best observed in a very low-power or wide-field eyepiece. Too high a power and you may not be able to see all of it at once.

The black patches of star-less sky about 3 degrees east of Mu Cep are the dark nebulae B 169/70 and B 171 [21:58;+58d45']. A few stars, mostly in the foreground, can be seen against the nebulae.

The area bounded by 21 Zeta, 23 Epsilon, and 27 Delta Cephei is a field rich in stars and open clusters. Unfortunately, many of the clusters in the area are very faint and you will need a scope capable of deep-sky observing to see them. Two of the easier open clusters to observe are NGC 7235 and NGC 7261. Each of these open clusters consists of about 30 members. NGC 7235 (III2p) [22:12;+57d17'] appears as a compact triangle with another grouping of about eight stars just southwest of the cluster. This smaller group is NGC 7234. Located about halfway between Zeta and Delta Cephei is NGC 7261 (III1p) [22:20;+58d05']. This cluster forms a rectangle in the sky with the hint of stars inside the box. You may even be able to spot this 8.4-magnitude cluster in your finder.

Probably one of the most astronomically important stars is the multiple and variable 27 Delta Cephei (ADS 15987; β 702) [22:29;+58d24']. In 1924, this short-period pulsating variable star was used by Edwin Hubble to discover the distance scale for the universe. The road to Hubble's discovery began in 1782 with John Goodricke's discovery of variable stars. Goodricke noticed that Algol in Perseus would change in magnitude on a regular basis. In 1784, Goodricke noticed that Delta Cephei also changed in brightness on a regular cycle lasting 5 days 8 hours and $37\frac{1}{2}$ minutes (5.366341 days). In his report to the Royal Society, Goodricke noted that these stars showed a quick and sharp rise in brightness while their decline in brightness took a much longer period than the rise, as shown in Figure 13.2.

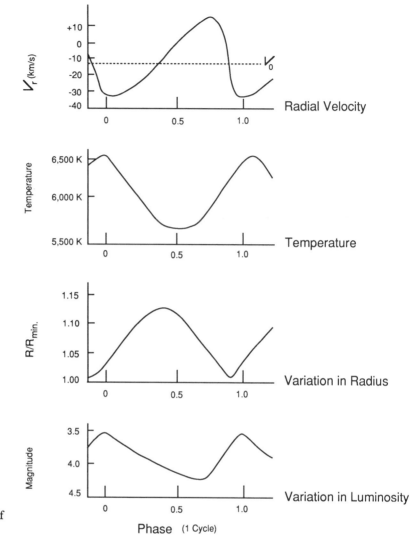

Figure 13.2
Variable properties of
Delta Cephei.

While studying the variables of the Small Magellanic Cloud, Harvard astronomer Henrietta Swan Leavitt (1886–1921), in 1912, discovered a relationship between a pulsating star's period of variability and its apparent luminosity. The longer the cycle, she found, the greater the apparent brightness of the star. She found 25 cepheids in the Small Magellanic Cloud with periods of 2 to 40 days. Four years after her death, she was nominated for the Nobel Prize.

Leavitt's "period–luminosity relation" work and the formulae of "stellar gas pressure versus stellar radiation" published in 1919, by Sir Arthur Eddington, led Harlow Shapley to conclude that cepheid variables also had a relationship between their period and apparent magnitude. Shapley figured that a pulsating star, rather than an eclipsing binary, would act this way.

Building upon the work of Leavitt, Shapley, and others, Hubble used the cepheid variables in the Andromeda nebula (M31) to calculate their distance from us. He proved that M31 was actually a separate galaxy. Until the time Hubble released his findings, people believed the patches of hazy light, like M31, were part of our own Galaxy. Using Hubble's formulae along with other techniques, such as spectroscopy, stellar distances can now be roughly calculated for almost every object in the universe.

Delta Cephei is the MK standard for spectrum F5 Ibv yellow supergiants. This star is about 1,200 times more luminous than the Sun. Delta's magnitude 7.5, spectrum A0, blue companion can easily be separated in a small telescope. This "C" star was discovered by F. G. W. von Struve, in 1835. The "B" star in this system is a 13.0-magnitude star first measured by S. W. Burnham, in 1878. This is located 20.4 arcsec from the primary in position angle 284°. The "B" star does not appear to be a physical member of this system. Delta is 410 pc from us and has a radial velocity of 17 km/s in approach.

To the southwest of Delta Cep are the tiny low-luminosity red dwarfs of Krüger 60 (ADS 15972; Kr 60) [22:28;+57d42']. This rapid binary system is 4.02 pc distant and we view it face on. Apparent magnitude 11.3, spectrum dM4e, Kr 60B orbits around its magnitude 9.8, spectrum dM2, primary Kr 60A in 44.46 years and is about as distant to 60A as Saturn is to the Sun. Kr 60B is one of the smallest stars we know of, being only 0.14 the mass of the Sun. The absolute magnitude for Kr 60A is +11.8 and that of the "B" star is estimated to be +13.4. It has a habit of dramatically increasing in brightness in a matter of minutes. The current theory is that a solar flare erupts on it, but, with the star being so small, the flare can actually increase the luminosity of the entire star. Finnish astronomer Adalbert Krüger (1832–96) discovered the "A" and "C" stars and thought them to be a double system. It now appears the "C" star is only a visual double as it does not have the same space motion as the "AB" pair. Burnham found the "B" star in 1890. Seven other faint stars make up the entire Kr 60 system.

East of Kr 60 is the asterism of NGC 7281 (IV4p) [22:24;+57d50']. Though NGC 7281 appears to be a cluster, astronomers are not quite sure if it actually is a cluster or just a line-of-sight grouping of maybe 20 to 35 faint stars.

I have led you on a path along the southern end of Cepheus; many other objects should however find their way into your eyepiece before you leave this constellation. NGC 188 [0:44;+85d20'] is a rich cluster of about 120 stars located near Polaris. At an estimated age of between 12 to 14 billion years old, NGC 188 is the oldest known open cluster. Most open clusters are less than half this age.

Star-hop in Cassiopeia

To the west of Cepheus, his queen Cassiopeia sits on her throne in a Milky Way lode of young stars and open clusters. The newer stars have condensed from pockets of gas and dust in our Galaxy's spiral arm. Being from the same material, many are therefore close together

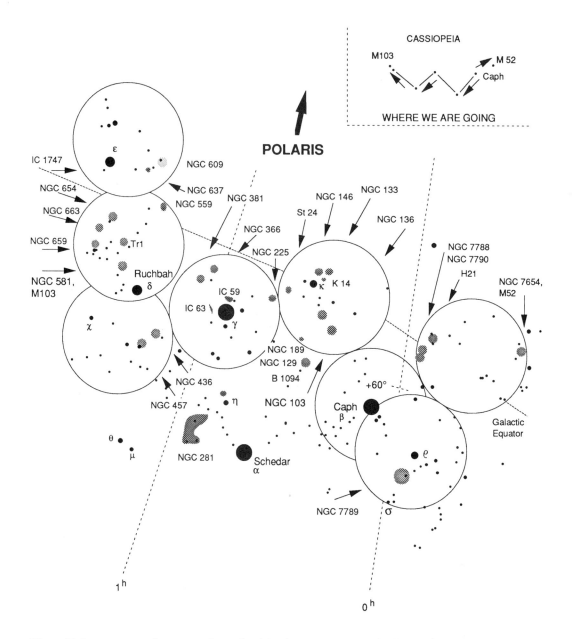

Figure 13.3
Star-hop in
Cassiopeia.

in approximately 30 clusters scattered about Cassiopeia. At 598 square degrees, the Queen ranks 25th in constellation size.

If you have a telescope and binoculars, observe this wondrous constellation with both of them. Some of the clusters will be easier to see with binoculars as the tight clusters will stand out from the haziness of the Milky Way. In a telescope, use a wide field of view eyepiece, otherwise your view may be too small to be able to differentiate the clusters from the background stars. You will also get by with a low-power (less than 100x) eyepiece for most of this star-hop.

During November evenings, Cassiopeia appears as a flattened "W" in the northwestern sky, and in the spring it is an "M" in the northeastern sky. The magnitude 2.3, MK standard spectrum F2 III-IV, blue-white star at the western end of the "W" is the open double and spectroscopic binary star Caph (11 Beta Cas; ADS 107) [0:09;+59d08']. Caph (pronounced kaff) is near the zero-hour line, which is also known the "prime meridian of the sky." This line runs from the North Pole, passes near Caph, Alpheratz (21 Alpha Andromedae), Algenib (88 Gamma Pegasi), then intersects the Ecliptic in Pisces. Caph will be the starting point on our hop around the galactic clusters of Cassiopeia, as shown in Figure 13.3.

Aim your optical aid at Caph. You should see an arrow-like grouping of stars. The brightest star near the tip of the arrow is Caph. Forming the shaft of the arrow is a line of stars, most of which are doubles. The shaft points away from Polaris. At a distance of 13 pc from us, Caph is the closest short-period pulsating Delta Scuti variable star. Caph fluctuates between magnitude 2.25 and 2.31 in 0.1043 day. A 13.6-magnitude binary companion orbits around Caph in a period of about 27 days.

Place Caph on the southern edge of your finder scope view. You should be able to see a grouping of three small open star clusters, NGC 7788, NGC 7790, and Harvard 21. NGC 7788 (I2p) [23:56;+61d24'] is 2.4 kpc away and appears as a hazy patch of nebulosity and five stars centered around a ninth-magnitude star. NGC 7790 (III2p) [22:58;+61d13'] is also a hazy patch of nebulosity with about 40 stars. These stars make up a 78 million-year-old eighth-magnitude triangle. Harvard 21 (IV2p) [23:54;+61d46'] contains six stars and is difficult to separate from the background stars.

At magnitude 6.9, both open clusters NGC 7654 and NGC 7789 are just beyond the naked-eye range, but are easy to locate in binoculars. Center NGC 7788 on the eastern edge of your finder scope. The open cluster NGC 7654 (M52) (I2r) [23:24;+61d35'] will be on the opposite edge of the view. The 193 stars of M52 are tightly compressed and form an easy to locate, glowing "V" filled with points of light. Messier discovered this young cluster, on September 7, 1774. The cluster is 924 pc from us.

The "V" and trail of stars south of M52 will point you in the direction of the open cluster NGC 7789 (III1r) [23:57;+56d44'], the richest in the area with about 600 stars. Caroline Herschel discovered this cluster. Many of its stars are red giants. About half of the stars are fainter than tenth magnitude and appear as a hazy fog around the brighter stars. Situated between 7 Rho and 8 Sigma Cas, this cluster fills an area of the sky almost as large as the Moon.

Rho Cas [23:54;+57d02'] is an SRD-type of pulsating variable. This slow and unpredictable semiregular suspected supergiant fluctuates between magnitude 4.1 and 6.2 during a period that has been as short as 100 and as long as 320 days. Its spectrum can be anywhere between F8 and K5 during a pulsation cycle. Because of its changing nature and the uncertainty as to its distance from us, exactly what type of star Rho is remains under discussion. A rapidly expanding

Photo 13.2 Open clusters NGC 7654 (M52) and Cz 43.

gaseous shell has been detected. Parallax and its estimated absolute magnitude of +0.4 suggest this star may be as close as 61.3 pc, but its spectrum indicates that it is 100 times brighter and may be as distant as 920 pc. Other estimates place Rho Cas at 1,500 pc.

Return to Caph then hop north-northest one finder view past it to an area rich with open clusters. Forming a triangle pointing south are the clusters NGC 103, NGC 136, and NGC 129. Use a low-power eyepiece to view the 30+ stars of NGC 129 (IV2p) [0:29;+60d14'] as it is large and very open. South of NGC 129 is a double star (ADS 412; β 1094) [0:30;+59d59'], which appears larger and brighter than any of the stars in NGC 129. At the northern corners of this cluster triangle are two faint clusters. To the west is NGC 103 (II2p) [0:25;+61d21']. NGC 136 (II2p) [0:31;+61d32'] is $\frac{3}{4}$ degree northeast of NGC 103. Both of these clusters are hard to locate because they are small, contain few stars, and are faint at magnitude 9.8 for NGC 103 and magnitude 11.3 for NGC 136. They blend in with the rich Milky Way star fields in the area.

Your finder field will include the tight grouping of open clusters King 14, NGC 133, and NGC 146 on the northern edge. Included in your view will be blue-white, MK standard spectrum B1 Ia, supergiant 15 Kappa Cas [0:32;+62d55'], and open clusters NGC 225, NGC 189, and Stock 24 (St 24). These six open clusters range from magnitude 7 to 9.4, but each are composed of stars of fainter magnitudes. King 14 (III2p) [0:31;+63d10'] contains about 20 stars located 2,600 pc from us. NGC 133 (VI1p) [0:31;+63d22'] contains seven stars spread out in a line about 7 arcsec long. East of NGC 133 is NGC 146 (IV3p) [0:33;+63d18']. This cluster of 20-plus stars seems almost circular in shape with a dark center. Embedded in a wide field of a faint nebula are the 76 stars of NGC 225 (III1p) [0:43;+61d47']. At 630 pc, this magnitude 7 cluster is one of the closest to us of Cassiopeia's clusters and is ancient at an estimated age of 140 million years. Use a low-power eyepiece or binoculars to view this loose cluster. NGC 189 (III2p) [0:39;+61d04'] consists of about 15 stars. The 20 stars of eighth-magnitude Stock 24 (IV2p) [0:39;+61d47'] are about 1 degree southeast of Kappa.

Over the centuries we have seen blue giants exhaust their fuel and die in massive explosions. One such explosion was noticed by Tycho Brahe in November 1572. The supernova, known as Tycho's Star 1572, is $1\frac{1}{2}$ degrees northwest of Kappa Cas. Only a rapidly expanding (9,333 km/s) very faint glow of nebulosity remains where the star was. You will not be able to see it in any amateur telescope; even on the best night of seeing. This is the "mysterious star" Poe referred to in his 1831 poem.

At the midpoint of the "W" is the close multiple and erratic variable 27 Gamma Cas (ADS 782) [0:56;+60d43']. While studying this 2.47-magnitude blue-white, spectrum B0.5 Ive, subgiant, in 1866, Father Pietro Angelo Secchi (1818–78) observed for the first time in any star the bright emission lines of hydrogen. Gamma is ejecting a spherical shell of this gas. In 1937, Gamma glowed at magnitude 1.6 and has unsteadily dimmed since then. This star is the prototype for the GCAS-type of rapidly rotating eruptive variables. This type of star spews material from its equatorial zones. Brightness changes are irregular and can be as much as 1.5 magnitudes. Two faint companions are associated with Gamma, but due to Gamma's brightness, these secondary stars are extremely difficult to locate visually.

To the north and east of Gamma Cas are the emission/reflection nebulae IC 59 [0:56;+61d04'] and IC 63 [0:59;+60d49']. Look carefully in a large scope, you may be able to see dark triangular shapes in these nebulae. The current theory holds that these shapes were caused by the solar wind from Gamma. In small scopes and binoculars, these nebulae get lost in Gamma's glow, so you may not be able to see them at all.

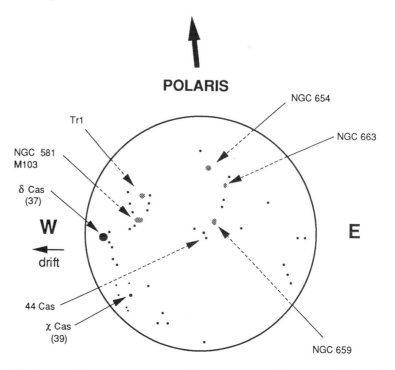

Figure 13.4 The clusters near Delta Cas, finder scope view, 9x60mm, with diagonal mirror.

With your finder centered on Gamma, you will see a field dominated by double stars. Also in the view will be two very faint clusters, NGC 381 and NGC 366. The 50 stars of NGC 381 (III2p) [1:08;+61d35'] appear as a hazy lollipop on a stick. NGC 366 (III2m) [1:06;+62d14'] is harder to locate since it is a tightly compressed group and is faint at tenth magnitude.

As the queen sits on her throne, her knees are bent at the open multiple and eclipsing binary star Ruchbah (37 Delta Cas) [1:25;+60d14']. Ruchbah (pronounced RUCK-bar) comes from the Arabic *Al Rukbah* which means "The Knee" or "The Knee of the Woman in the Chair." In 1669, French astronomer Jean Picard (1620–82) wanted to survey the size and shape of the Earth. In the first use of a telescope in geodesic research, he used Ruchbah as one of his reference points and accurately measured its movement along a north-south line. Using the distance it took for the star to move one degree of arc along the line, he was able to calculate the arc of the Earth's meridian. Delta Cas is a blue-white, 2.7-magnitude, MK standard spectrum A5 III-IV, star located somewhere between 13 and 27 pc from us. Ruchbah's proper motion and radial velocity of 7 km/s in recession indicate that it may be an outlying member of the Hyades open cluster in Taurus. The binary companion causes a dip in Delta's magnitude for a period lasting 759 days. A magnitude 11.4 secondary is located 131 arcsec from the primary in position angle 66°, as measured in 1925.

Slightly to the east of the line between Delta and 45 Epsilon Cas [1:54;+63d40'] is a U-shaped grouping of five open clusters, as shown in Figure 13.4. This group of clusters consists of NGC 581 (M103), Trumpler 1 (Tr1), NGC 654, NGC 663, and NGC 659. Except for M103 at 2,600 pc, these clusters are about 2,200 pc from us.

Photo 13.3 Open cluster NGC 581 (M103) in Cassiopeia.

The closest of these open clusters to Delta Cas is NGC 581 (M103) (III2p) [1:33;+60d42']. About 60 stars make up this loose magnitude 7.5 cluster discovered by Méchain prior to March 27, 1781. There is some disagreement as to whether this is an actual cluster or an accidental line-of-sight grouping. M103 is about 22 million years old and is 2,600 pc distant. The multiple star Σ 131 (ADS 1209) lies on the southwestern edge of M103. The brightest members of this multiple appear yellowish and blue.

The open cluster Trumpler 1 (I3p) [1:35;+61d17'] appears as two almost parallel lines of 10th- and 11th-magnitude stars embedded in a hazy patch of about 100 stars. This 26 million-year-old cluster is racing toward us at a speed of 65 km/s.

At the northern end of the "U" is the magnitude 6.5 open cluster

NGC 654 (II3m) [1:44;+61d53']. This easy to locate in binoculars cluster appears as a bright tiny spot of light surrounded by a dimmer loose field of 60-plus scattered stars. This cluster is estimated to be 15 million years old.

Forming a rough circle of about 105 stars is the large open cluster NGC 663 (III2m) [1:46;+61d15']. Glowing at an apparent magnitude of 7.1, this cluster has several 10th-, 11th-, and 12th-magnitude stars in its outer reaches. These 30 stars appear to be herding the other 50 or so fainter stars into a tight ball of haze.

To me, open cluster NGC 659 (III1p) [1:44;+60d42'] looks somewhat like the letter X with one short leg. On the southwestern fringe of NGC 659 is the magnitude 5.8 very close double star 44 Cas (ADS 1344; β 1103) [1:43;+60d33']. Golden 44 Cas, along with the two fainter stars flanking it, appear to be pointing right at the 40-plus members of NGC 659.

Back on the trail of Cassiopeia's principal stars, we hop over to magnitude 3.38, spectrum B3 III, Epsilon Cas. Depending on the way the mythological picture of the queen is drawn, Epsilon marks either her legs or a leg of the throne. At 160 pc, Epsilon is not a part of the Cassiopeia cluster. If it were at the same 27 pc distance as the balance of the constellation, Epsilon would be the brightest of them all with its absolute magnitude of -2.9. About $\frac{1}{2}$ degree southeast of and partly obscured by the glow of Epsilon is the visual magnitude 12 planetary nebula IC 1747 (PK 130+01.1) [1:57;+63d20'].

Another nice triangle of open clusters is formed by the clusters NGC 637, NGC 609, and NGC 559. These clusters will be in the western side of your finder if you place Epsilon Cas on the eastern edge of your field. Observing these clusters will be a test of your skills in locating difficult objects. With about 20 members, NGC 637 (I3p) [1:42;+64d00'] is one of the smallest open clusters in the area and the 2,100 pc of space between us reduces its reddish glow to a magnitude of 8.2.

About $\frac{1}{2}$ degree northwest of NGC 637 is the smaller open cluster NGC 609 (II3r) [1:37;+64d33'], which is many times fainter than NGC 637 at magnitude 11 and appears like a dim red bulb out alone in the wilderness. Its dimness is due partly to its greater distance, being 3,100 pc from us.

Of these three clusters, NGC 559 (II2m) [1:29;+63d18'] is probably the easiest to locate. I see it as a misshaped five- pointed star with a hole in the center of it. This is an old open cluster with an estimated age of 1.3 billion years.

We have hopped around the main galactic clusters in the vicinity of the principal stars which make up the "W" of Cassiopeia. Many more clusters and nebulae, as well as double and multiple star systems, await your discovery within the boundaries of the Queen of Ethiopia and King Cepheus.

14 *December*

Pisces, Triangulum, and Aries: Of fishes, a triangle, and a ram

Oh palpitante plata de pez pulido y puro,
cruz verde, perejil de la sombra radiante,
luciérnaga a la unidad del cielo condenada,

"Pure fish of space, quicksilver, incandescent,
green cross or parsley light in the dark,
firefly fixed in the singleness of heaven,"

Pablo Neruda (1904–73),
"Cien sonetos de amor:
Noche"; Canto LXXVI,
("A Hundred Love Sonnets:
Night"; Canto LXXVI)
(lines 9–11),
1960

At the beginning of winter, we venture into the night to explore our last three constellations: Pisces the Fishes, Triangulum the Triangle, and Aries the Ram. Pisces and Aries are ancient zodiac constellations and have been known as representing fishes and a ram since the earliest days of celestial mythology. Triangulum was also well known to the Greeks and Egyptians, but they called it Deltoton, because of its resemblance to the fourth letter in the Greek alphabet and the Nile delta region of Egypt.

Star-hop in Pisces

We will start this month's star-hop, as shown in Figure 14.1, in Pisces, hop to Triangulum, and finish in Aries. As the 14th largest constellation, Pisces (Psc) covers 889.147 square degrees of sky as it stretches almost 40 degrees of declination from its northern border with Andromeda to its border with Cetus the Sea Monster and Aquarius the Water Pourer. The Winged Horse Pegasus flies along the western side of the fishes. Aries and Triangulum are to the east of Pisces. Centuries ago, when the concept of the zodiac and celestial coordinates were first devised, the vernal equinox occurred with the Sun crossing the celestial equator, on March 21, while in Aries. Due to the westerly precession of the sky, Pisces is now the first sign in the zodiac. This 0° position on the ecliptic, known as the "First Point of Aries" is now near the southwestern border between Pisces and

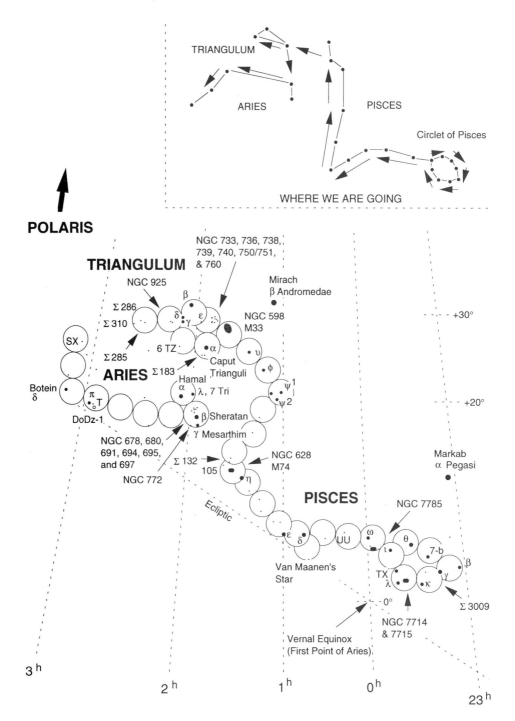

Figure 14.1
Star-hop in Pisces, Triangulum, and Aries.

Aquarius. This "First Point of Aries" slipped into Pisces over 2,000 years ago. The vernal equinox marks the intersection of the 0 hour of right ascension, the 0° circle of declination, and the ecliptic.

Pisces is depicted as being two fishes. Their origin is murky, but most scholars agree that they come from the Egyptian/Greek myth of a time when Aphrodite and Eros disguised themselves as fish and swam up the Nile to escape from the half man/half animal monster Typhon (also known as Typheus). Typhon had 100 dragon heads for fingers. Bursts of fire raged from his eyes. He was so tall that his head touched the stars. As the monster attacked Olympus, all the gods, except Zeus and Athena, fled. After a long and arduous battle, Zeus defeated Typhon with thunderbolts. To honor Aphrodite and Eros, their images as fishes, tied together by a cord, were placed in the sky.

The concept of a fish or a pair of fishes as celestial figures was also known in other ancient civilizations and was ascribed to the faint stars of this constellation. The Romans venerated the constellation as they felt it was their beloved goddess Venus. In the Christian faith, Pisces came to represent the loaves and fishes that Jesus used to feed the hungry. The early Chinese saw the stars as a pig and as the home of the Northern Emperor. After the introduction of western civilization into China, they changed their perception of this constellation to that of the fishes.

Pisces is poor in bright stars, but is rich in doubles and galaxies. Spectral types F, G, K, and M are the predominant stars in the area. Most of these stars are situated along the H–R diagram main-sequence line. Several of the brighter stars are giants. We will start our hop through Pisces at the head of the Southern Fish near the western boundary of Pisces and work our way to the knot in the cord at 113 Alpha Piscium, then north along the backbone of the Eastern Fish.

Locate the Great Square in Pegasus high overhead in your southern sky. Look for magnitude 4.53, spectrum B6 Ve, blue- white 4 Beta Psc at coordinates [23:03;+3d49'] 12 degrees south of Markab (54 Alpha Pegasi) [23:04;+15d12'], the southwestern bright white star in the Great Square of Pegasus. To the Arabs, Beta Psc was known as *Fum al Samakah* meaning the "Fish's Mouth." Spectral analysis of this star shows that it has a radial velocity of less than 1 km/s with a very slight easterly proper motion.

The head of the Southern Fish consists of an oval known as the "Circlet of Pisces" and is delineated by seven stars. The westernmost star is 6 Gamma Psc [23:17;+3d16']. Magnitude 3.69, MK standard for spectrum G9 III: Fe-2 stars, Gamma Psc is about 3 degrees east of Beta. With Beta on the western edge of your finder, you may be able to observe Gamma near the eastern side. The distance of 48 pc between us and Gamma is shrinking at a rate of 14 km/s.

Going clockwise from Gamma Psc, we hop 2 degrees southeast to the optical double, multiple, and variable star 8 Kappa Psc (S 830) [23:26;+1d15']. The primary for this system is a blue-white spectrum A0p:Cr; Si; Sr, magnitude 4.94 star. When first measured, in 1824 by James South, the magnitude 11.9 "B" star was separated from the primary by 150 arcsec in position angle 345°. By 1921, the star had moved 13 arcsec farther away and had shifted by 1 degree in position angle southwest. The magnitude 13.1 "C" star is located 76.1 arcsec from the "A" star in position angle 154°. These stars are located 30 pc from us and have a radial velocity of 3 km/s in approach. Kappa Psc is an example of Alpha[2] Canum Venaticorum-type (ACV) rotating variable stars. The magnitude change for Kappa is from 4.91 to 4.96 during a period of 0.05805 day.

The axial rotation of stars in the ACV-type variable class, combined with spots or other changes in the star's atmosphere, give them a non-uniform brightness. This type consists of main- sequence stars within the spectral classes of B8p to A7p and have strong magnetic fields. The magnetic field varies in conjunction with the rotation of the star. These stars have strong spectra lines for silicon (Si), strontium (Sr), chromium (Cr), and rare earths. Most stars in this type have periods of 0.5 to 160 days and their magnitudes fluctuate from 0.01 to 0.1.

The magnitude 6.25, yellow gG7 star southeast of Kappa is 9 Psc [23:27;+1d07']. This star is four times farther away (120 pc) from us than Kappa Psc and has a radial velocity of 11 km/s in approach.

About two-thirds of the way to 18 Lambda Psc is magnitude 5.68, spectrum DF0, yellow dwarf 16 Psc [23:36;+2d09']. This star leads us to the interacting galaxies NGC 7715 (VV51; Arp 284) and NGC 7714 [23:36;+2d09']. The irregular galaxy NGC 7715 [23:36;+2d09'] is the southern, closer, and fainter of these two objects. NGC 7714 is an SBbp spiral located 36.9 Mpc from us and appears slightly smaller than NGC 7715, which is located 36.3 Mpc from us. NGC 7714 is slightly more massive than its companion and about one absolute magnitude brighter at -20.03 versus -19.00.

The standard double star Σ 3009 (ADS 16730) [23:24;+3d43'] forms a right triangle with Kappa and Gamma Psc. The magnitude 6.8, spectrum K2 III, red giant primary is located about $2\frac{1}{2}$ degrees north of Kappa and 2 degrees northeast of Gamma. The absolute magnitude for the primary is greatly decreased from its estimated -0.1 by the tremendous distance of 230 pc between us. Since its first measure in 1829, the 8.8-magnitude comes has remained 7 arcsec from the primary in position angle 230°.

One degree southeast from 16 Psc is the magnitude 4.50, blue-white, spectrum A7 V, main-sequence star 18 Lambda Psc [23:42;+1d46']. This star marks the southeastern edge of the Circlet of Pisces.

About $1\frac{1}{2}$ degrees northeast from Lambda Psc is the pulsating slow irregular (LB-type) variable 19 Psc (TX Psc) [23:46;+3d29']. On the old spectral classification system, this star would be an N-type and is now classified as C5 II. TX Psc ranges from magnitude 6.9 to 7.7. This star is one of the very few carbon giant stars visible in small scopes or binoculars. The rare metallic element technetium (Tc) has been detected in this star's spectrum. This star is located 63 pc from us and has a radial velocity of 11 km/s in approach.

Head about 2 degrees north-northwest to the magnitude 4.13 open double star 17 Iota Psc [23:39;+5d37']. This spectrum F7 V, yellow system was first measured by S. W. Burnham, in 1879. The magnitude 12.8 comes is located 70 arcsec from the primary at position angle 294°. Iota Psc marks the northeastern edge of the Circlet. Of the stars forming the Circlet, Iota is the closest at 13 pc. This star has a radial velocity of 5 km/s in recession.

Three degrees northwest from Iota and forming the hump of the Fish's head is 10 Theta Psc [23:27;+6d22']. Theta is a yellowish-orange giant of spectral type K1 III. This star shines at magnitude 4.28. The radial velocity of Theta has been measured at 6 km/s in recession.

The last star in the Circlet of Pisces is another K2 III red giant, 7-b Psc [23:20;+5d22']. This star's apparent magnitude of 5.05 is reduced from its absolute magnitude of -0.1 by the 100 pc distance between us. Notice the 1-magnitude difference between the giants Σ 3009 (at 240 pc) and 7-b (at 100 pc), yet both stars have the same absolute magnitude. By comparing these stars, you can see the dimming effects of interstellar absorption.

From the Circlet of Pisces, we head east along the back of the fish to the yellow, spectrum F3 V, 4.01-magnitude, 28 Omega Psc [23:59;+6d51']. This star is located about $\frac{1}{4}$ degree west of the 0 hour of right ascension. The radial velocity for Omega is 2 km/s in recession.

One degree south-southwest of Omega Psc is the E5 elliptical galaxy NGC 7785 [23:55;+5d55']. At magnitude 11.59, this galaxy is one of the brightest in the area. It appears as an irregularly round object with a bright center.

About $3\frac{1}{2}$ degrees northeast of Omega is the ellipsoidal-type (ELL) rotating variable UU Psc (35 Psc; ADS 191; Σ 12) [0:14;+8d49']. Whitish subgiant UU Psc fluctuates from magnitude 6.0 to 6.3 during a period of 0.841678 day.

About 8 degrees to the east is the spectrum K5, magnitude 4.4, open double 63 Delta Psc [0:48;+7d35']. The magnitude 13 comes is located 132.2 arcsec from the primary in position angle 16°.

Van Maanen's Star (Wolf 28) [0:49;+5d22'] is a difficult to locate, but interesting, target located about 2 degrees south of Delta Psc. At a distance of 4.2 pc, this star is the 26th closest star to us. Dutch astronomer Adriaan van Maanen (1884–1946) discovered the nature of this star, in 1917. Being less than 10 pc from us, its visual magnitude is brighter at 12.4 than its absolute magnitude of 14.2, because this star is a low producer of energy. This degenerate white dwarf has one of the largest proper motions at 2.96 arcsec per year. Studies show that this star is about the diameter of the Earth (12,756.28 km; 7,930 miles), but contains 0.68 as much mass as the Sun. This star fits into a rare spectral class designated DG. With a temperature estimated at 6,000 K, this star was the first of the cool "late-type" dwarfs discovered. Van Maanen's star is one of the smallest stars we know of.

Follow the trail of faint stars northeast to magnitude 3.62, spectrum G7 IIIa, very close double 99 Eta Psc (ADS 1199; β 506) [1:31;+15d20']. The secondary star is difficult to separate being 1.0 arcsec from the primary and is faint at magnitude 10.6.

About 1 degree northeast of Eta is the face-on Sc spiral galaxy, NGC 628 (M74) [1:36;+15d47']. Méchain found this object in late September 1780. Messier logged it in as another "Nébuleuse sans étoiles" on his growing list. M74 is about midway between Eta Psc and reddish, 5.97-magnitude, spectrum K0 III, 105 Psc [1:39;+16d24']. Though its visual magnitude of 9.17 would indicate that this is a bright object, its light is spread out, which makes it difficult to see. Use a low-power eyepiece and averted vision at a dark site to observe this roundish galaxy. In deep-sky photographs, you can see thin dust lanes along the spiraling arms and can follow the arms into the core of the galaxy. You will not be able to see these dust lanes through your telescope. M74 is about 9.7 Mpc from us. Its absolute magnitude has been calculated at -20.32. Every second, M74 speeds 793 km away from us. This will be either our first or second target on the Messier Marathon.

About $1\frac{1}{2}$ degrees north of Eta Psc is the beautiful open multiple Σ 132 (ADS 1202) [1:32;+16d56']. Orbiting near the magnitude 6.9, spectrum G5, primary are five fainter stars that range from 9th to 14th magnitude. In 1921, the "B" star was 43 arcsec from the primary in position angle 348°.

Our last target in Pisces is 74 Psi^1 Psc (ADS 899; Σ 88) [1:05;+21d28'], which is an easy to separate binary triple system. It consists of a nice matched pair of blue-white stars and a faint second companion. The spectrum B9 V primary shines at magnitude 5.6. The magnitude 5.8, spectrum B9.5 IV, secondary is 30 arcsec from the primary in position angle 159°. The "C" star shines at magnitude 11.2 and is about 92 arcsec from the "A" star.

Star-hop in Triangulum

Encompassing 131.847 square degrees of sky, Triangulum (Tri) ranks as the 78th constellation in size. The area is populated with dozens of fine color-contrast double stars and bright galaxies. The Romans knew this area of the sky as Sicilia, since it reminded them of their island of Sicily. Hevelius tried to create the constellation Triangulum Minor from the stars 6 Iota, 10, and 12 Trianguli. His small constellation was not universally accepted, so the IAU dropped it when the official constellations were established in the 1920's. On January 1, 1801, Giuseppe Piazzi discovered the first and largest asteroid and named it "Ceres Ferdinandea" after Ceres, the patron goddess of Sicily, and Ferdinand I, King of Naples (1751–1825).

We will begin our tour of Triangulum at the last Messier object that we will observe in the regular star-hops: the late-type Sc galaxy NGC 598 (M33) [1:33;+30d39']. Messier came upon this glowing patch of light on August 25, 1764. The "Pinwheel Galaxy" gives the appearance of a sweeping "S" with a bright core and fading arms. Being only 0.7 Mpc (2.4 million light-years) from us, this member of the "Local Group" has undergone a tremendous amount of study. The galaxy contains many cepheid and irregular variables. The outer reaches of its four massive arms are dominated by hot young blue stars. Over 80 bright hydrogen II (H II) emission nebulae have been detected with several being given their own NGC number. The bright knot of the H II emission nebula NGC 604 can be seen on the northeast edge of the galaxy. The galaxy contains about 8 billion solar masses with 1.6 billion of that being neutral hydrogen. Spectroscopic

Photo 14.1 Spiral Galaxy NGC 598 (M33); the Pinwheel Galaxy.

study shows that the galaxy has a clockwise rotation and a blueshifted radial velocity of 2 km/s in approach.

M33 is a difficult object to locate the first time, due to its very low surface brightness and large angular size, which cause it to blend into the background sky glow. To locate M33, line up Mirach (43 β And)

and 37 Mu And. M33 is about double the distance southeast from Beta And as the distance between these two stars. The galaxy is also about 5 degrees west and 1 degree north of 2 Alpha Tri [1:53;+29d34']. Use a wide-field and very low-power eyepiece to get the best view of this large and diffuse object. Appearing about as big as the Moon, yet many times fainter, sweep the area where M33 is and locate it by noticing the change in the darkness of the sky in your eyepiece. You will have M33 when you can detect its foggy-like gray glow against the darker sky. Take a long look at M33 and try to pick out the individual gas clouds and knots of stars in its arms. These can be viewed in a large scope on a dark night.

The second brightest star in Triangulum is Caput Trianguli (2 Alpha Tri), which is Latin for "Head of the Triangle." To the Arabs, this 3.41-magnitude, spectrum F6 IV, open multiple star was *Rās al Muthallath*. In 1909, the magnitude 12.7 "B" star was 85 arcsec from the primary in position angle 309° and the magnitude 12 "C" star was in position angle 182°. The "C" star remained in the same position angle, but had moved from a separation of 228.3 arcsec, in 1879, to be 222.3 arcsec in 1909. The primary is also a spectroscopic binary. The unseen comes has an orbital period of 1.73565 days. Alpha is 18 pc from us and has a radial velocity of 13 km/s in approach.

About 1 degree southeast of Alpha Tri is the yellow and blue multiple Σ 183 (ADS 1522) [1:55;+28d48']. The yellow, spectrum F2, primary and the magnitude 8.7 blue "C" star are 5.6 arcsec apart. A magnitude 8.4 visual binary with an orbital period of 386 years is too close to the "A" to be seen in small scopes.

Hop about $3\frac{1}{3}$ degrees from Alpha Tri to the magnitude 5.5, spectrum A2 III, close double 3 Epsilon Tri (ADS 1621; Σ 201) [2:02;+33d17']. The blue-white primary and its white magnitude 11.4 companion are 3 arcsec apart. Look for the secondary in position angle 118°.

About one degree west of Epsilon is the yellowish, spectrum G5, magnitude 7.2 open double star at [1:59;+33d13'] and listed in Burnham's 1913 catalogue of double stars. The 12.2-magnitude comes is located 129 arcsec from the primary in position angle 139°. This double, along with another double (ADS 1548; A 819) [1:57;+31d01'] located about $\frac{1}{4}$ degree farther west, are visually within a small cluster of faint galaxies.

This galaxy cluster consists of NGC 733, NGC 736 (6Z 111), NGC 738, NGC 739, NGC 740, NGC 750/751 (VV 189; Arp 166), and NGC 760. These galaxies range from 12th to 14th magnitude and are a fine collection of elliptical to barred spirals. NGC 750/751 [1:57;+33d12'] consists of two magnitude 12, E-type galaxies in close contact. You can see the separate cores and the intermixing of their outer edges.

The brightest star in Triangulum is magnitude 3.0, spectrum A5 III, 4 Beta Tri [2:09;+34d59']. Together with Alpha, these stars were known by the Arabs as *Al Mīzān* which means "The Scale-Beam." Beta is a double-line spectroscopic with an orbital period of 31.3884 days.

Photo 14.2 Spiral galaxy NGC 925 in Triangulum.

The last of the trio of stars is 9 Gamma Tri [2:17;+33d50'], which is located about 2 degrees south-southeast of Beta. This bluish A0 IV-Vn main-sequence star shines at magnitude 4.01. Gamma has a radial velocity of 14 km/s in recession and is 46 pc from us.

About 2 degrees southeast of Gamma is the S(B)c-type galaxy NGC 925 [2:27;+33d35']. This magnitude 10.1 spiral galaxy is located in a field of tenth magnitude foreground stars, but appears as an elongated patch among them. The core is bright with a halo of nebulosity around it. This galaxy has a radial velocity of 716 km/s in recession. NGC 925 is 9.4 Mpc from us and has an absolute magnitude of -19.66.

Sweep $2\frac{3}{4}$ degrees east and about $\frac{1}{4}$ degree south to the very close double Σ 285 (ADS 2004) [2:38;+33d25']. The spectrum K0, magnitude 7.5 primary and 8.2-magnitude secondary are separated by 1.7 arcsec. Look for the secondary in position angle 167°. When F. G. W. von Struve first measured these stars, in 1832, the secondary was in position angle 178°.

Half a degree north-northeast of Σ 285 is the close double Σ 286 [2:39;+33d57']. The magnitude 7.9, spectrum G5 primary and its 10.2-magnitude secondary are in the same relative positions as when von Struve first measured them in 1830. Their separation is 2.9 arcsec with the secondary in position angle 254°.

For the last color-contrast multiple system we will hop to in Triangulum, return to Alpha Tri, then move about 4 degrees northeast to the close double and unclassified variable 6 Tri (ADS 1697; Σ 227; and TZ Tri) [2:12;+30d18']. Both the magnitude 5.3, spectrum G5 III, golden primary and the magnitude 6.9, spectrum F5 IV, bluish secondary are each double-line spectroscopic binaries. These two stars have a separation of 3.9 arcsec, which is only 0.1 arcsec farther apart than when von Struve first measured them, in 1836. For the primary, its spectroscopic companion has an orbital period of 14.732 days, which means the semimajor axis of the companion's orbit is about 11.27 million km. The companion of the "B" star is even closer at 2.89 million km. Its period is 2.236 days. Eighty-five parsecs separate us. This system has a radial velocity of 18 km/s in approach.

Star-hop in Aries

The last constellation we will visit on our regular star-hops is Aries (Ari). This asterism, which makes up the 39th largest constellation, has been known as a ram since ancient times. The Greeks knew it as such an animal and the Romans gave it the name we know it by. To many other ancient cultures, the ram was held in very high regard, so it seems fitting that this animal was used as the first sign of the zodiac.

During March and April, when the Sun is in Aries, Jews celebrate the festival of Passover. This holiday commemorates the release of the Hebrew slaves from Egypt about 4,000 years ago. As part of that first Passover, lambs were sacrificed and their blood was smeared on the door posts of the slave's homes to protect them from the Angel of Death. The slave's use of lambs may have been a way of mocking the greatest of Egyptian gods – the Sun god Amum-Re (also known as Ammon), a feathered-crowned ram. To the ancient Egyptians, the ram was the symbol of fertility and power.

Like eight other constellations, Aries is tied to the legend of Jason and the Argonauts. The Golden Fleece at one time had been Aries' hide. The legend begins when Athmamas, a king of Thebes, was tricked by his second wife, Ino (a daughter of Cadmus), into sacrificing the children of his first marriage. Ino had the women of her land roast their corn seeds before sowing the seeds. The crop naturally failed. Ino bribed the oracle of Delphi to inform Athmamas that his son Phrixus and daughter Helle had to be sacrificed to Zeus in order to rescue the crop. The god was upset by this and sent a ram with a golden fleece to rescue the children. They flew away from Thebes on the back of the ram. On the trip, Helle fell off the ram and drowned at the place now known as the Hellespont. Phrixus reached the safety of King Aeetes of Colchis. There, he sacrificed the ram and gave the fleece to King Aeetes. The fleece was hung on an oak tree limb in the Grove of Ares and was guarded by a terrible serpent. When Jason and his crew arrived in Colchis, he asked Aeetes for the fleece. The king refused. The king's daughter, Medea, fell in love with Jason. At the garden, she bewitched the serpent while Jason grabbed the fleece. Upon his return to Greece, Jason gave the fleece to the temple of Zeus at Orchomenus.

Hop about 6 degrees south from 6 Tri to magnitude 2.00, spectrum K2 IIIab: Ca-1, Hamal (13 Alpha Arietis) [2:07;+23d27']. Hamal (pronounced HAM-al) originally was the Arabic name for the entire constellation, but derives from their words *Al Rās al Ḥamal* and means "The Head of the Sheep."

About $2\frac{1}{4}$ degrees west of Hamal is the open multiple 9 Lambda Ari (ADS 1563; H V 12) [1:57;+23d35']. Sir William Herschel first measured the white 4.9-magnitude "A" and pale blue magnitude 7.7, "B" components in 1781. Their separation is 37.4 arcsec. The "B" star is in position angle 46°.

One half degree due west of Lambda is the EA-type variable 7 Ari (RR Ari) [1:55;+23d34']. This reddish-orange giant of spectral type K0 III fluctuates between magnitude 6.42 and 6.84 during a period of 47.9 days.

About 2 degrees south-southwest of 7 Ari is the color- contrast close double 1 Ari (ADS 1457; Σ 174) [1:50;+22d16']. The deep-yellow, spectrum F5 primary shines at magnitude 6.2. The white, spectrum A2, magnitude 7.4, comes is located 2.8 arcsec from the primary in position angle 166°. This system is 54 pc from us and has a radial velocity of 1 km/s in recession.

Forming a "V" to the east and south of 1 Ari is a cluster of faint galaxies consisting of NGC 678, NGC 680, NGC 691, NGC 694, NGC 695, and NGC 697. All of these galaxies are faint at 12th to 14th magnitudes. The Sbb-type spiral NGC 678 [1:49;+22d00'] and the Sop-type galaxy NGC 680 [1:49;+21d58'] are very close and can be seen in the same low-power eyepiece. NGC 678 appears slightly elongated and NGC 680 is round with a bright core. NGC 691 [1:50;+21d46'] is a Sb+ galaxy and is very diffuse, which makes it hard to see even though it has the brightest magnitude at 12.39 of this cluster of galaxies.

About 1 degree southeast of 1 Ari is the blue-white, 2.64-magnitude, spectrum A5 V, Sheratan (6 Beta Ari) [1:54;+20d48']. Along with Mesarthim (5 Gamma^2 Ari) [1:53;+19d17'], these stars marked the horns of the Ram and were known as *Al Sharaṭain* to the Arabs. In 1903, H. C. Vogel discovered that Sheratan (pronounced SHARE-ah-tan) is a spectroscopic binary. The comes has an orbital period 106.9973 days. The components of this pair have a mean separation of only 21.4 to 32.1 million km. This main-sequence star has about twice the mass as the Sun and its luminosity is 13 to 17 times brighter than our star. Sheratan is 14 pc from us and has a radial velocity of 2 km/s in approach.

Mesarthim or Mesartim (pronounced may-SHAR-thim) (5 Gamma^2 Ari) was one of the first double stars to be discovered. On February 8, 1665, British astronomer Robert Hooke (1635–1703) noticed the two components of Gamma Ari (ADS 1507; Σ 180) while tracking Comet Hevelius. The magnitude 4.68, spectrum A1 p:Si, northern star is Gamma^1. Magnitude 4.56, spectrum B9 V, Gamma^1, is an Alpha^2 Canum Venaticorum-type rotating variable. This magnetic star's brightness fluctuates by about 0.2 magnitude during a period of 2.6095 days. Another unusual feature of this pair of stars is that Gamma^1 is coming toward us at a rate of 4 km/s whereas Gamma^2 is receding at 1 km/s. These stars have an orbital period of five to ten thousand years. The data for these two stars have frequently been interchanged, so you may find them with the values for Gamma^1 being listed as those for Gamma^2.

Hop about $1\frac{1}{2}$ degrees southeast to locate the Sb-type galaxy NGC 772 (Arp 78) [1:59;+19d01']. In the same low-power eyepiece you may be able to see the 14th-magnitude, E3-type galaxy NGC 770 [1:59;+18d57'] to the south. NGC 772 glows softly at magnitude 10.33. We view this elongated spiral about three-quarters edge on. Bright H II regions near the bright core can be seen in large scopes. NGC 772 has a redshifted spectrum indicating that it is receding at a radial velocity of 2,570 km/s.

Sweep about 12 degrees east and 2 degrees south to magnitude 5.22, spectrum B6 IV, white subgiant, and triple system 42 Pi Ari (ADS 2151; Σ 311) [2:49'+17d28']. The primary is also a spectroscopic binary with an orbital period of 3.854 days. The three bright components for this system are almost in a straight line as the 8.7-magnitude "B" star is located in position angle 120° and the 10.8-magnitude "C" is at position angle 110°. The "A" and "B" stars have a separation of 3.2 arcsec. The "C" star is 25.2 arcsec from the primary. This system is 190 pc from us and has a system radial velocity of 9 km/s in recession.

To the west of Pi Ari is the pulsating SRA-type variable T Ari [2:48;+17d31']. This M6e to M8e spectral type star can be as bright as magnitude 7.5 or as faint as 11.3. Its period is 316.6 days. At a magnitude fainter than 14.5 you may be able to observe SU Tri-Nova 1854 [2:48;+17d22'] about $\frac{1}{8}$ degree south-southeast of T Ari.

About $\frac{1}{2}$ degree southwest of Pi Ari is the open cluster Dolidze-Dzimselejsvili 1 (DoDz-1) (III2p) [2:47;+17d12']. DoDz-1 is large in area, but contains only about 12 stars brighter than ninth magnitude. The stars are very loose and form a parallelogram-shaped asterism.

Magnitude 4.35, spectrum K0 III, Botein (57 Delta Ari) [3:11;+19d43'] marked Aries' belly on some old star charts, but today it would stand for the ram's tail. The name is derived from the Arabic *Al Buain* meaning "The Belly." This star is 78 pc from us and has a radial velocity of 3 km/s in approach.

The last stop on our regular star-hops is at the rotating variable 56 SX Ari [3:12;+27d15']. Look for this star $7\frac{1}{2}$ degrees due north of Botein. This star is the prototype for SX Ari-type (SXARI) variables also known as helium variables. The spectra for these high-temperature stars show intense lines of ionized helium (HeI), ionized silicon (SiIII), and other elements. These stars also show strong magnetic fields that vary as the star rotates. The intensity of the magnetic field and changes in the brightness of the star, which may be due to spots, coincide in a period lasting 0.7278925 day.

We have come full circle through the year and are now only a few degrees west of our starting point at the Pleiades in Taurus. It is my sincere hope that you have enjoyed all of the regular star-hops presented in this book and that you will continue to explore the wonders of the night sky and develop your own star-hops. Good luck and happy star-hopping.

15 *Messier Marathon*
A sundown to sunup hop across the skies

How lovely is this wildered scene,
  As twilight from her vaults so blue
Steals soft o'er Yarrow's mountains green,
  To sleep embalmed in midnight dew.

All hail, ye hills, whose towering height,
  Like shadows, scoops the yielding sky!
And thou, mysterious guest of night,
  Dread traveller of immensity!

Where hast thou roamed these thousand years?
  Why sought these polar paths again,
From wilderness of glowing spheres,
  To fling thy vesture o'er the wain?

James Hogg (1770–1835)
"To the Comet of 1811"
(lines 1–8 and 25–28),
1811

The work of comet-hunter and nebulae cataloguer Charles Messier comes alive in March of each year as amateur astronomers participate in a one night search for all of the objects in his catalogue of nebulae and star clusters. By a quirk of fate, we are fortunate that most of the objects Messier and Méchain took 24 years to discover, we can observe in one night around the time of the vernal equinox, on March 21.

All of the star-hops you have taken so far in this book have led you to this point in your quest up the mountain. We have already hopped to 87 of Messier's objects and we will look for the rest (23) tonight. Those objects already covered in the preceding chapters will only be mentioned in the suggested order you need to look for them. The other objects will be discussed in this chapter as we hop to them.

For this star-hop we will make a sundown to sunrise dash to see all of them, with the possible exception of the globular cluster M30 in Capricornus. The positions and search order are as shown in Figure 15.1, and as listed in Table 15.1.

Though I have mentioned their names in many places, this seems the most appropriate spot to tell you who Charles Messier and Pierre Méchain were. Charles Messier was born in Badonviller in Lorraine, on June 26, 1730. He was the tenth of 12 children. From an early age, Messier was interested in the night sky. At the age of 21, he left home to seek work in Paris.

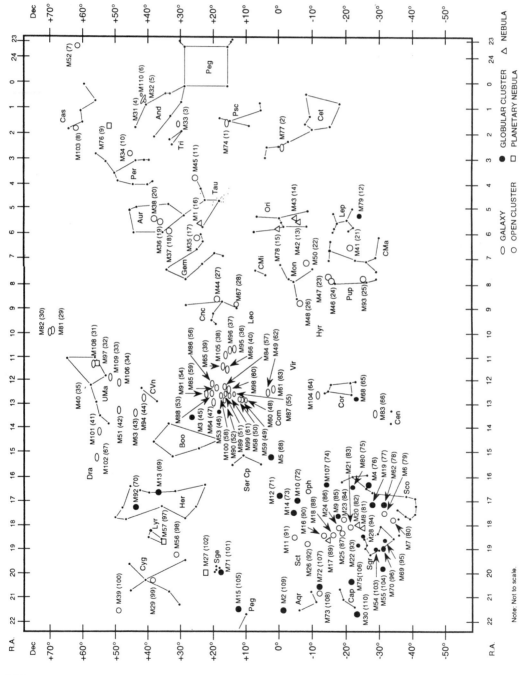

Figure 15.1 The Messier Marathon.

Astronomer of the Navy, Joseph-Nicholas Delisle (1688–1768), had set up an observatory in the Hôtel de Cluny in Paris and was in need of an assistant to take notes and to plot the objects as viewed through the observatory's telescopes. He hired Messier and began to instruct him in keeping accurate records on everything observed and to plot the object's exact celestial coordinates.

Sometime around 1754, Messier fell in love with hunting for comets. He saw the discovery of comets as his path to fame. Edmond Halley had predicted the famous comet would return in 1758. Messier began to search for it in 1757 and located a comet in early 1758, but it was not Comet Halley. On December 25, 1758, the German farmer and amateur astronomer Johann Georg Palitzch (also known as Baron de Zach) (1723–88) made the first sighting of the returning comet.

On January 21, 1759, Messier succeeded in locating the incoming Halley's Comet in the general area where it was scheduled to be. For unknown reasons, Delisle would not allow Messier to announce this sighting. Honoring his employer's orders, yet bitter toward him for the rest of his life, Messier withheld the news. He continued to observe the comet and plotted its course every clear evening.

While searching for Halley's Comet and tracking the 1758 comet, he came across a new hazy patch of light near 123 Zeta Tauri, now known as NGC 1952; M1; "the Crab Nebula." He plotted the object's coordinates and went on to other tasks.

In 1760, Delisle retired from his post and Messier was allowed to take over the observatory, but not Delisle's title or salary. Lacking formal scientific training, Messier struggled for academic recognition from professional astronomers. He plowed into the field of comet hunting and discovered several more during the next few years. Slowly, he gained the recognition he sought. His contacts outside of France got him memberships in prestigious scientific societies in Great Britain and Russia.

Though what is now listed as M1 was first seen by Messier, in 1758, the real genesis for his famous list came in May 1764. Messier came across another nebulous object, this time in Aquarius, and thought it was a new comet. After realizing the object was not moving like a comet, he decided to make a list of any more of these comet-like objects so that he would no longer be confused by them.

Messier decided to research the astronomical catalogues and atlases for any more of these objects and to add them to his new list. From the records of his predecessors, then by scanning the skies, he was able to locate 22 of their objects by January 1765. While searching for these objects, he also came across 17 new ones to add to the list. With his list set at 41 objects, he put the project aside for four years.

In 1769, Messier decided to publish his list of comet-like objects. For unknown reasons, he quickly added four more objectsto the list to make it an even 45. These four, M42 through M45, were already well known and it seems ridiculous they were put on a list of nebulous objects that could easily confused as being a comet. I doubt if anyone would mistake the Pleiades (M45), or the Great Nebula in Orion (M42) for comets. During this time, Messier also discovered another comet.

Table 15.1. *Messier Marathon search list.*

Date _____

Order	M#	Hop from	Object type	Mag. (V)	Size (arcmin)	Page	Time observed
1	M74	Eta Psc	S-gal	9.2	10 x 9	235	....
2	M77	Delta Cet	S-gal	8.8	7 x 6	253	....
3	M33	Alpha Tri	S-gal	5.7	62 x 39	236	....
4	M31	Mu And	S-gal	3.4	178 x 63	208	....
5	M32	M31	E-gal	8.2	8 x 6	209	....
6	M110	M32	E-gal	8.0	17 x 10	209	....
7	M52	Beta Cas	OC	6.9	13	225	....
8	M103	Delta Cas	OC	7.4	6	229	....
9	M76	Phi Per	PIN	10.1	2 x 1	211	....
10	M34	Gamma And	OC	5.2	35	207	....
11	M45	—	OC	1.2	110	62	....
12	M79	Beta Lep	GC	8.0	9	254	....
13	M42	Theta Ori	DN	4.0	66 x 60	77	....
14	M43	M42	DN	9.0	20 x 15	74	....
15	M78	Zeta Ori	DN	8.0	8 x 6	77	....
16	M1	Zeta Tau	SNR	8.4	6 x 4	67	....
17	M35	1 Gem	OC	5.1	28	254	....
18	M37	Theta Aur	OC	5.6	24	255	....
19	M36	M37	OC	6.0	12	256	....
20	M38	M36	OC	6.4	21	256	....
21	M41	Alpha CMa	OC	4.5	38	84	....
22	M50	Theta CMa	OC	5.9	16	257	....
23	M47	Gamma CMa	OC	4.4	30	87	....
24	M46	M47	OC	6.1	27	86	....
25	M93	Xi Pup	OC	6.2	22	85	....
26	M48	Zeta Mon	OC	5.8	54	257	....
27	M44	Delta Cnc	OC	3.1	95	93	....
28	M67	Alpha Cnc	OC	6.9	30	92	....

Table 15.1. (cont.)

29	M81	24 DK UMa	S-gal	6.8	26 x 14	177
30	M82	M81	Ir-gal	8.4	11 x 5	177
31	M108	Beta UMa	S-gal	10.0	8 x 2	116
32	M97	M108	PlN	9.9	3	117
33	M109	Gamma UMa	S-gal	9.8	8 x 5	118
34	M106	Chi UMa	S-gal	8.3	18 x 8	256
35	M40	Delta CMa	DS	8.0	—	119
36	M95	Alpha Leo	S-gal	9.7	7 x 5	98
37	M96	M95	S-gal	9.2	7 x 5	96
38	M105	M96	E-gal	9.3	4 x 4	99
39	M65	Theta Leo	S-gal	9.3	10 x 3	101
40	M66	M65	S-gal	9.0	9 x 4	101
41	M101	Eta UMa	S-gal	7.7	27 x 26	114
42	M51	Eta UMa	S-gal	8.1	11 x 8	51
43	M63	M51	S-gal	8.6	12 x 8	258
44	M94	M63	S-gal	8.1	11 x 9	258
45	M3	Beta Com	GC	6.4	16	126
46	M53	Alpha Com	GC	7.7	13	124
47	M64	35 Com	S-gal	8.5	9 x 5	125
48	M60	Epsilon Vir	E-gal	8.8	7 x 6	136
49	M59	M60	E-gal	9.8	5 x 3	136
50	M58	M59	S-gal	9.8	5 x 4	136
51	M89	M58	E-gal	9.8	4	136
52	M90	M89	S-gal	9.5	10 x 5	136
53	M88	M90	S-gal	9.5	7 x 4	131
54	M91	M88	S-gal	10.2	5 x 4	131
55	M87	M89	E-gal	8.6	7	137
56	M86	M87	E-gal	9.2	7 x 6	137
57	M84	M86	E-gal	9.3	5 x 4	137
58	M100	M84	S-gal	9.4	7 x 6	128
59	M85	11 Com	E-gal	9.2	7 x 5	130

Table 15.1. (*cont.*)

Order	M#	Hop from	Object type	Mag. (V)	Size (arcmin)	Page	Time observed
60	M98	6 Com	S-gal	10.1	10 x 3	128	....
61	M99	6 Com	S-gal	9.8	5	128	....
62	M49	20 Vir	E-gal	8.4	9 x 7	138	....
63	M61	17 Vir	S-gal	9.7	6 x 5	138	....
64	M104	Eta Crv	S-gal	8.3	9 x 4	109	....
65	M68	Beta Crv	GC	8.2	12	105	....
66	M83	Pi Hya	S-gal	7.6	11 x 10	260	....
67	M102	Iota Dra	E-gal	8.1	17 x 9	180	....
68	M5	5 SerCp	GC	5.8	17	141	....
69	M13	Eta Her	GC	5.9	17	260	....
70	M92	Pi Her	GC	6.5	11	262	....
71	M12	Delta Oph	GC	6.6	14	262	....
72	M10	M12	GC	6.6	15	262	....
73	M14	M10	GC	7.6	12	262	....
74	M107	Zeta Oph	GC	8.1	10	263	....
75	M80	Delta Sco	GC	7.2	9	153	....
76	M4	Alpha Sco	GC	5.9	26	152	....
77	M19	Alpha Sco	GC	7.2	14	154	....
78	M62	M19	GC	6.6	14	154	....
79	M6	Lambda Sco	OC	4.2	15	157	....
80	M7	M6	OC	3.3	80	156	....
81	M8	Lambda Sgr	DN	5.8	90 x 40	163	....
82	M20	M8	DN	8.5	29 x 27	164	....
83	M21	M20	OC	5.9	13	164	....
84	M23	M21	OC	5.5	27	165	....
85	M9	Eta Oph	GC	7.9	9	265	....
86	M24	Mu Sgr	SC	4.5	90	165	....
87	M25	M24	OC	4.6	32	165	....

Table 15.1. *(cont.)*

88	M18	M25	OC	6.9	9	165
89	M17	M18	DN	7.0	46 x 37	165
90	M16	M17	OC	6.0	7	166
91	M11	Beta Sct	OC	5.8	14	172
92	M26	Delta Sct	OC	8.0	15	171
93	M22	Lambda Sgr	GC	5.1	24	163
94	M28	Lambda Sgr	GC	6.9	11	163
95	M69	Epsilon Sgr	GC	7.7	7	198
96	M70	M69	GC	8.1	8	162
97	M57	Beta Lyr	PIN	9.0	1	197
98	M56	Beta Cyg	GC	8.2	7	198
99	M29	Gamma Cyg	OC	6.6	7	191
100	M39	Alpha Cyg	OC	4.6	32	190
101	M71	Gamma Sge	GC	8.3	7	200
102	M27	Gamma Sge	PIN	8.1	8 x 4	202
103	M54	Zeta Sgr	GC	7.6	9	162
104	M55	Zeta Sgr	GC	6.3	19	162
105	M15	Epsilon Peg	GC	6.4	12	265
106	M75	Beta Cap	GC	8.6	6	166
107	M72	Alpha-2 Cap	GC	9.4	6	266
108	M73	M72	A	9.0	3	266
109	M2	Beta Aqr	GC	6.5	13	266
110	M30	Gamma Cap	GC	7.5	11	267

A = asterism
DN = diffuse nebula
DS = double star
E-Gal = elliptical galaxy
GC = globular cluster

Ir-Gal = irregular galaxy
OC = open cluster
PIN = planetary nebula
S-Gal = spiral galaxy
SC = star cloud
SNR = supernova remnant

249

In hopes of gaining admission to more scientific societies, Messier sent word of the 1769 comet to his King Louis XV (1710– 74), and the Prussian King Frederick II (1712–86). Again he was turned down for a position in the French Académie des Sciences, but Frederick secured him a membership in the Berlin Academy of Sciences.

The recognition Messier had sought came after his discovery in 1770 of another comet. The Paris Academy finally admitted him as a member. The next year his list of 45 objects was published in the *Mémoirs de l'Académie des Sciences*. The king also appointed him to the post of Astronomer of the Navy, Delisle's old position, which had been denied to Messier when his mentor had retired.

During the 1770's Messier busied himself with his official duties and kept up his search for comets and those pesky non- comet things. He found several more comets and nebulae. Being poor in math, Messier sought out the assistance of others to do his astronomical calculations. In 1774, he was introduced to a mathematician in the Department of the Navy, Pierre Méchain. A strong bond of friendship quickly developed between these two men. Méchain discovered 28 of the objects on Messier's list.

After the publication of the first 45 objects, Messier slowly added more as he and Méchain discovered them. In 1780, they were up to 68 objects and the revised list was published in the 1783 edition of the French almanac *Connaissance des Temps*. This book was published three years in advance to allow time for its distribution to the far flung parts of the empire.

In 1781, Messier's list had reached 100 objects. He decided to publish this final list in that year's edition of the *Connaissance des Temps*, but just prior to submitting the list for publication Méchain discovered three more objects. These were quickly added to the list, and for the first time Messier did not verify his friend's discoveries. Later, it turned out that M102 was probably a rediscovery of M101 and then again maybe it was a new object. No one has ever been sure, since Méchain's coordinates were not checked at the time of discovery. Astronomers have speculated as to what really happened and if M101 and M102 are in fact the same object. The 1781 list was the final one published by Messier. The final seven objects (M104 through M110) were added later.

In April 1781, word reached Messier of the discovery of a strange object in the constellation Gemini. Nevil Maskelyne (1732–1811), the Astronomer Royal at Greenwich, asked Messier to observe this object and to if he could figure out what an amateur named William Herschel had discovered. Messier observed the object for several days and sent his notes to his friend and calculator Jean-Baptiste-Gaspard de Saron (also known as Bochart de Saron) (1730–94). The object's circular orbital elements showed that it was too far out from the Sun to be a comet. Messier and De Saron assisted in the confirmation of the discovery of the planet Uranus; the first such body to be discovered during historic times.

De Saron was a renowned instrument and optical glass maker. His skills at hunting comets had brought him in contact with Messier.

When the Reign of Terror had its firm grip on France during the French Revolution, De Saron was among the handful of scientists to die on the guillotine.

Messier continued his official work, until going into semi- retirement in 1808, at the age of 78. He received several honors, including the Legion of Honor, bestowed on him by Napoléon Bonapart (1769–1821) in 1806. Messier responded to this honor by naming a comet he had discovered in 1769 for the Emperor. The comet still also carries the name Messier 1769. In 1815, Messier suffered a partially paralyzing stroke. On the night of April 12, 1817, the "Ferret of Comets" passed away.

This man's accomplishments in the field of astronomy were many. Though he discovered or co-discovered 16 comets, he will be best remembered for his popular list of nebulae and clusters he did not want to see. A series of lunar craters in the Mare Fecunditatis were named for him. One of the craters has diverging streamers of material to its west side which gives the crater the appearance of being a comet heading toward the Moon's eastern equatorial limb.

The Hôtel de Cluny still stands at 24, rue du Sommerard, in the Latin Quarter of Paris (Arrondissement 5) and is now called Musée des Thermes et de l'Hôtel de Cluny. The museum is open to the public. Messier's observatory was removed sometime in the early 19th century.

Pierre Méchain was for a time a student of Joseph-Jérôme Le Français Lalande (1732–1807). In 1774, at the age of 20, he received a position as a calculator and surveyor in the Department of the Navy. Méchain discovered eight comets, several of which turned out to be periodic and now bear the names of the later observers who determined the periodic nature of these comets.

In 1792, Méchain and Jean-Baptiste-Joseph Delambre (1749–1822) were sent by the French Academy of Sciences to measure the arc of the meridian from Dunkirk to Barcelona. Their work was to establish the exact length of the meridian running north and south through Paris. A committee from the French Academy had decided the length of the new unit of measurement, to be known as a metre (meter), was to be set as being one-ten-millionth of the length of this meridian.

Delambre stayed in France and headed south from Dunkirk while Méchain traveled to Spain and began to work his way north. Méchain and his crew were arrested and held in a Spanish jail when war broke out between Spain and France. They were released after a few days, given their equipment back, and resumed work.

With his work done, Méchain left Spain, lived in Italy for about two years, and returned to Paris in 1795. The geodetic data collected by the Delambre–Méchain team were used by the French Academy to establish the length of a meter.

The length of a meter had to be fixed first as the metric system of weights and measures is based on this one measurement. The basic unit of mass, the gram, was established as being equal to the mass of a cubic centimeter (cm^3) of pure water at 4°C. The liter is the volume of a cube 10 cm on all sides. The entire metric system was officially adopted as the standard for France, in June 1799.

Worried that some of his measurements were incorrect, Méchain refused to publish his data. In 1803, he convinced Napoléon of the need for further surveys along the meridian, and was given permission to make the necessary measurements. Méchain contacted yellow fever during this work and died at Castillion de la Plana in Spain on September 20, 1804.

Outside the world of astronomy, Pierre Méchain is known only for his work with Delambre. Without his assistance and excellent observational and mathematical skills, Messier's famous list of 110 nebulae and star clusters would contain only 82 objects. No lunar feature has been named in Méchain's honor.

A Messier Marathon

To prepare for a Messier Marathon you need no additional equipment or observational skills than what you have already acquired as you progressed through the monthly star-hops. You will need to find a dark site that has a clear view of the southeastern horizon. The last objects you will be searching for rise just before dawn in the southeast, and, if you are going to have any chance to see them, you have to have an unobstructed view.

Plan to make your Messier Marathon as close to the vernal equinox (March 21) as possible. If you attempt your marathon early in March, the morning objects are difficult to find. When trying a marathon after the equinox, the evening objects are lower in the western sky, making them harder to find. March 21 is simply our target date as weather, the Moon, and your daily schedule may make it difficult for you to do a marathon on that date.

You will need to consult a lunar table to determine the nights of a New Moon. A thin crescent Moon a few days after or before New Moon will only affect your viewing for a short time after sundown or before sunrise. A Moon older than a few days will remain in the sky too long and will be too bright for you to see all of the objects.

Know the times of sunset and sunrise at your latitude, so you will know when to be ready and approximately how much time you will have near dawn to locate the last few objects. If you do not have lunar or solar tables, you can usually find this information in your local newspaper. On March 21, sunset at +40° latitude is around 6:10 p.m. with astronomical twilight ending about 90 minutes later. Morning astronomical twilight on the 22nd begins around 4:30 a.m. Sunrise will occur around 6:00 a.m.

Plan to arrive at your chosen site before sundown to allow plenty of time to set up your equipment. Be sure your finder and eyepiece are as closely aligned as possible. March nights are long and cold. Take plenty of refreshments and warm liquids with you and dress for cold weather.

The one basic thing to remember is to work your way easterly across the sky. In the event you are unable to locate an object, do not despair, just go on to the next one on the search list (Table 15.1). The objects are listed in the order you should attempt to find them in. The

"Hop from" column gives you the easiest star or Messier object to hop from to find the next object. In many cases, the target object will be in the same finder field as the "hop from" object. Consult your star atlas or maps for the exact locations of both objects. In Figure 15.1, the number in parentheses following the Messier number is the search order number. Record the time you positively identify the target object. The visual magnitudes listed in Table 15.1 are rounded up to the nearest decimal point.

You can search for M77 first and M74 second instead of the way I have listed them. Either way, you need to begin to search for them as soon as the first stars begin to appear.

As the Sun dips into the western horizon, be on the look out for Hamal (13 Alpha Arietis) at coordinates [2:07;+23d27'] and Sheratan (6 Beta Arietis) [1:54;+20d48']. These will be the first bright western stars to appear out of the twilight haze. Be aware that you may also encounter the hazy glow of the zodiacal light. This phenomenon is caused by dust in the ecliptic plane reflecting sunlight. The zodiacal light is more prominent in March than at other times of the year.

Use yellow, 2.0-magnitude Hamal and blue-white, magnitude 2.64 Sheratan to point you southwesterly to yellow, 3.63 magnitude 99 Eta Piscium [1:31;+15d20']. Our first Messier object is the 9.17-magnitude spiral galaxy M74, which is located about $1\frac{1}{3}$ degrees northeast of Eta Piscium. When you find M74, note the time on the search list and rush south to the constellation of Cetus the Sea Monster (also called the Whale) to find M77, another faint spiral galaxy. Time is very short as the first half dozen objects set into the western twilight and horizon soon after the Sun.

The Seyfert Sbp-type spiral galaxy M77 (NGC 1068) [2:42;–0d01'] is located about $\frac{2}{3}$ degree southeast of 82 Delta Ceti [2:39;+0d19']. M77 was first seen by Méchain on October 29, 1780. Messier verified the discovery on December 17 and added it to his list. The arms for this galaxy are very difficult to see, due to their low surface brightness and the vast strings of dust between the arms. The tiny core for M77 is almost star-like in brightness and appearance. This galaxy has strong emission lines in the green portion of the spectrum. Ejected from the nucleus is a rapidly expanding gas cloud, spreading perpendicular to M77's equatorial plane. The energy required to push this cloud is estimated to be equivalent to the explosive power of several million supernovae. M77 is the largest and most luminous galaxy on the list. Its absolute magnitude is estimated at -21.39. This island universe is located 14.4 Mpc from us.

The next nine objects on the list are ones we have already been to. Normally, they are easy to find, but early on March evenings they are beginning to enter the area of atmospheric extinction in the northwestern sky. The first of these nine is the Sc-type galaxy M33 in Triangulum. In Andromeda, we located the great spiral galaxy M31 and its companions M32 and M110. The open clusters M52 and M103 in Cassiopeia have also been covered. Perseus holds the planetary nebula M76 and the open cluster M34. The 11th object for the night is the Pleiades (M45) in Taurus.

Photo 15.1 Close-up of M45; the Pleiades, showing the wispy nebulosity around some of the stars.

The class 5 globular cluster M79 (NGC 1904) [5:24;124d33'] in Lepus the Hare is our next target. The discovery of this object is credited to Méchain, but in 1654 Hodierna had listed a nebulous object in the general area of M79. He did not give precise positional information, so it is not certain he saw this cluster. Méchain first saw this cluster on October 26, 1780, three days before he discovered M77. The Messier numerical order for M77, M78, and M79 is in the order Messier looked at them, on December 17, 1780, and not in the order Méchain found them. M79 is located 13.3 kpc from us and has a radial velocity of 185 km/s in recession. The luminosity for this cluster is equivalent to 90,000 Suns. High power or a large scope is needed to see individual stars near the center of the cluster. M79 is located about 4 degrees southwest of magnitude 2.84, spectrum G2 II, 9 Beta Leporis (ADS 4066) [5:28;−20d45'].

The next four objects we have already hopped to in previous chapters. Easy to spot are the three bright nebulae in Orion: M42, M43, and M78. In Taurus, we find the Crab Nebula, M1, near the tip of the Bull's eastern horn.

The 17th object on our list is the open cluster M35 (NGC 2168) (III2m) [6:08;+24d20'] in the constellation of Gemini the Twins. Credit for the discovery of this object is given to De Chéseaux. He first saw it in 1746. Messier observed it as a cluster of "very small stars" on August 30, 1764. At an age of 110 million years, M35 contains a mixture of main-sequence and giants in spectral types B through K. The brightest member of the cluster is a spectrum B3 V

Photo 15.2 Nebulae NGC 1976 (M42); the Great Nebula in Orion, NGC 1982 (M43), NGC 1973, NGC 1975, and NGC 1977.

star of magnitude 7.5. This cluster is located 870 pc from us and has a radial velocity of 5 km/s in approach. M35 is located about 1 degree northeast of 1 Geminorum (6:04;+23d15'].

About 8 degrees north-northwest of M35 is the open cluster M37 (NGC 2099) (II1r) [5:52;+32d33'] in the constellation Auriga the Wagoner. Hodierna included M37, along with neighboring open clusters M36 and M38, in his 1654 catalogue. Messier first observed M37 on September 2, 1764. This 300-million-year old cluster contains about 150 members. Its integrated magnitude is 5.6 and its brightest star is a spectrum B9, magnitude 9.21 white giant. The clus-

Photo 15.3 Open clusters NGC 1960 (M36), NGC 2099 (M37), and NGC 1912 (M38) in Auriga.

ter is 1,350 pc from us. The cluster stands out from the surrounding star field as the brighter half dozen stars form a cross. The cluster is about 5 degrees south-southwest of magnitude 2.62, spectrum A0p, 37 Theta Aurigae [5:59;+37d12'].

About $3\frac{1}{3}$ degrees north-northwest of M37 is the open cluster M36 (NGC 1960) (II3m) [5:36;+34d08']. This cluster is about half the apparent size as M37 and slightly dimmer at magnitude 6.0. Messier saw this cluster for the first time on September 2, 1764. The cluster contains about 60 members. The brightest star is a spectrum B2, blue-white giant of magnitude 8.86. The majority of the stars in M36 are young hot B-type giants and subgiants. There are no ancient red giants in M36. The cluster is 1,270 pc from us and is about 25 million years old. Look for a twisted "X" of eighth-magnitude stars in a field of fainter ones. The arm of stars to the northeast of the "X" extends out from the hazy center for about double the diameter of the central portion of the cluster. This cluster shows best under low power.

The third open cluster in this area is M38 (NGC 1912) (III2m) [5:28;+35d50']. M38 is fainter than its neighbors, but most of its 100 members still stand out as an irregular-shaped cluster from the rich star field around it. Messier added it to his list on September 25, 1764. The cluster contains a smattering of stars of various spectral types from white A-type main-sequence stars to yellow G0 giants. The brightest member is a G0 yellow giant of magnitude 7.9. The individual faint stars in this cluster are easier to resolve than in the other Messier Aurigae clusters.

Photo 15.4
Planetary nebula
NGC 3587 (M97);
the Owl Nebula and
spiral galaxy NGC
3556 (M108).

We head south to Canis Major to locate the open cluster M41, then head north into the land of the Unicorn Monoceros to add the open cluster M50 to our tally. M50 (NGC 2323) (II3m) [7:03;–8d20'] was first noticed by Italian astronomer Giovanni Domenico Cassini (1625–1712) sometime prior to 1711. In April 1669, he became the first director of the Paris Observatory. In 1771, Messier searched for, but was unable to locate, the nebula that Cassini had charted between Canis Major and Canis Minor. On April 5, 1772, he finally located Cassini's cluster and added it to his list. About 80 stars make up this middle-aged (78 million years) cluster. The distance to M50 is in dispute as sources list distances ranging from 800 to 7,000 pc. The brightest star is a spectrum B8 blue-white giant of magnitude 9.0. The integrated magnitude for M50 is 5.9. To find M50, hop north-northeast about 4 degrees from Sirius (9 Alpha CMa) to 14 Theta CMa [6:54;–12d02']. Continue in the same direction and hop another 4 degrees to the hazy patch of light; this is M50. Notice the nice color contrasts between the member of this loose cluster. Use lower power to see all of the cluster at once. You will need to have reached this point long before midnight as Sirius sets about that time on March 21.

South in Puppis, we re-observe the open clusters M47, M46, and M93. On the boundary between Hydra the Water Snake and Monoceros, we find the open cluster M48 (NGC 2548) (I2m) [8:13;–5d48']. M48 is located about 3 degrees southeast of magnitude 4.34, spectrum G2 Ib, bright yellow giant and double star, 29 Zeta Monocerotis (ADS 6617) [8:08;–2d59']. Messier first observed this 5.8-integrated magnitude cluster, on February 19, 1771. Until recently, it was considered as one of his lost objects as he gave the wrong coordinates for it. Where he said he saw a cluster, there is nothing. His description does match that of NGC 2548, so it is now considered to be the object he saw. The brighter members form a nice 300-million-year old triangle located 610 pc distant. At magnitude

8.8, the brightest member is a spectrum A0 main-sequence star. The cluster contains a mixture of G and K giants, with most of the members being spectrum A-type white main-sequence stars.

In Cancer we should have little trouble finding the large and bright open clusters M44 (the "Praesepe" or the "Beehive Cluster") and M67. We reach the quarter mark as we find these clusters and check them off the search list. After finding M67, you can take about an hour's break as we wait for the next batch of objects, mostly faint galaxies, to rise higher into the sky for easier viewing.

The next five targets, M81, M82, M108, M97, and M109, we have already hopped to in a previous chapter. The 34th target for tonight is the Sb+p spiral galaxy M106 (NGC 4258) [12:19;+47d18'] in Canes Venatici. Méchain found this 8.31-magnitude elongated hazy patch of light in July 1781. M106 is inclined 25 degrees to our line of sight, which affords us a splendid view of its sweeping, though faint, spiral structure. The arms are best viewed in large scopes. In a small scope, they disappear and only the core is easy to detect. M106 is 6.8 Mpc from us and has an absolute magnitude of -20.59. Hop south from Phecda (64 Gamma UMa) [11:53;+53d41'] to 63 Chi UMa then hop along a wavy line of stars the same distance (about 6 degrees) east to M106. The red main-sequence, magnitude 5.29, spectrum K4 V, star 3 CVn [12:19;+48d59'] is about 2 degrees north of M106.

The next eight targets: M40 in UMA; M94, M95, M105, M65, and M66 in Leo; M101 in UMa; and M51 in CVn, are ones we have explored in previous chapters. The Sb-type spiral M63 (NGC 5055) [15:08;+42d02'] in Canes Venatici is our 43rd target on the marathon. Méchain discovered this galaxy prior to June 14, 1779, for that is the evening Messier verified his colleague's find. In photographs of M63, massive sweeping multiple arms are visible, which gives this galaxy the seldom used name of the "Sunflower Galaxy." These arms extend out to touch the eighth-magnitude star to the west of the galaxy. Visually, the galaxy appears much smaller and you cannot see these extended arms. M63 is about 7.7 Mpc from us, which is about the same distance as its neighbor, M51. They appear about the same size and have almost the same absolute magnitude of -20.14 for M63 and -20.74 for M51. To find M63, hop about $5\frac{1}{4}$ degrees south-southwest from M51.

Méchain discovered the Sb-type barred galaxy M94 (NGC 4736) [12:50;+41d07'], our next object, on March 22, 1781. Messier verified its existence two nights later and added it to the list during his rush to get it ready for publication. We view this compact, near circular-shaped, galaxy face on. The multiple arms are tightly wound about the bright nucleus, so visually their structure appears only as a faint glow in small scopes. The apparent magnitude for M94 is 8.17 and its absolute magnitude is -19.37. M94 is located 4.3 Mpc from us and has a radial velocity estimated to be between 359 and 398 km/s in recession. To locate M94, hop about 5 degrees south-southwest to a triangle of seventh- and eighth-magnitude stars. M94 is near the northern point of the triangle. Alternatively, you can hop 4 degrees east from magnitude 4.26, spectrum G0 V, Chara (8 Beta CVn) [12:33;+41d21] to reach M94.

The globular clusters M3 and M53 are next on our list for tonight. M53 leads off our dash to again see the spring galaxies in Coma Berenices and Virgo. Using the marathon runners' old cliché, this segment is the "Heartbreak Hill" of the race. To have the best chance to view these faint galaxies, we have waited for this area of the sky to reach culmination on the southern meridian. In this segment of the marathon, we will look for M64, M60, M59, M58, M89, M90, M88, M91, M87, M86, M84, M100, M85, M98, M99, M49, M61, and M104. With the wealth of visible galaxies in the Coma–Virgo cluster, you have to be sure you are looking at a Messier object and not some other galaxy. Refer to the chapter covering these galaxies for data on them.

After viewing M53, locate magnitude 4.90, spectrum G8 III, yellow 35 Comae Berenicis [12:53;+21d14']. The Sb-type galaxy M64 is about 1 degree northeast of this star.

Magnitude 2.38, spectrum G9 III, Vindemiatrix (47 Epsilon Virginis) [13:02;+10d57'] leads us into the heart of the Coma–Virgo galaxies. In the same finder field as Epsilon Vir and about $1\frac{1}{2}$ degrees northwest of it are the galaxies M60 and M59. M58 is 1 degree west of M59. M60, M59, and M58 are usually the easiest galaxies to find in the area. If you have trouble finding them due to seeing conditions, you will probably have problems finding the rest of the Coma–Virgo targets. About $\frac{1}{4}$ degree west and $\frac{1}{2}$ degree north from M58 is M89. In the same low-power field as M89 will be M90. Look for M90 about $\frac{1}{2}$ degree north of M89. M88 is about $1\frac{1}{2}$ degrees north-northwest of M90. Slightly north and $\frac{1}{2}$ degree east of M88 is M91. Go back to M89 to then hop about $1\frac{1}{4}$ degrees southwest to M87. M86 and M84 are located about $1\frac{1}{2}$ degrees north-northwest of M87. M86 is the eastern of these two ninth-magnitude galaxies. Three degrees north of M84 is the Sc-type galaxy M100.

Line up magnitude 6.39, spectrum A2 V, 3 Comae [12:10;+16d48'] and magnitude 4.74, spectrum G8 III, 11 Comae [12:20;+17d47']. These two stars point you in the direction of M85. M85 is about 1 degree northeast of 11 Comae. Forming a triangle with 3 Comae and 11 Comae is 6 Comae [12:16;+14d53']. M98 is located about $\frac{1}{2}$ degree west of magnitude 5.10, spectrum A2 V, 6 Comae. Look for M99 about 1 degree southeast of 6 Comae. Head about $3\frac{1}{2}$ degrees south-southeast to the red giant 20 Virginis. This magnitude 6.26, spectrum K0 III, star is about $2\frac{1}{4}$ degrees north and $\frac{1}{4}$ degree east of M49. The last Messier galaxy in this densely packed area is M61. This Sc-type galaxy is about 1 degree south of the spectrum F7, magnitude 6.40, 17 Virginis [12:22;+5d18']. The last target in this segment of the marathon is the Sombrero Galaxy, M104, located at the southern tip of Virgo.

As a way of pacing yourself, you should be finishing up the Coma–Virgo segment about 2:00 a.m. The objects in Coma and Virgo will begin to dip into the western haze around 4:00 a.m., so you still have plenty of time to search for them if you have not finished by 2:00 a.m.

Number 65 on our list is another Messier object we hopped to earlier in this book: the class 10 globular cluster M68 in Hydra. Also in Hydra is the Sc-type galaxy M83 (NGC 5236) [13:37;–29d52']. While on his observational expedition to the Cape of Good Hope in 1751–2, De Lacaille discovered this southern galaxy. Thirty years later (February 17, 1781), Messier observed it and added it to the list. With an absolute magnitude of -20.31 and visual magnitude of 7.6, M83 is probably the brightest, intrinsically, of Messier's galaxies. It ranks as one of the 25 brightest galaxies we know of. One reason M83 is so bright is that only 4.7 Mpc of space separate us. Its closeness has also allowed us to observe and study in great detail many supernovae in its swirling arms during the last 50 years. Though originally classified as a Sc-type spiral, recent studies indicate that this galaxy should be reclassified to be an SAB (s)c barred spiral. In photographs, the bar structure is visible, but you cannot see it visually through a telescope. You can see some of the mottling and dark dust lanes in the diffuse arms of the galaxy in a low-power eyepiece. The galaxy appears large with a bright stellar-like core. M83 is in a sparse area of the sky almost directly on the border between Hydra and Centaurus. The closest bright star to M83 is magnitude 3.27, spectrum K2 III, 49 Pi Hydrae [14:06;–26d40']. M83 is about $3\frac{1}{4}$ degrees south and $6\frac{1}{3}$ degrees west of Pi Hya. Follow the trail of fifth-, sixth-, and seventh-magnitude stars until you come to a yellowish magnitude 5.83, spectrum F6, subgiant [13:38;–29d33'] and a magnitude 7.0 white A5 V main-sequence star [13:38;–29d50']. These stars are about $\frac{1}{2}$ degree northeast of M83.

The last object we must observe before we take a break is the controversial Messier galaxy M102. Many Messier scholars feel that this object should not be on the list, because of Méchain's letter claiming that he felt it was a relisting of M101. Other students of the Messier catalogue insist that M102 is the elliptical galaxy NGC 5866 in Draco (see Chapter 10). You can decide if you want to include it tonight or skip it. You have plenty of time to find it as we have to wait for the next segment of objects in the Serpens Caput/Hercules/Lyra/Cygnus/Sagitta area to rise out of the atmospheric-distorted eastern haze. M102 is the last galaxy on the marathon search list.

When you see the leading side of Serpens Caput nearing your southern meridian around 3:00 a.m., it is time to get running again. The first object for this segment is the class 5 globular cluster M5 (NGC 5904). In Chapter 8, we hopped to M5 as a side trip.

The most spectacular northern sky globular awaits us next: the "Great Globular Star Cluster in Hercules", also known as M13 (NGC 6205) [16:41;+36d28']. This 5.9-magnitude cluster was discovered by Edmond Halley, in 1714. Messier again was unable to resolve a globular cluster's stars and added this "Nébuleuse sans étoile" to his list on June 1, 1764. M13 is a naked-eye object and even in 20-power binoculars you should be able to begin to resolve some of the estimated one million members of the cluster. Old red giants are the dominant type of star in the cluster. No young B-type stars have been detected as members of this ten billion-year old cluster.

Photo 15.5 NGC 6205 (M13); the Great Globular Cluster in Hercules and spiral galaxy NGC 6207.

Positioned far away from the plane of the Milky Way, the stars have received little of the heavy elements that are spewed forth from super-novae in the densely populated arms of the galaxy. New stars form from gaseous clouds of this material and its absence contributes to the lack of young stars in M13. The presence of variables in M13 has allowed astronomers to calculate its distance to be about 7.2 kpc from us. Its redshift indicates that M13 has a radial velocity of 248 km/s in approach. M13 appears as a glowing cloud in binoculars or small (less than 4-inch) scopes. In larger instruments, individual stars can be resolved deep into the core. The cluster is almost on the line between magnitude 3.53, spectrum G8 III-IV, 44 Eta [16:42;+38d55'] and magnitude 2.81, spectrum G0 IV, 40 Zeta Herculis [16:41;+31d36']. Look about $2\frac{1}{2}$ degrees south of Eta for M13.

Another spectacular globular in Hercules is M92 (NGC 6341) [17:17;+43d08']. This rich class 4 cluster was discovered by Bode, on December 27, 1777. It became the 92nd object on Messier's list, on March 18, 1781. M92 appears about half as large as M13, but is almost as luminous at magnitude 6.5. M92 is a young cluster at about two to three billion years. Spectral analysis shows this cluster contains more younger stars than M13. Though red giants predominate, many A- and F-type stars are also present in the cluster. At 7.8 pc, M92 is slightly farther away than M13. The 14 known short-period pulsating RR Lyrae-type variables in the cluster were used to calculate its distance. M92 has a radial velocity of 121 km/s in approach. The stars in this cluster are not as tightly compacted as those in M13, so even in small scopes you can resolve many of the stars near the center of the cluster.

The next segment of the marathon takes us into the summer Milky Way in Ophiuchus, Scorpius, Sagittarius, and Scutum. In Ophiuchus, we will first search for the globular clusters M12, M10, M14, and M107 that we have not been to, then hop south to enter Scorpius.

On May 30, 1764, Messier saw M12 (NGC 6218) [16:47;–1d57'] for the first time. The stars in this 6.60-magnitude globular are not as compact as in most of the other Messier globulars. Stars deep into the central area in this class 9 cluster are easy to resolve, even in medium-size scopes. With its loose central region and streamers of stars, M12 looks almost like an open cluster. Just one variable has been detected in this medium-aged cluster. The cluster is estimated to be 30 pc in diameter and to be located 5.5 kpc from us. Its radial velocity is 44 km/s in approach. To locate M12, hop about $8\frac{1}{2}$ degrees north-northeast from the multiple system, magnitude 2.7, spectrum M1 III, Yed Prior (1 Delta Ophiuchi; Σ 2032) [16:14;–3d41']. You can also hop 4 degrees south-southeast from multiple system, magnitude 4.2, spectrum A1 V, Marfic (10 Lambda Ophiuchi; ADS 10087; Σ 2055) [16:30;+1d59']. M12 is at the right angle of a triangle formed by these stars and the cluster. A degree northwest of the cluster is a box of fifth- and sixth-magnitude stars.

In the same finder field as M12 is the class 7 globular cluster M10 (NGC 6254) [16:57;–4d06']. With an integrated magnitude of 6.57, this cluster is just slightly brighter than M12, but its more compact core makes it appear even brighter. The outer stars are easy to resolve in small scopes. The dense core is almost impossible to resolve, even in photographs. The true diameter for M10 is calculated to be 26 pc. It is interesting that M10 and M12 are so close, visually, yet Messier did not see them at the same time. (He saw this cluster the night before he saw M12.) Estimates have these two clusters separated by about 613 pc. Though close, these clusters are not bound gravitationally, since M12 is coming toward us and M10 is receding at a rate of 70 km/s. M10 is located 5.5 kpc from us. Look for M10 about 2 degrees southeast of M12.

Sweeping about 10 degrees east from M10, we come to another bright globular cluster, M14 (NGC 6402) [17:37;–3d15']. This class 8 cluster glows with an integrated magnitude of 7.56. In small to

Photo 15.6
Emission nebulae
NGC 6523 (M8);
the Lagoon Nebula,
and NGC 6514
(M20); the Trifid
Nebula.

medium scopes, the stars appear of uniform brightness from its loose core to outer reaches. You can resolve some of the central region stars. Though appearing smaller and dimmer than its neighbors M10 and M12, this cluster has an absolute visual magnitude of -9.34 compared to that of -7.48 for M10 and -7.70 for M12. M14 appears dimmer than these other two clusters, because at 10.2 kpc it is about twice as distant from us. Its radial velocity in approach of 123 km/s is almost three times that of M12. Messier discovered M14 on June 1, 1764.

In the southern reaches of Ophiuchus is the class 10 globular cluster M107 (NGC 6171) [16:32;−13d03']. Méchain discovered this

Photo 15.7
Emission nebula
NGC 6618 (M17) in
Sagittarius.

glowing patch of stars in April 1782. With its very loose core, you can resolve most of its stars. This cluster is small and being faint (magnitude 8.13, absolute visual magnitude -6.90) it gets lost in the rich Milky Way star field around it. M107 is located 5.9 kpc from us and has a radial velocity of about 60 km/s in approach. You will find M107 about $3\frac{1}{2}$ degrees southwest of magnitude 2.56, spectrum O9.5 V, 13 Zeta Ophiuchi [16:37-10d34'].

The 75th target is the globular cluster M80 in Scorpius, and with it we begin the downhill jog before the rush to beat the oncoming twilight at the end of the marathon. Except for M9, we have already observed the next 22 objects, located in Scorpius, Scutum, Sagittarius, Serpens Cauda, and Ophiuchus, in Chapter 9. Refer to that chapter for information on M80, M4, M19, M62, M6, M7, M8, M20, M21,

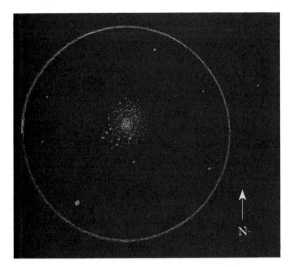

M23, M9, M24, M25, M18, M17, M16, M11, M26, M22, M28, M69, and M70. All of these objects will be low in your southeastern horizon.

After observing the open cluster M23 (NGC 6494), locate magnitude 2.43, spectrum A2 V, Sabik (35 Eta Oph; ADS 10374) [17:10;–15d43']. The class 8, magnitude 7.9, globular cluster M9 (NGC 6333) [17:19;–18d31'] is about 3 degrees southeast of Sabik. Messier found this cluster on May 28, 1764. The stars near the edge of the bright core are resolvable in small scopes. M9 has an absolute magnitude of -7.4. This cluster is about 6.9 kpc from us and has a radial velocity of 225 km/s in approach.

Just before we begin the mad dash to the finish, we turn to look high overhead in Lyra, Cygnus, and Sagitta for six very easy targets. During the summer, we hopped to the Ring Nebula (M57) as well as M56, M29, M39, M71, and the Dumbbell Nebula (M27).

Take a deep breath, for we are about to kick it into high gear in order to outrace the Sun as we hunt down the last eight objects. Astronomical twilight will be starting soon, so you have no time to waste at this stage in the race. Targets 103 and 104 on the search list are the globular clusters M54 and M55 in Sagittarius, objects we have already explored in Chapter 9.

Our next target is the class 4 globular cluster M15 (NGC 7078) [21:30;+12d10'] in Pegasus the Winged Horse. Italian- French astronomer Jean-Dominique Maraldi (1709–88) discovered M15 on September 7, 1746. He assisted in the publication of 25 volumes of the *Connaissance des Temps* and he published De Lacaille's *Coelum Australe Stelliferum*. Messier acknowledged Maraldi as the discoverer when he found M15 on June 3, 1774. Among the 250,000 members of this visual 6.35-magnitude cluster, is a rich assortment of variables and at least one planetary nebula (PK 065-27.1; Pease 1). The distance to M15 is estimated to be between 9.4 and 16 kpc with a radial velocity of 112 km/s in approach. The radial velocity for the planetary is 141 km/s in approach. This planetary is located near the northeast side of the cluster and is very difficult to see, because it is about 3 arc-sec in diameter and is very faint at magnitude 15.5. M15 is an old globular and contains mostly red giants. This cluster is a bright X-ray emitter, which leads astronomers to believe there are one or more supernovae within the cluster. The core is too dense for you to visually resolve any of its inner stars. Thousands of faint stars appear to swarm around the core. The cluster is about 4 degrees northwest from the bright red giant Enif (8 Epsilon Pegasi; ADS 15268) [21:44;+9d52']. Enif (pronounced ENN-if) is derived from the Arabic *Al Anif*, and as its name implies this magnitude 2.38, spectrum K2 Ib, star marks "The Nose" of the horse.

Except for the globular cluster M2 in Aquarius, the last five objects (M75, M72, M73, M2, and M30) will be very low in your southeastern horizon as morning twilight begins to brighten the eastern sky. Depending on any visual obstructions, such as trees, houses, or hills, you may want to go for these last objects in a different order from that presented here. You might, for example, consider looking for M2 before M75 and M73. The globular cluster M30 in Capricornus the Goat will still have to be the last target as it is the last to rise just before the Sun.

The class 1 globular cluster M75 in Sagittarius is first of the last rush hour targets on our list. We have already visited this object during the summer in Chapter 9.

Méchain discovered the loosely compacted class 9 globular cluster M72 (NGC 6981) [20:53;–12d32'] in Aquarius the Water Pourer, on August 29, 1780. He reported the find to Messier, who verified it on October 4, 1780. This cluster is faint at magnitude 9.35 and its brightest members are about 14th magnitude. Approximately 45 short-period variables are known in the cluster. The distance to M72 is calculated to be 17.3 kpc. The cluster displays a radial velocity of 278 km/s in approach. To find this object, locate magnitude 3.57, spectrum G9 III, yellow giant multiple star Algedi (6 Alpha^2 Capricorni; ADS 13645) [20:18-12d32'] and its visual companion magnitude 4.24, spectrum F5 V, main-sequence star 5 Alpha^1 Capricorni (ADS 13632) [20:17;–12d30']. Both of these stars are multiple systems totally unrelated to the other system. M72 is located about 9 degrees east of Algedi (pronounced al-JEE-dee).

About $1\frac{1}{2}$ degrees east, and in the same finder field as, M72 is the four-star open cluster (asterism?) M73 (NGC 6994) (IV1p) [20:58;–12d38]. Messier found these stars on October 4, 1780. To him, they appeared as a cluster with nebulosity. These stars are 10th, 11th, and 12th magnitudes and have an integrated magnitude of 9.0. No nebulosity is detectable in medium or larger scopes.

Like M15, the class 2 globular cluster M2 (NGC 7089) [21:33;–0d49'], in Aquarius, was discovered by Maraldi, on September 11, 1746, while he was searching for his friend De Chéseaux's comet. Messier came across it in his comet-hunting activities 24 years later to the day. To Messier, the cluster resembled a comet. You need a large-aperture scope to resolve any of the central stars. The outer stars are easy to pick out from the sparse

Photo 15.8
Globular cluster
NGC 7099 (M30) in
Capricornus.

background of fainter stars. M2 glows as a hazy patch of light at
integrated visual magnitude 6.50. The absolute magnitude for this
object is -8.95. Most of the 100,000-plus members of the cluster are
red and yellow giants. Of these stars, about 17 are short-period pulsating RR Lyrae-type variables. The lack of O and B stars indicates
this cluster is very old. Estimates place it in the 13 billion-year old
range. M2 is located about 11.3 kpc from us and has a radial velocity
of 6 km/s in approach. This cluster is in the northern portion of the
constellation, about 4 degrees north of magnitude 2.41, spectrum
G0 Ib, yellow giant Sadalsuud (22 Beta Aquarii; ADS 15050)
[21:31;–5d34'].

   If you are still in the dark, you may be able to search for the class 5
globular cluster M30 (NGC 7099) [21:40;–23d11'] in the eastern

portion of Capricornus. Usually, this object is hopelessly lost in the morning twilight and is therefore extremely difficult to find in March as the Sun is only about $2\frac{1}{4}$ hours of right ascension behind it on March 21. M30 was discovered by Messier on August 3, 1764. The core is dense and bright, but you cannot visually resolve any of the central stars. The radial velocity for this cluster is 172 km/s in approach. The cluster is located 8.2 kpc from us. The best time to see this cluster is actually in late summer or early fall. To find M30, hop either about 6 degrees south from magnitude 3.68, spectrum F0, Nashira (40 Gamma Capricorni) [21:40;–16d39'] or $2\frac{1}{2}$ degrees southeast from magnitude 3.74, spectrum G4 Ib, yellow giant, and wide double 34 Zeta Capricorni (ADS 14971; SEE 446) [21:26;–22d24']. In the same field as M30 is the magnitude 5.30, spectrum K0 III, red giant 41 Capricorni [21:42;–23d15'].

It has been a long and probably very cold night, but if you were able to locate 100 or more of the objects, consider it a successful marathon. I sincerely hope you enjoyed the challenge of the Messier Marathon and viewing all of the objects we hopped to along the way from the first stars on the first star-hop to this point.

A few additional notes about taking a Messier Marathon

In this chapter I have presented just one way in which you can hunt down Messier's list. I based it on making the least amount of telescope movements from one object to the next while taking into consideration the rising, culmination, and setting of the objects. Another method is to look for all of the objects contained on the same page of your star chart. Doing this may have you jumping all over the sky, but will ease the page flipping exercise. If you missed M31, M32, M110, M76, M52 or M103, you usually have a second chance to find them shortly before dawn as they have returned to the sky for the second time tonight.

I am fully aware that because of the limitations of your optical equipment, observing site, and so many other factors you may not have been able to locate some of the objects covered in this book. I hope that, as you become a better observer of the night sky, you will surmount any obstacle in your way up the mountain and will return to these pages to look again. I leave you with the words of Johannes Kepler from his letter of April 19, 1610, to Galileo and hope they inspire you to press on vigorously with your own observations of the night sky.

> It remains for me to make an urgent request of you, most illustrious Galileo. Press on vigorously with your observations, and let us know at the very earliest opportunity what results your observations have attained. Finally, forgive my diffuse and independent way of discussing nature. Farewell.

Appendix A

Classification tables

Table A.1. *Henry Draper spectral classification of stars.*

HD class	Apparent color	Surface temperature (K)	Key spectral line
O	Blue-white	25,000 to 40,000	Ionized helium
B	Blue-white	10,000 to 25,000	Atomic helium
A	White	7,500 to 10,000	Atomic hydrogen
F	White	6,000 to 7,500	Strong hydrogen lines
G	White/yellow	5,000 to 6,000	Ionized and atomic metals
K	Deep yellow/orange	3,500 to 5,000	Atomic metals
M	Red	2,000 to 3,500	Strong molecular lines

The colors are not absolute for each class as there is gradual shading at the ends of each class. You may also see slightly different shades than someone else. The temperature ranges are also estimates as many astronomical sources give different ranges for each spectral class.

Table A.2. *MK system of stellar luminosity.*

MK class	Description	Spectral classes
Ia-O	Very bright supergiants	O to G5
Ia	Bright supergiants	O to M3
Iab	Faint supergiants	B0 to M3
Ib	Very faint supergiants	O to M3
II	Bright giants	B0 to M4
III	Normal giants	B0 to M5
IV	Subgiants	B0 to K0
V	Main-sequence and dwarfs	O to M
VI	Subdwarfs	F to K5
VII	White dwarfs	B5 to F

Spectral class is dependent on the star's temperature which in turn affects the star's placement on the H–R diagram and its absolute visual magnitude.

Table A.3. *Classification of variable stars.*

Class/type	Designation
Cataclysmic variables	
AM Herculis stars	AM
Novae	N
Fast novae	NA
Intermediate speed novae	NAB
Slow novae	NB
Very slow novae	NC
Nova-like variables	NL
Recurrent novae	NR
Supernovae	SN
Type I supernovae	SNI
Type II supernovae	SNII
Dwarf novae	UG
SS Cygni stars	UGSS
SU Ursae Majoris stars	UGSU
Z Camelopardalis stars	UGZ
Z Andromedae stars	ZAND
Eclipsing binary variables[*]	
(a) Shape of light curve	
Eclipsing binary systems	E
Algol-type eclipsing variables	EA
Beta Lyrae-type eclipsing variables	EB
W Ursae Majoris-type eclipsing variables	EW
(b) Degree of contact between companions	
Detached systems of AR Lacertae type	AR
Detached systems	D
Detached main-sequence stars	DM
Detached with one subgiant component	DS
Detached like W Uma-type systems	DW
Contact systems with both Roche lobes filled	K
Contact system of spectral types O to A	KE
Contact systems of W UMa type in spectral classes F0 to K	KW
Semidetached systems	SD
(c) Physical characteristics	
Systems where one or both stars are giants	GS
Systems with at least one planetary nebula	PN
RS Canum Venaticorum systems	RS
Systems with a dwarf star component	WD
Systems with a Wolf–Rayet component	WR
Eruptive variables	
Stars with small-scale variations	BE
FU Orionis stars (Fuors)	FU
Gamma Cassiopeiae stars	GCAS
Poorly studied irregular variables	I
Poorly studied irregular variables of spectral types O to A	IA
Poorly studied irregular variables of spectral types F to M	IB

Table A.3 *(cont.)*

Class/type	Designation
Orion variables	IN
Orion variables of spectral types B to A or Ae	INA
Orion variables of spectral types F to M or Fe to Me	INB
T Tauri stars	INT
Rapid irregular variables	IS
Rapid irregular variables of spectral types B to A or Ae	ISA
Rapid irregular variables of spectral types F to M or Fe to Me	ISB
R Coronae Borealis stars	RCB
RS Canum Venaticorum stars	RS
S Doradus stars	SDOR
UV Ceti stars	UV
Flaring Orion variables	UVN
Wolf–Rayet variables	WR
Pulsating variables	
Alpha Cygni stars	ACYG
Beta Cephei stars	BCEP
Short-period Beta Cephei stars	BCEPS
Cepheid variables	CEP
Cepheids with two or more simultaneous pulsations	CEP(B)
W Virginis stars	CW
W Vir stars with periods longer than eight days	CWA
W Vir stars with periods shorter than eight days	CWB
Delta Cephei stars	DCEP
Delta Cephei stars with visual amplitudes less than 0.5 magnitude	DCEPS
Delta Scuti stars	DSCT
Delta Scuti stars with visual amplitudes less than 0.1 magnitude	DSCTC
Slow irregular variables	L
Slow irregular variables of spectral types K, M, C, and S	LB
Irregular variable supergiants	LC
Mira stars (long-period variables)	M
PV Telescopii stars	PVTEL
RR Lyrae stars	RR
RR Lyrae with two simultaneous pulsation modes	RR(B)
RR Lyrae with asymmetric light curves	RRAB
RR Lyrae stars with symmetrical or sinusoidal light curves	RRC
RV Tauri stars	RV
RV Tauri stars with constant mean magnitudes	RVA
RV Tauri stars with variable periodic mean magnitudes	RVB
Semiregular variables	SR
Semiregular giants of spectral types M, C, S, or Me, Ce, or Se with regular light curves	SRA
Semiregular giants of the same classes as SRA, but with irregular light curves	SRB
Semiregular supergiants of spectral types M, C, S, or Me, Ce, or Se	SRC
Semiregular yellow supergiants	SRD
SX Phoenisis stars	SXPHE

Table A.3 (*cont.*)

Class/type	Designation
ZZ Ceti stars	ZZ
Hydrogen-rich ZZ Ceti dwarf (DA) stars	ZZA
Helium-rich ZZ Ceti dwarf (DB) stars	ZZB
Very hot ZZ Ceti white dwarf (DO) stars	ZZO
Rotating variables	
Alpha^2 Canum Venaticorum stars	ACV
Rapidly oscillating Alpha-2 CVn stars	ACVO
BY Draconis stars	BY
Ellipsoidal variables	ELL
FK Comae Berenices stars	FKCOM
Optical variable pulsars	PSR
Reflection variables	R
SX Arietis stars	SXARI
X-ray variables	
X-ray bursters	XB
Fluctuating X-ray variables	XF
Irregular X-ray variables	XI
Binary stars with jets in X-ray or radio range	XJ
X-ray nova-like systems with a dwarf or subgiant	XND
X-ray nova-like system with giants or supergiant components	XNG
X-ray pulsars systems	XP
X-ray pulsars systems with a reflection effect	XPR
X-ray systems with a DK to MK dwarf star	XPRM
Miscellaneous variables	
BL Lacertae objects	BLLAC
Mislabeled, not a variable	CST
Optical variable galaxy	GAL
Optical variable quasi-stellar objects	QSO

[*] Individual eclipsing variables are classified by combining any of the three characteristics to classify the variable according to its light curve, the degree of contact between its component stars, and physical properties of the component stars. The third characteristic is not used if both stars are main-sequence objects.

Table A.4. *Classification of double stars by separation.*

Title	Angular separation (arcsec)
Very close	0.5 to 2.0
Close	2.1 to 5.0
Standard	5.1 to 10.0
Wide	10.1 to 30.0
Open	30.1 and wider

Table A.5. *Classification of planetary nebulae.*

Type	Description
I	Appears stellar, no shell visible
IIa	Oval shape with a bright concentrated center
IIb	Oval shape, bright non-concentrated center
IIc	Oval shape with trace of ring structure
IIIa	Oval, uneven brightness
IIIb	Oval, uneven brightness with traces of a ring structure
IV	Annular ring structure around dim core
V	Anomalous shape

Table A.6. *Trumpler classification system for open clusters.*

Component 1: Concentration of cluster members		
I	Detached from surrounding star field with strong central concentration	
II	Detached with little central concentration	
III	Detached with no central concentration	
IV	Not well detached	
Component 2: Range in magnitude		
1	Small range from brightest to faintest member	
2	Moderate brightness range	
3	Large range	
Component 3: Richness		
p	Poor, less than 50 stars	
m	Medium rich, 50–100 stars	
r	Rich, more than 100 stars	

The components are put together such as III2p for open cluster NGC 1891 (in Orion) and shown in this book as NGC 1891 (Cr 73) (III2p).

Table A.7. *Classification of galaxies.*

Class	Type	Description
Elliptical		
	cE	Compact elliptical (V)
	E	Elliptical (V)
	EO	Globular structure, appears round (H) (V)
	E1–6	Ellipticity increasing with E6 being the flattest (V) (Hubble's E7's were misclassified edge-on lenticulars)
	E/SO	Transitional elliptical/lenticular (V)
Lenticular		
	SO	Bright nucleus without spiral arms (H)(V)
	SAO	Ordinary (V)
	SBO	Barred structure (V)
	SABO	Mixed (V)
	S(r)O	Contains an inner ring of stars (V)
	S(s)O	Has the general appearance of a letter S (V)
	S(rs)O	Is a mixture of inner-ringed and spiral galaxies (V)
	SO/a	Transitional lenticular/spiral (V)
Spiral *		
	SA	Ordinary spiral (V)
	SAB	Mixed spiral (V)
	Sa	Large amorphous center with tightly wound weak, smooth arms (H & V)
	Sb	Center not as bright and arms not as tightly wound as in Sa (H)
	Sc	Arms are spread wide from core (H)
	Sd–Sm	Arms are spread wider and increasingly irregular from Sd through Sm (V)
Barred spiral		
	SB	Barred spiral (V)
	SBa	Barred with well-defined arms close to core (H)
	SBb	Barred with well-defined arms spread more than **SBa (H)**
	SBc	Barred with well-defined arms spread very wide from core (H)
	SBd–SBm	Barred with irregular arms becoming more asymmetric from SBd through SBm. An example of a SBm galaxy is the Large Magellanic Clouds (V)
Irregular		
	Irr	Irregular shape, no well-defined central core or spiraling arms (H)
	IA	Irregular (V)
	IB	Irregular with bar-like structure (V)
	IAB	Mixed irregular with some bar structure (V)

Table A.7 (*contd.*)

Class	Type	Description
	IO	Non-Magellanic galaxy (V)
	Im	Magellanic-irregular galaxy (V)
	cI	Compact irregular (V)
Peculiar		
	P	Can have any of a number of peculiar features (V)

(H) = Hubble system
(r) = An inner ring structure is present
(R) = Has outer ring structure
(s) = No ring structure; S-shaped spiral
(rs) = Transitional stage between (r) and (s) stages
(V) = De Vaucouleurs extension of the Hubble system
★ = In De Vaucouleurs' system, spirals are sub-classed as Sa, Sab, Sb, Sbc, Scd, Sd, Sdm, through Snm, and Im depending on how smooth or irregular the arms are; Sa is tightest, Sm most irregular.

Catalogues

Table A.8 lists the celestial catalogues mentioned in this book. It is not intended to be a complete listing of the hundreds of celestial listings that you may want to refer to later. Refer to the Hirshfeld/Sinnott *Sky Catalogue 2000.0* catalogue, as listed in the Bibliography, for more catalogues as Volume 2 of their work contains an extensive listing of celestial catalogues. The IDS code is used for identifying double stars as listed in various catalogues. The system was developed by the staff at Lick Observatory, Mount Hamilton, CA.

Table A.8. *Catalogue designations.*

Designation	IDS code	Catalogue or discoverer	Type
ADS	A	R. G. Aitken (1918, revised 1935)	Double stars
Abell		G. O. Abell	Galaxies, clusters, and planetary nebulae
AGC	AGC	Alvin G. Clark	Double stars
Arp		*Atlas of Peculiar Galaxies* (1966) by Halton Christian Arp (1927–)	Peculiar galaxies
β	BU	S. W. Burnham	Double stars

Table A.8 (*contd.*)

Designation	IDS code	Catalogue or discoverer	Type
B	BAR	E. E. Barnard	Dark nebulae and double stars
B	B	Willem H. van den Bos (1896–1974)	Double stars
Bas		Basel	Clusters
Berk		Berkeley	Clusters
Cr		P. Collinder (d. 1974)	Clusters
Dawes	DA	R. W. Dawes	Double stars
Do		Dolidze	Clusters
DoDz		Dolidze–Dzimselejsvili	Clusters
Σ	STF	F. G. W. Von Struve *Dorpat Catalogue* (1827)	Double stars
Σ II		Wilhelm Struve, 2nd supplement	Double stars
h	HJ	John Herschel	Double stars
Hdo	HDO	Harvard College Observatory and Stations	Double stars
HN	H	W. Herschel (1821)	Double stars
Ho	HO	George Washington Hough (1836–1909)	Double stars
Hu	HU	William Joseph Hussey (1862–1926) (1901)	Double stars
IC		*Index Catalogue* (1895, 1908)	Deep-sky objects
IDS	IDS	Lick Observatory *Index Catalogue of Visual Double Stars, 1961.0*	Double stars
K		Ivan Robert King (1927–)	Clusters
Kr	KR	A. Krüger	Double stars
M		Messier/Méchain (1781)	Deep-sky objects
Mel		Philibert Jacques Melotte (1880–1961)	Clusters
MGC		*Morphological Catalogue of Galaxies* (1962–74)	Galaxies
NGC		*New General Catalogue of Nebulae and Clusters of Stars* (1888)	Deep-sky objects

Table A.8 (*contd.*)

Designation	IDS code	Catalogue or discoverer	Type
OΣ	STT	Otto Wilhelm von Struve (Otto Vasilievich) (1819–1905) *The Pulkovo Catalogue* (1843, revised 1850)	Double stars
Pease		Francis Gladheim Pease (1881–1938) (1928)	Planetary nebulae
PK		Perek/Kohoutek (1967)	Planetary nebulae
Ru		J. Ruprecht	Clusters
See	SEE	T. J. J. See	Double stars
Sh2		Stewart Lane Sharpless (1926–)	Bright nebulae
s,h	SHJ	James South and John Herschel (1824)	Double stars
Stock		Jurgen Stock	Clusters
Tr		R. J. Trumpler	Clusters
UGC		*Uppsala General Catalogue of Galaxies* (1973)	Galaxies
vdB		G. van den Bergh (1890–1966)	Nebulae
VV		B. A. Vorontsov-Vel'iaminov (1959)	Interacting galaxies

Appendix B

The constellations

Table B.1. *The constellations.*

Constellation	Genitive	Abbr.	Description	Size ranking
Andromeda	Andromedae	And	The Chained Maiden	19
Antlia	Antliae	Ant	The Air Pump	62
Apus	Apodis	Aps	The Bird of Paradise	67
Aquarius	Aquarii	Aqr	The Water Pourer	10
Aquila	Aquilae	Aql	The Eagle	22
Ara	Arae	Ara	The Altar	63
Aries	Arietis	Ari	The Ram	39
Auriga	Aurigae	Aur	The Wagoner	21
Boötes	Boötis	Boo	The Ploughman	13
Caelum	Caeli	Cae	The Burin	81
Camelopardalis	Camelopardalis	Cam	The Giraffe	18
Cancer	Cancri	Cnc	The Crab	31
Canes Venatici	Canum Venaticorum	CVn	The Hunting Dogs	38
Canis Major	Canis Majoris	CMa	The Large Dog	43
Canis Minor	Canis Minoris	CMi	The Small Dog	71
Capricornus	Capricorni	Cap	The Goat	40
Carina	Carinae	Car	The Keel	34
Cassiopeia	Cassiopeiae	Cas	Cassiopeia	25
Centaurus	Centauri	Cen	The Centaur	9
Cepheus	Cephei	Cep	King Cepheus	27
Cetus	Ceti	Cet	The Sea Monster	4
Chamaeleon	Chamaeleontis	Cha	The Chamaeleon	79
Circinus	Circini	Cir	The Pair of Compasses	85
Columba	Columbae	Col	The Dove	54
Coma Berenices	Comae Berenicis	Com	The Hair of Berenice	42
Corona Australis	Coronae Australis	CrA	The Southern Crown	80
Corona Borealis	Coronae Borealis	CrB	The Northern Crown	73
Corvus	Corvi	Crv	The Crow	70
Crater	Crateris	Crt	The Cup	53
Crux	Crucis	Cru	The Southern Cross	88
Cygnus	Cygni	Cyg	The Swan	16
Delphinus	Delphini	Del	The Dolphin	69
Dorado	Doradus	Dor	The Swordfish	72
Draco	Draconis	Dra	The Dragon	8
Equuleus	Equulei	Equ	The Foal	87
Eridanus	Eridani	Eri	The River	6
Fornax	Fornacis	For	The Furnace	41

Table B.1 (*cont.*)

Constellation	Genitive	Abbr.	Description	Size ranking
Gemini	Geminorum	Gem	The Twins	30
Grus	Gruis	Gru	The Crane	45
Hercules	Herculis	Her	Hercules	5
Horologium	Horologii	Hor	The Clock	58
Hydra	Hydrae	Hya	The Water Snake	1
Hydrus	Hydri	Hyi	The Sea Serpent	61
Indus	Indi	Ind	The Indian	49
Lacerta	Lacertae	Lac	The Lizard	68
Leo	Leonis	Leo	The Lion	11
Leo Minor	Leonis Minoris	LMi	The Small Lion	10
Lepus	Leporis	Lep	The Hare	51
Libra	Librae	Lib	The Balance Scales	29
Lupus	Lupi	Lup	The Wolf	46
Lynx	Lyncis	Lyn	The Lynx	28
Lyra	Lyrae	Lyr	The Lyre	52
Mensa	Mensae	Men	The Table	75
Microscopium	Microscopii	Mic	The Microscope	66
Monoceros	Monocerotis	Mon	The Unicorn	35
Musca	Muscae	Mus	The Fly	77
Norma	Mormae	Nor	The Square	74
Octans	Octantis	Oct	The Octant	50
Ophiuchus	Ophiuchi	Oph	The Serpent Bearer	11
Orion	Orionis	Ori	The Hunter	26
Pavo	Pavonis	Pav	The Peacock	44
Pegasus	Pegasi	Peg	The Winged Horse	7
Perseus	Persei	Per	Perseus	24
Phoenix	Phoenisis	Phe	The Phoenix	37
Pictor	Pictoris	Pic	The Easel	59
Pisces	Piscium	Psc	The Fishes	14
Piscis Austrinus	Piscis Austrini	PsA	The Southern Fish	60
Puppis	Puppis	Pup	The Poop Deck	20
Pyxis	Pyxidis	Pyx	The Compass	65
Reticulum	Reticuli	Ret	The Net	82
Sagitta	Sagittae	Sge	The Arrow	86
Sagittarius	Sagittarii	Sgr	The Archer	15
Scorpius	Scorpii	Sco	The Scorpion	33
Sculptor	Sculptoris	Scl	The Sculptor	36
Scutum	Scuti	Sct	The Shield	84
Serpens Caput	Serpentis Caput	SerCp	The Serpent (head)	23
Serpens Cauda	Serpentis Cauda	SerCa	The Serpent (tail)	23
Sextans	Sextantis	Sex	The Sextant	47
Taurus	Tauri	Tau	The Bull	17
Telescopium	Telescopii	Tel	The Telescope	57
Triangulum	Trianguli	Tri	The Triangle	78

Table B.1 (*cont.*)

Constellation	Genitive	Abbr.	Description	Size ranking
Triangulum Australe	Trianguli Australis	TrA	The Southern Triangle	83
Tucana	Tucanae	Tuc	The Toucan	48
Ursa Major	Ursae Majoris	UMa	The Great Bear	3
Ursa Minor	Ursae Minoris	UMi	The Little Bear	56
Vela	Velorum	Vel	The Sails	32
Virgo	Virginis	Vir	The Virgin	2
Volans	Volantis	Vol	The Flying Fish	76
Vulpecula	Vulpeculae	Vul	The Little Fox	55

Appendix C

The Greek alphabet

In 1603, Johann Bayer introduced the system of assigning a Greek letter to the brighter stars in a constellation. Since classical times, the shapes for several of the Greek letters have been changed. You will quickly discover that some modern sky mapmakers employ either the letter designs of the Classical Greek alphabet or of the Modern Greek style. The first letter in the "Lower case" column in Table C.1 is the Classical Greek style, whereas the second letter is the Modern Greek or a variant design that you may find used on your charts.

Table C.1. *The Greek alphabet.*

Letters		Name	Letters		Name
Capital	Lower case		Capital	Lower case	
A	α	alpha	N	ν	nu
B	β	beta	Ξ	ξ	xi
Γ	γ	gamma	O	o	omicron
$\varnothing$	δ, ∂	delta	Π	π	pi
E	ϵ, ε	epsilon	Π	ρ, ϱ	rho
Z	ζ	zeta	Σ	σ	sigma
H	η	eta	T	τ	tau
Θ	θ, ϑ	theta	Y	υ	upsilon
I	ι	iota	Φ	ϕ, φ	phi
K	$\kappa, \varkappa$	kappa	X	χ	chi
Λ	λ	lambda	Ψ	ψ	psi
M	μ	mu	Ω	ω	omega

Appendix D

Decimalization of the day

In calculating time intervals during the day, the system of dividing the day into its decimal parts is used. Each minute is 0.00069444444 day based on the astronomical unit of time consisting of 86,400 seconds per day (1,440 minutes). Table D.1 lists the decimal equivalent for each ten minute period during the day.

Table D.1. *Decimalization of the day.*

Hours	00 min	10 min	20 min	30 min	40 min	50 min
0	0.000000	0.006944	0.013889	0.020833	0.027778	0.034722
1	0.041667	0.048611	0.055556	0.062500	0.069444	0.076389
2	0.083333	0.090278	0.097222	0.104167	0.111111	0.118056
3	0.125000	0.131944	0.138889	0.145833	0.152778	0.159722
4	0.166667	0.173611	0.180556	0.187500	0.194444	0.201389
5	0.208333	0.215278	0.222222	0.229167	0.236111	0.243056
6	0.250000	0.256944	0.263889	0.270833	0.277778	0.284722
7	0.291667	0.298611	0.305556	0.312500	0.319444	0.326389
8	0.333333	0.340278	0.347222	0.354167	0.361111	0.368056
9	0.375000	0.381944	0.388889	0.395833	0.402778	0.409722
10	0.416667	0.423611	0.430556	0.437500	0.444444	0.451389
11	0.458333	0.465278	0.472222	0.479167	0.486111	0.493056
12	0.500000	0.506944	0.513889	0.520833	0.527778	0.534722
13	0.541667	0.548611	0.555556	0.562500	0.569444	0.576389
14	0.583333	0.590278	0.597222	0.604167	0.611111	0.618056
15	0.625000	0.631944	0.638889	0.645833	0.652778	0.659722
16	0.666667	0.673611	0.680556	0.687500	0.694444	0.701389
17	0.708333	0.715278	0.722222	0.729167	0.736111	0.743056
18	0.750000	0.756944	0.763889	0.770833	0.777778	0.784722
19	0.791667	0.798611	0.805556	0.812500	0.819444	0.826389
20	0.833333	0.840278	0.847222	0.854167	0.861111	0.868056
21	0.875000	0.881944	0.888889	0.895833	0.902778	0.909722
22	0.916667	0.923611	0.930556	0.937500	0.944444	0.951389
23	0.958333	0.965278	0.972222	0.979167	0.986111	0.993056

Glossary

aberration: The speed of light coming toward us from a celestial body is constant, but, on taking account of the orbital motion of the Earth about the Sun, the observed position of the object appears to be displaced from its actual position as plotted on a celestial map. This relatively small angular displacement (up to 23 arcsec) is called aberration. You are actually seeing the object where it was at the time the light left it, not where it is at the instant you see its light.

absolute brightness: How bright a celestial object would shine if we were viewing it from the standard distance of 10 parsec. This term is interchangeable with *absolute luminosity*.

absolute luminosity: The actual amount of light measured from a star over a specific range of wavelengths, assuming the star is 10 pc from us. Like *absolute brightness*, absolute luminosity is a measure of how bright a star is intrinsically.

absolute magnitude: Using the logarithmic magnitude scale, this is an object's visual magnitude as if it were at a distance of 10 pc. Comparing the absolute magnitudes of objects allows astronomers to compare their brightness regardless of their distance from us.

absolute zero: The temperature of -273 °C which is 0 K. This is the lowest possible temperature.

absorption spectra: The black bands on a continuous spectrogram caused by the absorption of specific wavelengths of energy by a particular element. Absorption spectra are produced when starlight passes through relatively cooler gases in stellar atmospheres or interstellar space. If the energy (and wavelength) of the incoming light precisely equals the difference in energy levels for an element's electrons, the light can be absorbed, and a shadow is left on the spectrogram. Absorption spectra reveal both the chemical composition and the physical state of the intervening gas.

achromatism: The ability of a lens to produce a clear sharp image without colored fringes. As white light passes through a lens, it is bent into a rainbow of color by the refracting effect of the glass. This phenomenon is called *chromatic aberration*. A properly designed set of lenses will bend the light so the final image is free of the rainbow at the focal point of the image. The use of crown glass for the primary lens and denser flint glass for a secondary lens will usually eliminate most, but not all of the chromatic aberration of a lens. Lenses designed to eliminate chromatic aberration are called achromatic. Achromatic means "without color."

Airy disk: The apparent disk of a star at the focus of a telescope caused by *diffraction* of the light. When your scope is out of focus, you can see the concentric rings of light around bright stars. In 1834, the mathematical formula to explain this was devised by British astronomer Sir George Biddell Airy (1801–92).

altazimuth mount: This type of telescope mount allows the instrument to move up and down (in *altitude*) and along the horizon (in *azimuth*). Most telescopes with this type of mount do not easily track the stars as they move across the sky.

altitude: The angular distance of a celestial object above your horizon, measured in degrees from zero at the horizon up to 90° at the *zenith*.

angstrom (Å): An angstrom is a unit of length equal to one ten-billionth of a meter (100,000,000 Å = 1 cm) and is used in this context to measure the distance between *spectra* lines.

angular diameter: The diameter of a celestial object, as we view it, given in degrees, minutes, or seconds of arc.

angular separation: The distance, expressed in degrees of arc, between two celestial objects as seen from Earth. This term is usually expressed in either *arcminutes* or *arcseconds*.

angular size: This term is used to express, in degrees of arc, how big a celestial object appears to be. Angular size and *angular diameter* are interchangeable terms.

aperture: The diameter of the opening through which light enters the telescope. Aperture size is usually given in inches, centimeters, or millimeters. The larger the aperture, the more light the telescope can gather, allowing fainter objects to be visible with greater detail.

apparent brightness: This is one way of expressing how bright a celestial object appears to be as viewed from Earth from a dark viewing site. An object could be close, but be intrinsically dim, or it could be far away and be actually very luminous, and yet have the same apparent brightness rating.

apparent luminosity: This term is one way of expressing how bright a celestial object appears to be when viewed from Earth.

apparent magnitude: Apparent magnitude and *magnitude* mean the same thing; namely how bright a stellar object appears to us on Earth ranked on the historic logarithmic magnitude system.

arcminute: Sixty arcminutes equals one degree of arc. Arcminute is abbreviated as "arcmin" or by an apostrophe ('). One arcmin can also be expressed as being 1/360 degree of arc.

arcsecond: Sixty arcseconds equals one arcminute. Arcsecond is abbreviated as "arcsec" or by a double quote mark ("). One arcsec can also be expressed as being 1/60 arcmin or 1/3600 degree of arc. This term is used extensively to indicate the apparent distance between members of a double or *multiple star system*.

asterism: A recognizable pattern or geometric shape created by a group of stars.

astrometric binary: The slight changes in the *proper motion* of what appears to be a single star revealing the presence of a second massive object near to or orbiting around the primary star. This object (often a dimmer companion star or a dark, massive stellar corpse) disrupts the primary star's straight line proper motion and causes the star to appear to be wobbling when compared to background stars.

astronomical photometry: The study and measurement of the brightness of celestial objects usually viewed in narrow, defined wavelength bands. Photoelectric instruments are employed to measure stellar magnitudes. These newer techniques have better accuracy, reliability, and sensitivity than the visual and photographic methods used prior to the Second World War.

astronomical unit (AU): A shorthand method of measuring solar system distances. One astronomical unit equals the nominal (mean) distance (about 149,597,000 km or 93,000,000 miles) from the Earth to the Sun.

atmospheric extinction: As a celestial object gets closer to your horizon, you have to look through a thicker amount of air to see it. The more air you have to look through, the dimmer the object becomes. Your best view of an object will be when you can observe it at or near your *zenith*, for in that position you are looking through the thinnest amount of atmosphere.

autumnal equinox: The autumnal equinox occurs when the Sun, heading south, crosses the *celestial equator* at 12 hours of *right ascension* in the constellation Virgo. This position is 180 degrees of arc along the celestial equator from the *vernal equinox*. The autumnal equinox marks the first day of autumn (September 22 or 23) in the northern hemisphere.

average density: The average density of an object is calculated by dividing the object's mass by its volume.

averted vision: The technique of looking indirectly at a very faint object in your eyepiece.

azimuth: Azimuth refers to the *angular distance* of objects along the horizon. It also refers to how many degrees an object is horizontally from another object.

Barlow lens: A telescope magnification multiplying system which uses a concave lens or a doublet in a removable tube that you place between the telescope's secondary mirror and *eyepiece*. Using a Barlow lens allows you to increase the power of your eyepieces, saving you the cost of buying lots of additional eyepieces.

barred spiral galaxy: A spiral galaxy with a bar-like structure connecting the galaxy's arms to its core.

blueshift: The shift of absorption or emission spectra lines toward the blue end of the spectrum. This is caused by the shortening of the object's wavelengths as the distance between the object and the viewer decreases.

binary stars: Two or more gravitationally bound stars revolving around a common center of mass.

binoculars: A matched pair of low-power refractor-type telescopes. Binoculars are lightweight, compact, and employ prisms to correct the image and keep it right side up.

black hole: An object so dense and with so strong a gravitational pull that it allows no radiation near it to escape. Black holes are believed to be the cores of tightly collapsed stars which exploded in a *supernova*. Much larger black holes may be at the heart of most galaxies.

bolometric magnitude: The brightness of a star based on the total measurement of its *electromagnetic radiation* over all wavelengths.

brown dwarf: A very cool star with a mass too low to cause nuclear reactions to begin in its core.

Cassegrain: A Cassegrain telescope reflects the incoming light from a primary mirror, to a secondary mirror, then back through a hole in the center of the primary mirror. The result is a very compact telescope with a long *focal length*. Focusing is accomplished by moving the primary mirror.

catadioptric: A telescope design which uses a combination of mirrors and lenses to direct the light path to an *eyepiece*. This type of telescope is sometimes referred to as a compound telescope.

cataclysmic variable: A class of *variable star* which includes *novae* and *dwarf novae*. The stars in this class may also be losing stellar matter to a companion star or undergoing violent outbursts from the conversion of hydrogen to helium.

celestial equator: The celestial equator represents the projection of the Earth's equator into the sky.

celestial pole: The north and south celestial poles are the points in the sky directly over the north and south terrestrial poles. The night sky appears to slowly revolve around them.

celestial sphere: As you look up at the night sky, all of the objects appear to be at the same distance from us as if they were attached to some giant dome. This dome is referred to as the celestial sphere, and the Earth is located at the center of it.

center of mass: The central point of the mass of an object, like the center of the Earth, or the gravitational center of a system of objects.

cepheid variable: A type of radially *pulsating star* with a high *luminosity* and periods of 1 to 135 days. Their visual amplitudes can range from a few hundredths of a *magnitude* to as much as two magnitudes. The *radial velocity* curve of the star's surface layer expansion usually matches its light curve.

chromatic aberration: The wavelengths in a beam of light bend at different angles as the light passes through a lens. This causes the focal points for the colors of light to occur at different distances from the lens. The use of two or more lenses with different focal points can overcome this problem.

circumpolar stars: Stars located in the vicinity of either the celestial north pole or celestial south pole. These stars always appear above the nighttime horizon for people in the same hemisphere as the stars. Which of the particular stars or constellations are circumpolar depends on the latitude of the observer.

cluster: A group of stars bound together by their mutual gravitational pull, and exhibiting as a group a similar *proper motion*.

cluster of galaxies: A grouping of dozens (or up to thousands) of gravitationally bound galaxies.

colatitude: Your colatitude is 90° minus your latitude. You can determine how high an object will rise above your southern horizon by subtracting your latitude from 90 degrees then adding the object's *declination*. If an object's declination is below the *celestial equator* and is equal to or greater than your colatitude, the object will not be visible above your southern horizon:

colatitude = 90° – your latitude + object's declination.

collimation: The alignment of an optical instrument's mirrors or lenses. The optics must be perfectly aligned to give the best views possible with the instrument. A telescope out of collimation will give you out of focus stars and blurred views of everything else.

color: The color of a star is dependent on the its mass, surface temperature, and chemical composition. The star's surface temperature determines its most intense color, yet all colors are visible when the star's light is recorded on a spectrogram. The hottest stars we see as blue, slightly cooler stars appear white, mid-range stars are yellow, and cool stars are red. Stars emit energy at practically all wavelengths.

color index: The difference in the magnitudes of a star when measured at different wavelength bands.

comes: The secondary star or stars orbiting a primary star in a double or *multiple star system*.

compact galaxy: A tightly compressed galaxy that appears to be like a ball of stars. These galaxies contain many young hot stars and are therefore considered to be the primary stage in the evolution of a galaxy. On photographic plates, they appear as smudged stars.

constellation: The *celestial sphere* is divided into 88 areas known as constellations. These areas were defined by the International Astronomical Union, in 1928. Each constellation consists of an *asterism* which generally forms a very recognizable shape. Many of the constellation names came from mythological tales and characters. Constellations serve as major guideposts.

continuous spectrum: A spectrum of light lacking spectral lines.

culmination: Culmination occurs when a celestial object crosses your southern *meridian*. This event is also described as a *transit*.

dark adaptation: The ability of your eyes to see fainter objects as your pupils open wider in the dark. It generally takes about $\frac{1}{2}$ hour for your pupils to fully open.

dark nebula: Dark nebulae are un-illuminated dust and gas clouds. We know of their existence simply because they block the light from brighter objects beyond them. All we see is the black silhouette of the dark nebula against the illuminated background.

declination (dec.): The north or south position of an object, as measured from the *celestial equator*. This position is given in degrees, minutes, and seconds of arc.

deep-sky object: Any stellar object located beyond the solar system.

degree (angular): A degree of arc as measured against the *celestial sphere* equal to 1/360 of the circumference of the celestial sphere.

density: This term describes how packed together the atoms of an object are or the number of stars in a globular cluster.

dewshield: A tubular device attached to the objective end of a telescope to slow down the formation of dew on the objective lens. Using a dewshield will extend your stargazing on cold moist nights.

diffraction: The thin bands of light or shadows caused by the bending of light as it passes by an object, passes through a narrow slit, or reflects off the ruled groves.

diffraction grating: A system of very closely spaced slits or grooves used to disperse light passing through or over the grating in order to produce a spectrum.

diffuse nebula: This type of nebula appears wispy as if it was almost transparent. Diffuse nebulae often have low surface brightness.

Doppler effect: As an object moves toward or away from you, the frequency of its light or sound shifts slightly. On a spectrogram, this *blueshift* or *redshift* is noticed as a change in the placement of emission or absorption lines on a spectra.

doublet: An objective lens consisting of two pieces of glass.

double-lined spectroscopic binary: This type of *spectroscopic binary* occurs when the two stars have approximately the same magnitude. The *spectra* of the stars is superimposed on a single spectrum, thereby revealing the presence of the secondary star.

dwarf nova: A type of *cataclysmic variable* sometimes called a U Geminorum star. Dwarf novae are close binary systems that consist of a *dwarf star* or a *subgiant* filling its inner *Roche lobe*, and a *white dwarf* surrounded by an accretion disk.

dwarf stars: Normal stars, fusing hydrogen to helium in their cores, are known as dwarf stars. They are found on the H–R diagram main-sequence. They are smaller stars near the end of their life and as they have consumed most of their hydrogen fuel they begin to collapse and their cores can reach very high densities. These stars are known as *white dwarfs*.

eclipse: This event occurs when one object or its shadow passes in front of another object and either fully or partially blocks our view of the background object.

eclipsing binary: In a *binary star* system, when one of the members transits and eclipses the other star, or it passes behind the other member and is eclipsed.

eclipsing binary variable: The brightness variation of these *variable stars* occurs when one of the stars in the pair transits or goes behind the other star. A variation can also occur if the stars are close to each other and matter passes back and forth between the stars. These are called either contact binary stars or a mass-transfer pair.

ecliptic: The ecliptic is the plane of the Earth's orbit around the Sun. Seen from Earth, the ecliptic is the portion of the sky where the Sun, the planets, and the Moon pass through. The constellations which comprise the *zodiac* straddle the ecliptic.

electromagnetic radiation: The radiation produced by oscillating coupled electric and magnetic fields which is detectable in various spectra.

electromagnetic spectrum: The range of electromagnetic wavelengths from short-wavelength gamma rays, X-rays, ultraviolet light, visible light, infrared light, to long-wavelength radio waves.

elliptical galaxy: This type of galaxy appears round or elliptical. They do not have a central bulge and their *luminosity* is evenly distributed throughout the galaxy. These galaxies contain some young and mostly older stars.

emission nebula: An irregular gas cloud where stars are forming. The young hot *supergiants* embedded in the cloud emit a tremendous amount of short-wave radiation. This radiation excites the gas atoms in the cloud and causes them to glow. Emission nebulae are generally brighter than *reflection nebulae*.

emission spectra: The spectral lines emitted by atoms when they are heated to very high temperatures. High temperatures cause the atoms to collide forcing electrons into excited states. As the electrons fall back to less energetic levels, they give off radiation. This radiation appears on a *spectrogram* as colored spectral lines. Each emission line is produced by a specific element undergoing a specific energy transition. Variations in the temperature of a gas cause it to emit different emission spectrum lines. A lower case "e" is added to a star's spectral class to indicate the presence of emission lines in its spectrum.

emission spectrum: A spectrum that contains *emission spectra*.

epoch: Due to the processes of *precession* and *proper motion*, the stars gradually shift coordinates as plotted on celestial charts and in catalogue listings. Stars are usually plotted and listed based on their positions for a certain year. The year of the chart or listing is its epoch. In general use today are charts and catalogues with the epochs of 1950.0 and 2000.0 (for January 1 of each year).

equatorial coordinate system: The system of *right ascension* and *declination* coordinates applied to the entire *celestial sphere*. The equatorial coordinate system uses the plane of the Earth's equator and axial poles, as if projected out into space, as its main reference points.

equatorial mount: This type of telescope mount has its up and down axis (*declination*) tilted to be parallel with the Earth's polar axis. The horizontal axis (*right ascension*) is perpendicular to the polar axis. As the instrument rotates in right ascension (driven at the same speed [sidereal rate] as the stars appear to move), it tracks the stars and gives them the appearance of standing still in your *eyepiece*.

equinox: A term used to define the *epoch* of a star chart or catalogue.

eruptive variable: An unpredictably variable star. Flares and other violent processes in a star's chromosphere and corona can cause the brightness of the stars in this class to vary.

exit pupil: The diameter of the beam of light exiting the *eyepiece*. The maximum useful exit pupil is 7.0 mm:

$$\text{exit pupil} = \frac{\text{aperture (in millimeters)}}{\text{magnification}}.$$

extragalactic: Objects located beyond the *Milky Way Galaxy*.

eyepiece: The lens at the end of the telescope where the image is magnified and focused, also known as the ocular. Eyepieces generally consist of two lenses. The lens closest to your eye is called the eye lens. The lens closer to the telescope *aperture* is the field lens. Some eyepiece designs use several lenses to correct and sharpen stellar images. The *focal length* of an eyepiece determines its *magnification* for a given focal length telescope. The shorter the eyepiece focal length (measured in millimeters) the greater the magnification. To increase magnification, you lose image brightness and clarity.

field of view: The amount of sky you can see through the *eyepiece*. An eyepiece's field of view is given in degrees of arc. Sometimes you will see a low-power, wide-field eyepiece or telescope referred to as a rich field.

finder scope: A small, usually refracting, telescope mounted on a larger telescope. Because of its larger *field of view* than that of the telescope it is mounted on, a finder scope is used to locate the general area of a target object.

First Point of Aries: Thousands of years ago, when the Sun crossed the *celestial equator* in the constellation of Aries the Ram, this *vernal equinox* position on the *ecliptic* was called the First Point of Aries. The Sun now crosses the celestial equator in Pisces.

focal length: The distance, usually given in millimeters, that light travels from the *aperture* to the *eyepiece* of a telescope.

focal ratio: The focal ratio "f" number of a telescope is obtained by dividing the *focal length* by the *aperture*. The resulting number (f/number) determines how a telescope will function. The lower the focal ratio, the wider its *field of view* will be, but its *magnification* will also be weaker. Focal ratios below f/8 are best suited for astrophotography of large areas of the sky. Mid-range (f/8 to f/11) focal ratios are good for observing galaxies, nebulae, and star clusters. Focal ratios larger than f/12 are more suited to planetary work and splitting close double stars for they have low image brightness and small fields of view:

$$\text{focal ratio} = \frac{\text{focal length (in millimeters)}}{\text{aperture (in millimeters)}}.$$

focus (pl., **foci**): The point where light rays converged by a mirror or lens meet.

galactic cluster: Can be either an open cluster or globular cluster of stars located within our Galaxy.

galactic coordinate system: Uses the plane of the *galactic equator* and *galactic poles* as the reference points. This system ignores the stars of our Galaxy and plots only the positions of other galaxies.

galactic equator: The great circle on the *celestial sphere* marking the centerline of the *Milky Way Galaxy*.

galactic poles: The north and south axial pole for the *Milky Way Galaxy*. The poles are 90 degrees from the *galactic equator*.

galaxy: A massive assemblage of stars, nebulae, clusters, and interstellar matter.

giant stars: Either very large *main-sequence stars*, or old-age stars that have burned off enough fuel to lose some of their gravitational pull and have therefore inflated in size with a resultant rise in *luminosity*. These old stars have begun to fuse elements heavier than hydrogen at their cores, creating the enormous pressures that swell the stars to gigantic diameters.

globular cluster: A tight grouping of old stars which looks like a huge fuzzball of stars with a few stragglers nearby.

grating spectroscope: This type of spectroscope uses a grooved piece of flat glass to reflect and break up the light beam into its spectrum. Grating-type instruments give a spectra in which the lines are equally spaced at the scope's focal plane. Being reflective, they can also work at wavelengths outside the range of *prism spectroscopes.*

H I region: An interstellar region consisting of a cloud of neutral hydrogen.

H II region: An interstellar region consisting of a cloud of ionized hydrogen.

H-Beta filter®: This LUMICON brand eyepiece filter is used to enhance your view of diffuse and extremely faint nebulae. It allows the emissions of the beta level of hydrogen at 486.1 nm to pass through it, while blocking most of the other wavelengths.

Hertzsprung–Russell diagram: The H–R diagram graphically plots stars according to the four components of spectral type, luminosity, surface temperature, and magnitude. Stellar mass and age are also determining factors for a star's placement on the H–R diagram. All these components are interdependent. As stars age, these stellar components gradually change, thereby causing a shift in a star's H–R diagram position.

High-Contrast filter®: This LUMICON brand eyepiece filter is designed to darken the sky around faint objects, such as planetary and emission nebulae. This gives you a visually sharp contrast between the faint object and the sky around it.

horizon system of coordinates: This system is based on where you are. Its main reference plane is through your observing point parallel to your horizon. The angular distance of an object above the horizon is its *altitude. Azimuth* is the object's angular distance from north and is measured from north toward the east. The main problem with this system is that the coordinates for any object are in constant motion as the Earth rotates. This system is used with altazimuth-mounted telescopes.

hour circle: One of the 24 great circles on the *celestial sphere* which passes through the *celestial poles* and intersects the *ecliptic.* Each of the hour circles marks an hour of *right ascension.*

integrated magnitude: When viewing objects other than a single star, the varying luminosities of the entire object must be calculated by totaling them in order to establish a single integrated magnitude for the object.

interstellar dust: Tiny particles of matter between the stars.

interstellar gas: Gas located between the stars. The most prominent gas in space is hydrogen.

interstellar medium: The cosmic rays, dust, gas, and magnetic fields located between the stars.

irregular galaxy: Irregular galaxies show a lack of the uniform structure found in spiral and elliptical galaxies. They usually appear as a hazy blob of stars, both young and old. They often contain significant amounts of gas.

Julian date: A system of calculating the time interval in days and fractional days since 1 January 4713 B.C. at noon at Greenwich, UK. Each day, since the starting day, has a sequential Julian Day Number assigned to it. This allows the timing of or the interval between astronomical events (past, current, and future) to be calculated no matter what civil or religious calendar is used.

kilometers per second (km/s): Everything in the universe is moving at tremendous speeds relative to everything else. This speed is usually given in kilometers per second. To convert from miles to kilometers, multiply the miles by 1.609. To convert from kilometers to miles, divide the kilometers by 1.609.

kiloparsec: One kiloparsec (1 kpc) equals 1,000 *parsec*. To convert kpc to light-years, multiply the kpc by 3,261.631.

lenticular galaxy: Galaxies having the shape of a double-convex object. They have a *nuclear bulge* and flattened disk, but no spiral structure is visible in the disk.

light curve: The change in brightness of a *variable star* or *multiple star system* as plotted over time on a graph.

light pollution: City lights throw massive amounts of light into the night sky. The glow from these lights pollute our view of the night sky as they overpower the faint glow from dim sky objects. A dark viewing site is any place you can get to where the glow from city lights is reduced. If possible, try to observe from a site at least 50 miles from a large city.

light-pollution reduction filters: These LPR filters are designed to block most of the wavelengths of light emitted by typical city lights while allowing the wavelengths of light emitted by stellar objects to pass through them. LPR filters enhance your view of nebulae, galaxies, and faint clusters. These filters are usually mounted on the field end of your *eyepiece*. Some are designed to be mounted prior to a *star diagonal* at the eyepiece end of the telescope.

light-year: One light-year is the distance that a photon of light energy will travel in one Earth year. At 186,000 miles per second in a vacuum, the photon travels about 5,878,000,000,000 miles (give or take a few miles and not counting the extra day in a leap year) during a year. A light-year equals 9.460536×10^{15} meters. The term light-year is used more by amateurs and the popular press than the term *parsec*.

limiting magnitude: The level of *magnitude* of the faintest objects that can theoretically be seen in a particular scope. Local *seeing* and *transparency* can limit the visible magnitude to objects brighter than a scope's theoretical limiting magnitude.

Local Group of Galaxies: The group of 20 galaxies which are within 1 Mpc (three million light-years) from the center of our Galaxy. The group includes the *Milky Way Galaxy* and the Magellanic Clouds.

local sidereal time: Your local sidereal time is the same as the *right ascension* coordinate for any star on your southern *meridian*.

Local Supercluster of Galaxies: The supercluster of galaxies which consists of our Galaxy, the Local Group, and the Virgo Cluster.

lucida: Latin for brightest.

luminosity: How bright a star shines by measuring the rate at which it emits radiant energy. Luminosity is generally given as a comparison of how bright or dim a celestial object is in relation to the Sun's luminosity.

luminosity classification of stars: A way of classifying stars based on their *absolute luminosity*. The more mass a star contains, the hotter its surface temperature and the brighter it shines. The largest stars are the hottest and brightest (blue *supergiants*). The *luminosity* of a star helps to position it on the H–R diagram. The MK classes of stellar luminosity are listed in Table A.2 in Appendix A.

magnification: Magnification and power refer to a telescope's ability to increase in size the image of an object.

magnitude: The unit of brightness of a stellar object and the system of classifying stars according to their apparent or *absolute brightness*.

main-sequence stars: About 90 percent of the stars we see have *luminosity* and *spectral classes* which place them on a downward sloping line, called the main-sequence on the H–R diagram. These stars are in the process of converting their hydrogen fuel core to helium through thermonuclear fusion.

Maksutov–Cassegrain: In this type of *Cassegrain* telescope, the corrector plate is highly curved with a circular patch of aluminum centered on the inside surface of the plate. This patch of aluminum serves as the secondary mirror. Limited adjustment of the *collimation* of the optics is available.

mass: The measure of how much material an object contains.

megaparsec: One megaparsec (1 Mpc) equals one million *parsec*.

meridian: The lines in the celestial coordinate system running from the north *celestial pole* to the south celestial pole. Your local meridian is the imaginary line rising from the point on your horizon due south of you, up through your *zenith*, then down to the celestial north pole.

Milky Way Galaxy: The proper name for the *spiral galaxy* we are located in.

multiple star system: A primary star with two or more companion stars.

nanometer (nm): One nanometer is one-billionth of a meter. The placement of spectral lines on a spectrum of light are sometimes given in nanometers. One nanometer equals 10 angstroms.

nebula (pl., **nebulae**): Nebula is the Latin word meaning cloud, mist, or smoke. In astronomy, it was applied to the hazy patches of light observed by early astronomers and is still in use today.

neutron star: The rapidly spinning and extremely dense core of a *supernova* remnant, made of neutrons formed when the star collapsed and crushed the electrons into the nuclei. Neutron stars are typically about 10 km in diameter and can emit a rotating beacon of light producing a star known as a *pulsar*.

nova (pl., **novae**): When hydrogen gas is pulled from one star in a binary system onto its companion *white dwarf*, a luminous thermonuclear explosion can result. This flare-up in brightness of the white dwarf in called a nova. Nova derives from the New Latin word *novus* meaning new, fresh, recent, or strange.

nuclear bulge: The central region of a galaxy.

objective: The principal mirror or lens of an optical instrument.

obscuration: The intervening *interstellar dust* and gas can absorb the light energy from stars farther away. This absorption causes the stars to appear dimmer than they would otherwise.

occultation: This event occurs when one object passes in front of another. This term is interchangeable with *eclipse*.

open cluster: A loose grouping of young gravitationally bound stars which is usually closer to us than *globular clusters*. The stars in open clusters are somewhat scattered, yet all members of the cluster have approximately the same *proper motion*.

optical double: The stars in an optical double system are not related and are not gravitationally bound. They appear as a double only because of our line of sight to them.

Oxygen-III filter®: This LUMICON brand filter allows the bandwidths for doubly ionized oxygen (495.9 nm and 500.7 nm) to pass through it, while blocking most other wavelengths. An O-III filter enhances your view of faint planetary and diffuse *emission nebulae*.

parallax: The apparent displacement of a nearby object relative to more distant ones when viewed from different points. Stellar parallax is used to determine the distance to nearby celestial bodies. The two points used are positions along the Earth's orbit six months apart. Knowing the distance between these points (2 astronomical units), and measuring the angle of the shift (parallax), the distance to the star can be calculated. Parallax is useful for stars closer than 100 pc (300 light-years). Beyond this distance, the angle becomes too small to measure with precision.

parsec (pc): A shorthand way of referring to the incredible distances of space. One parsec is the distance at which a star would have an annual *parallax* of one second of arc and equals 3.261.631 *light-years*, which is about 19,200,000,000,000 miles. One parsec is also equal to 206,265 *astronomical units* or 3.085678×10^{16} meters. The word is derived from PAR-allax of one SEC-ond. Professional astronomers and publishers are increasingly using the terms pc, kpc, and Mpc instead of "light-years" in technical papers and articles.

peculiar galaxy: These galaxies do not fit into any of the other classes of galaxies. Each one has something odd about it. Some have only one arm; some have three arms; some have ill-defined arms; or one heavy and one thin arm. They can also be two galaxies interacting with each other.

period-luminosity relation: The relationship between the period and the average density of a pulsating star.

photographic magnitude: The *magnitude* of a stellar object as determined by how bright it appears on a photographic plate or film. The photographic emulsion often determines its greatest sensitivity to a narrow band of wavelengths of light. A blue star would appear to be brighter on blue-sensitive emulsion than would a red star of equal magnitude, so a star's photographic magnitude refers to the magnitude of the star in that particular region of the star's spectrum based upon the sensitivity of the emulsion.

photovisual magnitude: The technique used to give a *magnitude* rating that is close to how bright an object appears to your eyes.

planetary nebula: The remains of typically a *red giant* star which has expelled large amounts of its outer envelope of material. The remaining hot *white dwarf* star is surrounded by a shroud of glowing gas, which is speeding away from the star. This envelope of expanding material gives the star either a halo appearance or leaves an irregular patch of glowing gas.

population I stars: Younger stars, usually located in the galaxy's disk and arms. These stars are rich in heavy elements gained from earlier generations of stars.

population II stars: Older stars located in the halo or core of a galaxy. These older stars either lack or have low amounts of the heavy elements commonly found in *population I stars*.

position angle: The angular distance of the *comes* in its orbit around the primary star. The angular distance in degrees is measured counterclockwise from the north pole of the primary star, as viewed from Earth (0°, 90° at east, 180° at south, and 270° at west).

precession: The Earth wobbles slightly on its polar axis. This causes the poles to gradually shift their positions relative to the stars. This wobble takes about 25,725 years to complete one revolution.

prism spectroscope: This instrument uses a glass prism to refract and separate the starlight into its spectrum for analysis.

prime meridian of the sky: The celestial meridian that runs from the north *celestial pole*, bisects the *ecliptic* at 0 hour of *right ascension* and ends at the south celestial pole. This *meridian* is also known as the zero meridian or the zero hour of right ascension in the celestial coordinate system.

proper motion: A star's apparent motion across our line of sight. The apparent distance is given in *arcseconds*. Proper motion is a measure of the sideways, or transverse, motion of a star as we view it. Combined with a star's radial motion determined by its Doppler shift, we can determine a star's true motion through three-dimensional space.

protostar: The earliest stages of a star when its energy output is detectable. These stars are found in regions of *interstellar gas* and dust clouds where star formation is taking place.

pulsar: A rapidly spinning *supernova* remnant (believed to be a *neutron star*) which is spewing flashes of high-energy radio waves.

pulsating variable: A star whose diameter periodically expands and contracts. This size change affects the *luminosity* of the star.

radial velocity: The speed of a celestial object which is moving in our line of sight, toward or away from the Sun. When the distance between the Sun and the object is expanding, the radial velocity of the object is "in recession" and is given in catalogues as a positive number. When the object is moving closer to us, its radial velocity is "in approach" and is given as a negative number.

radiation: One method by which energy is transmitted.

recurrent nova: A *nova* that has erupted more than once.

red giant: In the upper right-hand portion of the H–R diagram are the red giants. These older stars have converted their hydrogen fuel into helium at their core, and have expanded in size at a slow pace, therefore avoiding the explosive *supernova* stage. The core of these stars becomes inert as hydrogen burning begins in its shell. This hydrogen burning causes the shell to expand. These stars are generally hundreds of times greater in diameter than the Sun, but may contain only a few times the amount of matter. Red giants appear in various shades from yellow to yellow-orange to pale red.

redshift: The shift of absorption or emission spectra lines toward the red end of the spectrum. This is caused by the lengthening of the object's wavelengths as the distance between the object and the viewer increases.

reflection nebula: A cloud of dust or gas reflecting the light from nearby stars or cool stars embedded in it. The cloud is not emitting its own light like an *emission nebula*.

reflector: A type of telescope that uses one or more mirrors to gather light and direct it to the telescope's *eyepiece*.

refraction: As light passes from one medium (a vacuum or the atmosphere) into another denser medium (glass, or water) the speed of the light is decreased, causing the wavelengths of the light to bend at different angles. Refraction by the Earth's atmosphere of the light coming from a celestial body also causes the object to appear in a slightly shifted position than it actually is.

refractor: Refractor-type telescopes use a glass lens or combination of lenses to bend the light and direct it to the *eyepiece*.

resolution: Defines how sharply an optical instrument can focus an object.

resolving power: The useful *magnification* of an optical instrument, which is limited by its ability to increase the size of an image while keeping that image sharp and bright.

right ascension (R.A.): The east or west position of a celestial object on the *celestial sphere* and is measured from the zero hour of right ascension semicircle. The hours of right ascension are divided into minutes and seconds of time.

rotating variable: Stars in this class have non-uniform surface brightness or they may be ellipsoidal in shape. The variability of these stars depends on when the dimmer and brighter portions of the star's surface are facing us. Spots on these stars may also account for the fluctuation in their magnitudes.

Schmidt–Cassegrain (SCT): Schmidt–Cassegrain telescopes are *catadioptric*-type instruments which use a glass corrector plate at the *aperture* with an adjustable secondary convex mirror mounted on the corrector plate.

secondary spectrum: The faint violet halo around bright celestial objects when viewed through a *refractor*. This effect is more noticeable around a planet than other celestial objects.

seeing: Seeing refers to the steadiness of the sky and therefore how clearly an object will be visible. Wind-tossed air causes the image in your telescope to blur, flicker, and dance about as the light rays are bent, reflected off water droplets and dust particles, and are scattered about in the atmosphere. Thermal currents in your optical equipment can also affect seeing.

separation: The angular distance, as measured in *arcseconds*, between the centers of the *Airy disks* for the primary and its comes. Doubles are classified in terms of their separation, as listed in Table A.4 in Appendix A.

Seyfert galaxy: Seyfert galaxies have very bright nuclei caused by charged particles being accelerated in strong magnetic fields within the galaxy's core. This synchrotron radiation, which requires very large total energies from the center of the galaxy to operate, is detected by the polarization of portions of the galaxy's spectra. This energy drives hot gases within the nuclei at speeds of several thousand km/s.

sidereal day: The interval between two consecutive transits of the 0 hour of *right ascension* is a sidereal day. The mean sidereal day is 23 hours 56 minutes and 4 seconds. The shorter sidereal day equals 86,164.093 seconds whereas the civil (solar) day equals 86,400 seconds.

sidereal hour: A sidereal hour is 9.833 seconds shorter than a civil clock hour and equals 59 minutes 50.167 seconds.

sidereal time: Sidereal time is the measure of the Earth's rotation with respect to the stars instead of the Sun. Due to our movement around the Sun, a star will appear at the same place in the sky about four minutes earlier tomorrow than tonight at the same civil clock time. Sidereal time is equal to the hours of *right ascension* in the *equatorial coordinate system*, so the right ascension coordinate of an object is also its sidereal time.

single-lined spectroscopic binary: A *spectroscopic binary* system in which one star is more than two magnitudes fainter than the other star, which causes the fainter star's spectrum not to be visible. The star's presence is revealed by the periodic shift in the brighter star's *spectra*.

solstice: The two points along the *ecliptic* at which the Sun reaches its highest position north (*summer solstice*) and south (*winter solstice*) of the *celestial equator*.

spectra: All energy can be measured by its *wavelength* and frequency on the *electromagnetic spectrum*. A single component of the spectrum is called a spectra. When energy is dispersed, such as by a prism, its component parts can be seen as they arrange themselves in the order of their wavelengths.

spectral class: This classification system for stars is based on a comparison of certain spectral lines and decreasing surface temperatures. The stars are designated in classes O, B, A, F, G, K, and M. Each of these seven classes is divided into ten subdivisions. The system has been modified with the addition of three minor classes, N, R, and S and a subclass called Wolf–Rayet (WR) stars. The N, R, and S classes are rare and are noted by their deep red tint. A few years ago, the N and R classes were replaced by a single new class. Stars in this new C class of stars show intense bands of carbon molecules and are cool at 2,500 K. The WR subclass consists of extremely hot white stars with expanding gas shells.

spectrogram: A photograph of the *spectra* of light.

spectrograph: A telescope set up to take *spectrograms*.

spectroscope: A type of telescope designed to break up a light beam into its component spectrum so the *spectra* can be recorded on a *spectrogram*, and analyzed to see what elements and their temperatures combined to form the light beam.

spectroscopy: The study of the *spectra* of light.

spectroscopic binary: Like a *visual binary*, the stars in a spectroscopic binary system rotate around a common center of gravity, but one or more of the stars are observable only with the use of spectroscopic equipment. The star's Doppler shift and spectra reveal its motion and position.

spiral galaxy: A galaxy with arms of stars that appear to sweep out and gently curve away from the core. The sweep of the arms can vary from being tightly wound around the core to being spread wide from it. Spiral galaxies contain a mixture of young and old stars.

spiral arms: The regions of galaxies that contain dust, gas, and young stars. These regions sweep out with varying degrees of curvature from the core of a galaxy.

star: An intensely hot glowing ball-shaped condensation of matter in which thermonuclear fusion is taking place which in turn causes a tremendous amount of radiation to be spewed into space.

star diagonal: A flat mirror placed at a 45 degree angle to the light path in a *refractor* or *Cassegrain* telescope just prior to the *eyepiece*. A disadvantage of using a star diagonal is that when reflecting the image it becomes inverted, thereby giving you a mirror image of the object you are viewing.

subdwarf: A star with a lower *luminosity* than *main-sequence stars* of the same spectral type.

subgiant: A star with a *luminosity* between that of *main-sequence stars* and giants of the same spectral type.

summer solstice: The most northerly point the Sun reaches along the *ecliptic* and marks the first day of summer (June 21 or 22) in the northern hemisphere.

supercluster of galaxies: A vast assemblage of clusters of galaxies.

supergiant: A giant star with extremely high *luminosity* and diameter.

supernova: The cataclysmic explosion of a massive star. This can happen after the star has converted most of its hydrogen fuel to helium or after a star has fused elements in its core up to iron and suffers a rapid inward collapse of its outer shell. Unable to slow the inward rush of material, the star's core temperature rises by several thousand kelvin. The temperature rise causes an increase in the star's internal pressure, which in turn causes an overheated outward rush of material. The star's weakened gravitational field is unable to overcome this outward rush of material and the star therefore blows itself apart in a luminous explosion. This collapse and explosion can take place in a matter of seconds. Another type of supernova can result from a *white dwarf* star in a *binary star* system collecting so much new matter from its companion that it collapses suddenly. Supernovae can leave behind *neutron stars* or *black holes*, as well as a surrounding cloud of debris.

telescope: An instrument designed to collect and magnify the radiation from distant objects. The basic principle of an optical telescope is to gather photons of light energy by bending the light with a lens, bouncing the light beams off a mirror, or a combination of both methods, then passing the intensified light through a focusing lens (*eyepiece*). Gamma ray, X-ray, ultraviolet, infrared, and radio telescopes gather and magnify radiation outside the visible wavelengths of the *electromagnetic spectrum*. Computers and other electronic equipment are used to focus and display the gathered image.

temperature: Stellar temperatures are measured on the kelvin (K) scale. On this scale, *absolute zero* (0 K) equals -273.16° celsius (C) and -460° fahrenheit (F). There is no upper limit on the kelvin scale.

transit: Transit occurs when an object crosses your southern *meridian* or passes in front of another celestial body.

transparency: Transparency is a measure of how easily you can see dim objects through the atmosphere. The less light pollution, dust, smoke, haze, or moisture over your head, the better the transparency. The transparency of the sky above you can change in a matter of minutes since the atmosphere is in constant motion. When the transparency is good, "*seeing*" is good and you can observe fainter objects than when their light is blocked by junk in the atmosphere.

triplet: An *objective* lens made of three individual pieces of glass is a triplet and is used to reduce *secondary spectrum*.

true space motion: The direction in which a star is actually moving.

UBV system of stellar magnitude: This system uses photoelectric photometry in three broad wavelength bands: (U) ultraviolet, (B) blue, and (V) visual spectral regions. By comparing the star's light in these three bands and making corrections for *interstellar dust* absorption (which reddens a star's B-V *magnitude* and is called color excess), *temperature* (which affects a star's B and V magnitudes to a different degree than color excess), and spectral type, the UBV system gives a more accurate magnitude for a star than most other techniques. The U-B and B-V indices have their zero magnitude point set as that for A0 spectral-type stars of normal *luminosity*.

universal time (UT): Universal time is the standard worldwide civil timekeeping system which uses the solar time on the prime meridian (which runs through the Royal Observatory in Greenwich, UK) as its base. All civil time is then calculated by how far you are east or west of Greenwich. The zero hour of UT is midnight at the Greenwich Observatory. A UT time signal is given each second on short-wave radio station WWV in the United States, CHU in Canada, and worldwide on station MSF from the UK.

variable stars: A variable star is any star that exhibits changes in its *apparent magnitude*. Variables are divided into six main classes (*cataclysmic, eclipsing binary, eruptive, pulsating, rotating,* and *X-ray*). Each class is subdivided into one or more types.

vernal equinox: The vernal equinox occurs when the Sun, heading north, crosses the *celestial equator* at 0 hours of *right ascension* in the constellation Pisces. When the vernal equinox and *autumnal equinox* were established, the Sun crossed the celestial equator in Aries at a position called the *First Point of Aries*. Due to *precession*, the vernal equinox now occurs while the Sun is in Pisces. The vernal equinox marks the first day of spring (March 20 or 21) in the northern hemisphere.

visible spectrum: Visual light consists of the radiation emitted or reflected by its source and makes up a small portion of the *electromag-*

netic spectrum. By passing the light through a prism or reflecting it off a grating, the light is broken down into a rainbow-like pattern. This rainbow is actually the light separated and arranged in the order of its wavelengths from ultraviolet (short wavelengths at 4,000 Å) to infrared (long wavelengths at 7,000 Å) in the visual part of the spectra. This is the narrow portion of the entire spectrum that our eyes are sensitive to, so we are able to see it. We see different colors because every shade of color is fixed at a certain wavelength. We have developed machines that can "see," "hear," or "produce" the rest of the electromagnetic spectrum.

visual binary: The stars in a visual binary system are physically related as they rotate around a common center of gravity with both the primary and *comes* visible from Earth.

visual magnitude: The brightness of an object as seen by your eye. You can generally see, on a good night of *seeing*, at a dark site, stars down to sixth magnitude.

wavelength: The distance between the repeated points in any wave of vibrations in individual particles of matter or light. Frequency and wavelength are interrelated, because the wavelength multiplied by the frequency equals the speed of light (300,000 km/s or 186,000 miles/s).

winter solstice: The most southerly position the Sun reaches along the *ecliptic* and marks the first day of winter (December 20 or 21) in the northern hemisphere.

white dwarf: After consuming most of its fuel, and depending on its initial mass, an old-age low to medium mass star can collapse into a dense, highly luminous and extremely hot white dwarf. Due to their small size, generally about that of a small planet, these stars have low *visual magnitudes*. White dwarfs are rather rare and are usually associated with larger companion stars.

X-ray variable: X-ray type variables occur in close *binary star* systems in which one of the stars is a hot compact star, like a *white dwarf*, a *neutron star*, or a *black hole*, and the other star in the system is cooler. As matter from the cooler companion star is pulled into the compact star by its greater gravity, X-rays are given off. These high energy bursts are then reflected off the cooler star as high-temperature optical radiation, which causes a variation in the brightness of the entire binary system.

zenith: The point in the sky directly over your head.

zero hour of right ascension: The hour semicircle that runs from the *celestial poles* and bisects the *ecliptic* at the *vernal equinox*. This *meridian* marks the starting point for measuring an object's *right ascension* position.

zodiac: The 18° wide path through the 13 constellations the Sun and all the planets (except Pluto) follow during the year. The *ecliptic* is the centerline for this path. The word zodiac comes from the Greek ὸ Ζωδιακός Κύκλος meaning "circle of animals." In Latin, it is *zodiacus*. The zodiac constellations are: Pisces, Aries, Taurus, Gemini, Cancer, Leo, Virgo, Libra, Scorpius, Ophiuchus (though not considered a zodiac constellation for astrological purposes), Sagittarius, Capricornus, and Aquarius.

Bibliography

Atlases

Audouze, Jean, and Israel, Guy, eds. *The Cambridge Atlas of Astronomy.* Cambridge University Press, ISBN 0-521-26369-7, 1985.

Cox, John, and Monkhouse, Richard. *Philip's Color Star Atlas: Epoch 2000.* London: George Philip Limited, ISBN 0-91315-08-9, 1991.

Dickinson, Terence, Costanzo, Victor, and Chaple, Glenn. *The Edmund Mag 6 Star Atlas.* Barrington, NJ: Edmund Scientific Co., 1982.

Ridpath, Ian, ed. *Norton's 2000.0: Star Atlas and Reference Handbook.* 18th edn. New York: John Wiley & Sons, ISBN 0-470-21460-0, 1989.

Sandage, Alan R. *The Hubble Atlas of Galaxies.* Washington, D.C.: Carnegie Institution of Washington, ISBN 0-87279-629-9, 1961.

Snyder, George Sergeant. *Maps of the Heavens.* New York: Abbeville Press, ISBN 0-89659-456-4, 1984.

Tirion, Wil. *Sky Atlas 2000.0.* Cambridge, MA: Sky Publishing, ISBN 0-933346-33-6, 1981.

Tirion, Wil, Rapport, Barry, and Lovi, George. *Uranometria 2000.0.* 2 vols. Richmond, VA: Willmann-Bell, ISBN 0-943396-14-X (vol. 1) and 0-943396-15-8 (vol. 2), 1987. *Volume I-The Northern Hemisphere to -6 Degrees* and *Volume II- The Southern Hemisphere to +6 Degrees.*

Tully, Brent R., and Fisher, Richard J. *Nearby Galaxies Atlas.* Cambridge University Press, ISBN 0-521-30136-X, 1987.

Books

Allen, David A., Barker, Edmund S., and Jones, Kenneth Glyn, eds. *Webb Society Deep-Sky Observer's Handbook.* Vol 2: *Planetary and Gaseous Nebulae.* Hillside, NJ: Enslow Publishers, ISBN 0-89490-028-5, 1979.

Allen, Richard Hinckley. *Star Names: Their Lore and Meaning* (formerly titled *Star Names and Their Meaning.* New York: G. E. Stechert, 1899). Revised version New York: Dover Publications, ISBN 0-486-21079-0, 1963.

Barker, Edmund S., and Jones, Kenneth Glyn, eds. *Webb Society Deep-Sky Observer's Handbook.* Vol 3: *Open and Globular Clusters.* Hillside, NJ: Enslow Publishers, ISBN 0-89490-034-X, 1980.

Barker, Edmund S., and Jones, Kenneth Glyn, eds. *Webb Society Deep-Sky Observer's Handbook.* Vol 4: *Galaxies.* Hillside, NJ: Enslow Publishers, ISBN 0-89490-050-1, 1981.

Berry, Arthur. *A Short History of Astronomy: From Earliest Times Through The Nineteenth Century.* John Murray, 1898. Reprint version New York: Dover Publications, SBN 486-20210-1, 1961.

Berry, Richard. *Discover the Stars.* New York: Harmony Books, ISBN 0-517-56529-3, 1987.

Beyer, Steven L. *The Star Guide: A Unique System For Identifying The Brightest Stars in the Night Sky.* Boston: Little, Brown and Company, ISBN 0-316-09267-3, 1986.

Boorstin, Daniel J. *The Discoverers.* New York: Random House, ISBN 0-394-40229-4, 1983.

Bulfinch, Thomas. *Bulfinch's Mythology*. New York: Crown Publishers, Inc., ISBN 0-517-274159, 1979.

Burnham, Robert, Jr. *Burnham's Celestial Handbook: An Observer's Guide to the Universe Beyond the Solar System*. 3 vols. New York: Dover Publications, ISBN 0-486-24063-0, 1966, revised 1978.

Claiborne, Robert. *The Summer Stargazer: Astronomy for Absolute Beginners*. Nature Library, LCCCN 74-30616, 1975.

Consolmagno, Guy, and Davis, Dan M. *Turn Left At Orion: A Hundred Night Sky Objects to See in a Small Telescope - and How to Find Them*. Cambridge University Press, ISBN 0-521-34090, 1989.

Daintith, John, gen. ed. *Biographical Encyclopedia of Scientists*. New York: Facts on File, 1981.

Doig, Peter. *A Concise History of Astronomy*. London: Chapman & Hall Ltd, 1950.

Edmonds, Margot, and Clark, Ella E. *Voices of the Winds: Native American Legends*. New York: Facts on File, ISBN 0-8160-2067-1, 1989.

Eicher, David J. *Deep-Sky Observing With Small Telescopes*. Hillside, NJ: Enslow Publishers, Inc., ISBN 0-89490-075-7, 1989.

Egan, Edward W., Hintz, Constance B., and Wise, L.F., eds. *Kings, Rulers, and Statesmen*. New York: Sterling Publishing Company, ISBN 0-8-69-0050-4, 1976.

Evslin, Bernard. *Gods Demigods & Demons: An Encyclopedia of Greek Mythology*. New York: Scholastic Book Services, 1975.

Freeman, Lenore. *A Starhopper's Guide to Messier Objects*. Oakland, CA: Everything in the Universe, ISBN 0-913399-57-4, 1981.

Galilei, Galileo. *Discoveries and Opinions of Galileo*. Translated by Stillman Drake. New York: Doubleday Anchor Books, 1957.

Goldstein, Bernard R., trans. *Al-Bitrūjī: On the Principles of Astronomy*. Vol 1. New Haven: Yale University Press, ISBN 0-300-01387, 1971.

Graves, Robert. *The Greek Myths*. Mt. Kisco, NY: Moyer Bell Limited, ISBN 0-918825-80-6, 1988.

Grimal, Pierre, and Kershaw, Stephen, eds. *A Concise Dictionary of Classical Mythology*. Oxford: Basil Blackwell Ltd, ISBN 0-631-16696-3, 1990. Originally published in French as *Dictionnaire de la Mythologie Grecque et Romaine*. Paris: Presses Universitaires de France, 1951.

Hamilton, Edith. *Mythology: Timeless Tales of Gods and Heroes*. New York: New American Library, 1969.

Harrington, Philip S. *Touring the Universe Through Binoculars: A Complete Astronomer's Guidebook*. New York: John Wiley & Sons, ISBN 0-471-51337-7, 1990.

Hawking, Stephen W. *A Brief History of Time: From the Big Bang to Black Holes*. New York: Bantam Books, ISBN 0-553-34614-8, 1988.

Haynes, Steven. *Planetary Nebulae: A Practical Guide and Handbook for Amateur Astronomers*. Richmond, VA: Willmann-Bell, ISBN 0-943396-30-1, 1991.

Hirshfeld, Alan, and Sinnott, Roger W., eds. *Sky Catalogue 2000.0*. 2 vols: Cambridge, MA: Sky Publishing Corporation, ISBN 0-933346-34-4 (vol. 1) and 0-933346-39-7 (vol. 2), 1982. *Volume 1 Stars to Magnitude 8.0* and *Volume 2 Double Stars, Variable Stars and Nonstellar Objects*.

Hoffleit, Dorrit. *The Bright Star Catalogue*. 4th rev. edn. New Haven, CT: Yale University Observatory, 1982.

Holyoke, Edward A. *Observe: A Guide to the Messier Objects, An Observing Program For Use With Small Instruments*. Washington, D.C.: The Astronomical League, 1962, reprinted 1987.

Isles, John E., and Jones, Kenneth Glyn, eds. *Webb Society Deep-Sky Observer's Handbook*. Vol 8: *Variable Stars*. Hillside, NJ: Enslow Publishers, ISBN 0-89490-208-3, 1990.

Jastrow, Robert, and Thompson, Malcolm H. *Astronomy: Fundamentals and Frontiers*. New York: John Wiley & Sons, ISBN 0-471- 440752, 1972.

Jones, Brian, and Edberg, Stephen, eds. *The Practical Astronomer*. New York: Simon and Schuster, ISBN 0-671-69303-2, 1990.

Jones, Kenneth Glyn. *The Search for the Nebulae*. Cambridge: Alpha Academic, 1975.

Jones, Kenneth Glyn. *Messier's Nebulae and Star Clusters*. 2nd edn. Cambridge University Press, ISBN 0-521-37079-5, 1991. First edition published by Faber & Faber, Ltd, 1968.

Jones, Kenneth Glyn, ed. *Webb Society Deep-Sky Observer's Handbook*. Vol 1: *Double Stars*. Hillside, NJ: Enslow Publishers, ISBN 0-89490-027-7, 1979.

Krupp, Edwin C. *Echoes of the Ancient Skies: The Astronomy of Lost Civilizations*. New York: Harper & Row, ISBN 0-06-015101-3, 1983.

Leeming, David Adams. *The World of Myth: An Anthology*. New York: Oxford University Press, ISBN 0-19505601-9, 1990.

Mallas, John H., and Kreimer, Evered. *The Messier Album*. Cambridge, MA: Sky Publishing, LCCCN 78-63243, 1978.

Mayer, Ben. *Astrowatch*. New York: Perigee Books, ISBN 0-399-51431-7, 1988.

Mitton, Jacqueline. *A Concise Dictionary of Astronomy*. Oxford University Press, ISBN 0-19-853967-3, 1991.

Mitton, Simon, ed. *The Cambridge Encyclopedia of Astronomy*. New York: Crown Publishers, ISBN 0-517-52806-1, 1977.

Moore, Patrick. *The Sky at Night*. New York: W. W. Norton & Company, ISBN 0-393-30390-X, 1985.

Moore, Patrick. *Exploring the Night Sky With Binoculars*. Cambridge University Press, ISBN 0-521-30756-2, 1986.

Morales, Ronald J. *The Amateur Astronomer's Catalog of 500 Deep-Sky Objects*. Vol. 1. Tucson, AR: Aztex Corporation, ISBN 0-89404-076-6, 1986.

Motz, Lloyd, and Nathanson, Carol. *The Constellations: An Enthusiast's Guide to the Night Sky*. New York: Doubleday, ISBN 0-385-17600-7, 1988.

Muirden, James. *How to Use an Astronomical Telescope: A Beginner's Guide to Observing the Cosmos*. New York: Linden Press, ISBN 0-671-4774-7, 1985.

Nakayama, Shigeru. *A History of Japanese Astronomy*. Cambridge, MA: Harvard University Press, LCCCN 68-21980, 1969.

North, Gerald. *Advanced Amateur Astronomy*. Edinburgh University Press, ISBN 0-7486-0253-4, 1991.

Ogilive, Marilyn Baily. *Women In Science: Antiquity Through the Nineteenth Century*. Cambridge, MA: Massachusetts Institute of Technology Press, ISBN 0-262-15031-X, 1986.

Petit, Michel. *Variable Stars*. New York: John Wiley & Sons Ltd, 1987. Originally published in French as *Les Etoiles Variables*. Paris: Masson, Editeur, ISBN 0-471-90920-3, 1982.

Ptolemy. *The Almagest*. Translated by R. Catesby Taliaferro in *Great Books of the Western World*. Chicago: Encyclopedia Britannica, LCCCN 55-10325, 1952.

Radice, Betty. *Who's Who in the Ancient World*. New York: Stein and Day/Publishers, ISBN 0-8128-1338-3, 1971.

Raymo, Chet. *365 Starry Nights: An Introduction to Astronomy for Every Night of the Year*. Englewood Cliffs, NJ: Prentice-Hall, ISBN 0-13-920512-8, 1982.

Ridpath, Ian. *Star Tales*. New York: Universe Books, ISBN 0-87663-694-6, 1988.

Riordan, Michael, and Schramm, David N. *Shadows of Creation: Dark Matter and the Structure of the Universe*. New York: W. H. Freeman and Company, ISBN 0-7167-2157-0, 1990.

Room, Adrian. *Dictionary of Astronomical Names*. New York: Routledge, Chapman & Hall, Inc., ISBN 0-415-012988, 1988.

Rosen, Edward, trans. *Kepler's Conversation With Galileo's Sidereal Messenger*. New York: Johnson Reprint Corporation, LCCCN 64-20952, 1965.

Sanford, John. *Observing the Constellations: An A-Z Guide For the Amateur Astronomer*. New York: Simon and Schuster, 1989.

Schroeder, Daniel J. *Astronomical Optics*. San Diego: Academic Press, ISBN 0-12-629805-X, 1987.

Shapley, Harlow. *Source Book in Astronomy: 1900-1950*. Cambridge, MA: Harvard University Press, LCCCN 60-13294, 1960.

Snow, Theodore P. *Essentials of the Dynamic Universe: An Introduction to Astronomy*. St. Paul, MN: West Publishing Company, ISBN 0-314-30403-7, 1987.

Staal, Julius D. W. *Stars of Jade: Calendar Lore, Mythology, Legends and Star Stories of Ancient China*. Decatur, GA: Writ Press, ISBN 0-914653-008, 1984.

Tenney, Merrill C., gen. ed. *Handy Dictionary of the Bible*. Grand Rapids, MI: Zondervan Publishing House, 1965.

Thompson, Malcolm J., Morales, Ronald J., and Jones, Kenneth Glyn, eds. *Webb Society Deep-Sky Observer's Handbook*. Vol 6: *Anonymous Galaxies*. Hillside, NJ: Enslow Publishers, ISBN 0-89490-027-7, 1987.

Tully, Brent R. *Nearby Galaxies Catalog*. Cambridge University Press, ISBN 0-521-35299-1, 1988.

Vico, Giambattista. *The New Science*. Abridged translation of the 3 rd edn. (1744), translated by Thomas Goddard Bergin and Max Harold Fisch in *The New Science of Giambattista Vico*. Ithaca, NY: Cornell University Press, ISBN 0-8014-9099-5, 1948, revised 1970.

Webb, William L. *Brief Biography and Popular Account of the Unparalleled Discoveries of T.J.J. See*. Lynn, MA: Thos. P. Nichols & Sons, Co., 1913.

Whiston, George O., and Jones, Kenneth Glyn, eds. *Webb Society Deep-Sky Observer's Handbook*. Vol 5: *Cluster Galaxies*. Hillside, NJ: Enslow Publishers, ISBN 0-89490-066-8, 1982.

Zombeck, Martin V. *Handbook of Space Astronomy and Astrophysics*. Cambridge University Press, ISBN 0-521-34787-4, 1982, revised 1989.

Filters

LUMICON, 2111 Research Drive, Suite 5, Livermore, CA 94550, USA, (510) 447-9570.

Orion Telescope Center, 2450 17th Avenue, P.O. Box 1158, Santa Cruz, CA, 95061, USA, 800-447-1001 (CA 800-443-1001).

Miscellaneous books

Belitt, Ban, ed. and trans. *Pablo Neruda Five Decades: A Selection (Poems: 1925-1970)*. New York: Grove Weidenfeld, ISBN 0-8021-3035-6, 1974.

Ciardi, John, trans. *The Inferno: Translated in Verse*. New Brunswick: Rutgers University Press, LCCCN 54-9668, 1954.

Duncan, A. M., trans. *Copernicus: On the Revolutions of the Heavenly Spheres*. New York: Barnes & Noble Books, ISBN 0-06-491279-5, 1976.

Gardner, W. H., ed. *Poems and Prose of Gerard Manley Hopkins.* Baltimore, MD: Penguin Books, Inc., 1953.

Gibb, A. M., ed. *Sir William Davenant: The Shorter Poems, and Songs from the Plays and Masques.* Oxford University Press, 1972.

Johnson, Thomas H., ed. *The Complete Poems of Emily Dickinson.* Boston, MA: Little, Brown and Company, LCCCN 60-11646, 1960.

Jowlett, Benjamin, trans. *Plato: The Republic.* New York: The Heritage Press, 1944.

Lounsbury, Thomas R., ed. *Yale Book of American Verse.* New Haven: Yale University Press, 1912.

Mabbott, Thomas Ollive, ed. *Collected Works of Edgar Allen Poe.* Vol. 1: *Poems.* Cambridge, MA: Harvard University Press, ISBN 0-674-13935-6, 1969.

Porte, Joel, ed. *Ralph Waldo Emerson: Essays & Lectures.* Cambridge University Press, ISBN 0-940450-15-I, 1983.

Rich, Adrienne. *The Will to Change.* New York: W. W. Norton & Co, Inc. LCCCN 78-146842, 1979.

Ross, W. D. ed., Stocks, J. L., trans. *The Works of Aristotle.* Vol II: *De Caelo.* Oxford University Press, 1930.

Singleton, Charles S., trans. *Dante Alighieri: The Divine Comedy; Inferno.* Princeton University Press. ISBN 0-691-098557, 1970.

Whitman, Walt. *Leaves of Grass.* New York: Alfred A. Knopf, 1945.

Wilson, James Grant, ed. *The Poets and Poetry of Scotland: From the Earliest to the Present Time.* London: Blackie & Sons, 1877.

Periodicals

Astronomy. Kalmbach Publishing Co., P.O. Box 1612, Waukesha, WI, 53187, USA: published monthly.

Deep Sky. Kalmbach Publishing Co., P.O. Box 1612, Waukesha, WI, 53187, USA: ceased publication in 1992, back issues only.

Mercury: The Journal of the Astronomical Society of the Pacific. Astronomical Society of the Pacific, 390 Ashton Avenue, San Francisco, CA 94112, USA: published bimonthly.

Observer's Handbook. The Royal Astronomical Society of Canada, 136 Dupont Street, Toronto, Ontario, Canada M5r 1V2. ISSN 0080-4193: published annually.

Science News. 231 West Center Street, Marion, OH, 43305, USA: published weekly.

Sky Calendar. Abrams Planetarium, Michigan State University, East Lansing, MI, 48824, USA: published monthly.

Sky & Telescope. Sky Publishing, 49 Bay State Road, Cambridge, MA, 02138, USA: published monthly.

The Astronomical Almanac. United States Naval Observatory, Nautical Almanac Office, Washington: Superintendent of Documents, U.S. Government Printing Office, Washington, D.C., 20402, USA, and Royal Greenwich Observatory, Nautical Almanac Office. London: Her Majesty's Stationery Office, P. O. Box 276, London, SW8 5DT, UK: published annually.

The Griffith Observer. Griffith Observatory, 2800 East Observatory Road, Los Angeles, CA, 90027, USA: published monthly.

The Observer's Guide. Astro Cards, P.O. Box 35, Natrona Heights, PA, 15065, USA: ceased publication in 1992, back issues only.

Planispheres

Griffith Observatory Astrorama. Griffith Observatory, 2800 East Observatory Road, Los Angeles, CA 90027, USA.

Star and Planet Locator. Edmund Scientific Company, 101 East Glouchester Pike, Barrington, NJ, 08007-1380, USA.

The Miller Planisphere. Datalizer Slide Charts, Inc., 501 Westgate Street, Addison IL, 60101-4524, USA.

Index of Arabic titles

Index of celestial objects

Mythological characters are listed here in their celestial context. See the Index of mythological characters and place names for these names when used in their mythological context.

Index of mythological characters and place names

Index of names

General index